AF837691

Recent Trends in Environmental Pollution and its Impact on Human Health

Pollution refers to the negative impact of human activity on the environment, posing risks to human health, ecosystems, and economic growth. Environmental sciences, an interdisciplinary field, investigates the influence of humans on the environment and solutions to solve problems.

The present book, encompassing ten chapters, covers the impacts of aerosol, and bioaerosol pollution, ozone and black carbon, micro plastics on marine life, rare earth elements, the environmental impact of energy production, heavy metal pollution, wastewater treatment, and the influence of biogenic volatile organic compounds (BVOCs) on health and environment in a single overarching scientific theme.

Features:

- The book provides up-to-date information on various types of pollution and methods of their mitigation/remediation.
- For anyone with a curiosity for environmental biology, environmental science, or environmental studies, this is a desk reference.
- New innovations and data are accounted for and included.
- This book is intended not only for research scientists, engineers, biologists, and chemists but also for laymen and policy makers.
- The book not only engages the fundamentals of environmental science but also demonstrates to the general public how environmental science studies show methods of sustainability, environmental policies, economics, and personal choices.
- Diversity of the chapters chosen to represent this book is one of its distinguishing features.

Recent Trends in Environmental Pollution and its Impact on Human Health

R. Ramakrishna Reddy and T. Pullaiah

CRC Press
Taylor & Francis Group
Boca Raton London New York

CRC Press is an imprint of the
Taylor & Francis Group, an **informa** business

Designed cover image: Getty Images (Image ID: 899102868)

First edition published 2026
by CRC Press
2385 NW Executive Center Drive, Suite 320, Boca Raton, FL 33431

and by CRC Press
4 Park Square, Milton Park, Abingdon, Oxon, OX14 4RN

CRC Press is an imprint of Taylor & Francis Group, LLC

© 2024 R. Ramakrishna Reddy and T. Pullaiah

Reasonable efforts have been made to publish reliable data and information, but the author and publisher cannot assume responsibility for the validity of all materials or the consequences of their use. The authors and publishers have attempted to trace the copyright holders of all material reproduced in this publication and apologize to copyright holders if permission to publish in this form has not been obtained. If any copyright material has not been acknowledged, please write and let us know so we may rectify it in any future reprint.

Except as permitted under U.S. Copyright Law, no part of this book may be reprinted, reproduced, transmitted, or utilized in any form by any electronic, mechanical, or other means, now known or hereafter invented, including photocopying, microfilming, and recording, or in any information storage or retrieval system, without written permission from the publishers.

For permission to photocopy or use material electronically from this work, access www.copyright.com or contact the Copyright Clearance Center, Inc. (CCC), 222 Rosewood Drive, Danvers, MA 01923, 978-750-8400. For works that are not available on CCC, please contact mpkbookspermissions@tandf.co.uk

For Product Safety Concerns and Information please contact our EU representative GPSR@taylorandfrancis.com. Taylor & Francis Verlag GmbH, Kaufingerstraße 24, 80331 München, Germany.

Trademark notice: Product or corporate names may be trademarks or registered trademarks and are used only for identification and explanation without intent to infringe.

ISBN: 9781041022091 (hbk)
ISBN: 9781041022107 (pbk)
ISBN: 9781003618164 (ebk)

DOI: 10.1201/9781003618164

Typeset in Times
by Newgen Publishing UK

Contents

Preface

Environmental risks continue to have an impact on public health despite numerous advancements. Around the world, factors like air pollution, noise, ozone, BVOC, heavy metal emissions, plastics, heat waves, and cold spells continue to cause illnesses and deaths. Numerous human activities have an impact on the environment; thus, the relationship between people and their physical surroundings has been thoroughly researched. The biotic (living things and microbes) and abiotic (atmosphere, lithosphere, and hydrosphere) combine to form the environment. Because human activity pollutes the soil where plants grow, the water we drink, and the air we breathe, it hurts the ecosystem. Environmental pollution is a worldwide issue that has drawn human attention due to its serious long-term effects and potential to have a significant impact on human health. Numerous factors, including mining, deforestation, urbanization, transportation, household waste, industrial operations, and agricultural practices, can contribute to pollution. The definition of pollution is changing as more individuals from all backgrounds become aware of its consequences. Environmental pollution is a widespread issue that is likely to have a significant impact on population health.

New environmental issues that are typically not included in the majority of published publications in this field are covered in this book. The topics chosen for this book are what make it special. The intricate connections between environmental contaminants and their significant impacts on human health are examined. The book, which is divided into ten in-depth chapters, starts with aerosols, Black Carbon, Ozone, BVOCs, heavy and rare earth elements, Microplastics, and the impact of power generation on pollution. It then goes on to examine the health consequences of these factors. This book will comprise several unique chapters that are rarely collected in one volume.

With its insights into the intricate relationship between environmental health and human well-being and its advocacy for further study and preventative measures to safeguard public health, it is a valuable resource for all professions. There is growing evidence in the literature that pollution is a significant risk factor for human illness. Numerous common exposures can affect various organ systems and either directly or indirectly contribute to a wide range of illnesses and disorders, which presents a challenge to environmental health science. We hope that the reader will get an understanding of the wide-ranging health effects of pollution that are discussed in this book. Increased knowledge of the negative effects of pollution on human health will help to better understand the pathophysiology of disease and may open up new avenues for improving human health.

For professionals, researchers, scholars, engineers, and businessmen and businesswomen interested in the most recent advancements in environmental science, this book is a perfect resource. We appreciate the assistance of several scientists in allowing their papers to enhance the caliber of the book. The first author would like to thank his wife, C. Ramathulasamma, for her perseverance, as well as his children, R. Sreevani and R. Vamsydhar Reddy.

R. Ramakrishna Reddy
T. Pullaiah

About the Authors

R. Ramakrishna Reddy, PhD, is a former Professor at Sri Krishnadevaraya University in Andhra Pradesh, India. With over 30 years of teaching and research experience, he has authored numerous books and published around 152 popular science articles. His research interests include spectroscopy, molecular structure, solid-state and materials chemistry, and atmospheric science. He has received numerous awards, including the UGC research award in 1999, and has been a visiting scientist to IUCAA and IISc. He is a reviewer for international scientific journals and a life member of scientific societies.

T. Pullaiah, a former Professor at Sri Krishnadevaraya University in Andhra Pradesh, India, has held various positions, including Dean, Faculty of Biosciences, Head of Department of Botany, and Biotechnology. He has authored 80 books, edited 40 books, and published over 340 research papers. Under his guidance, 54 students have obtained doctoral degrees. He received his PhD from Andhra University and attended Moscow State University in Russia. He was also a member of the Species Survival Commission of the International Union for Conservation of Nature (IUCN).

1 Aerosol Contamination and a Threat to Plant Vegetation and Health

1.1 HISTORY OF AEROSOL POLLUTION

About 78% of the atmosphere is made up of nitrogen (N_2), 21% of the atmosphere is made up of oxygen (O_2), and the remaining 2% is made up of other gases. Particulate matter (PM), which includes both liquid droplets and solid particles, can also be found in the atmosphere with the gases just mentioned. Although they have a significant impact on human health and climate, these particles are typically too small for the naked eye to detect.

The term "aerosol" refers to a system of finely divided particles of liquid or solid that are distributed uniformly throughout a gas, often air. For precipitation to occur, nuclei for condensation and freezing must first be present, and aerosol particles like dust perform this role (www.britannica.com/science/aerosol). They modify the weather by changing the amount of incoming solar energy that is reflected or absorbed, and by making clouds brighter and more reflective. In addition to playing a role in chemical processes, they also affect the electrical characteristics of the environment. Particles of condensed matter are suspended in a gaseous medium, forming an aerosol. Because of its effects on the weather, visibility, biogeochemical cycle, atmospheric reactivity, and human health, atmospheric aerosol, or particulate matter (PM), is of paramount importance (Griffin, 2013).

The energy balance of the Earth is affected by aerosols, which can be anything from microscopic droplets to dust particles to fragments of fine black carbon. The sky is made up of more than simply empty space. Aerosols, often known as particulates, are millions upon billions of microscopic floaters. Hundreds or even thousands of aerosols can occupy the same volume of air as one human hair. Aerosols significantly alter the global climate. Airborne solid particles and liquid droplets are collectively referred to as particulate matter (PM) or particle pollution. Dust, dirt, soot, and smoke are all examples of particles that are either large enough or dark enough to be visible to the naked eye. Others are undetectable without the use of an electron microscope (Hinds, 1999; Twomey, 1987). Black and brown carbon (Li et al., 2023a) are warming the Earth's atmosphere, whereas sulphate droplets are cooling it. The overall budget of aerosols, according to scientists, somewhat cools the Earth. It is yet unclear how much and how frequently that impact can change over days, years, or centuries.

Floating in the air are tiny particles called aerosols. There are a lot of natural ones, but their numbers haven't changed significantly throughout the years. While greenhouse gases are mostly to blame for climate change, human-caused aerosols have increased rapidly due to smokestacks, car exhaust, wildfires, and even clothes dryers (https://news.mongabay.com/2022/03/aerosol-pollution-destabilizing-earths-climate-and-a-threat-to-health/). Some estimate that annually one million individuals are lost worldwide due to aerosol pollution, including 200,000 in the United States. They should be quickly contained. However, there is a problem with that, as the aerosols humanity now

DOI: 10.1201/9781003618164-1

releases into the atmosphere help to keep the climate cool. Therefore, they shield us from some of the warming caused by the persistent emissions of greenhouse gases.

Greenhouse gases like carbon dioxide and methane are the first things that come to mind when discussing the causes of climate change because their compositions are pretty uniform and their effects, the warming of the Earth, are well understood. Greenhouse gases have been studied extensively, but scientists are beginning to realize that they only tell part of the picture. Aerosols also provide a significant insight into the climate. These materials are a hodgepodge of liquids and solids that are significantly different from their gaseous counterparts. Aerosols can have a wide range of paradoxical effects on the climate, including cooling and warming, causing drought and intense rainfall, and so on, and they tend to linger in the atmosphere near their source or migrate as localized or regional masses via air currents. The effects of aerosols are difficult to define and characterize, but they may help climate scientists fill in many gaps.

PM10 refers to inhalable particles with a diameter of 10 micrometers or less; PM2.5 refers to fine inhalable particles with a diameter of 2.5 micrometers or less; both PM10 and PM2.5 are considered to be particulate pollution. In what ways is 2.5 micrometers minute? Consider one of the hairs on your head. To put that in perspective, the diameter of a human hair is roughly 70 micrometers, making it 30 times larger than the greatest fine particle.

1.1.1 Pollutants

These particles can be formed from a wide variety of substances and can have a wide range of sizes. Some are released immediately from their points of origin, such as building sites, dirt roads, fields, factories, and fireplaces. Pollutants from sources including power plants, industry, and vehicles, such as sulphur dioxide and nitrogen oxides, are responsible for the formation of most particles in the atmosphere. Aerosols are a harmful component of air pollution. These minute particles cause harm to lung tissue and can even cause lung illnesses if inhaled. In many places of the world, aerosols contribute to reduced vision by creating haze.

The amount of sunlight that is reflected away from Earth can be altered by atmospheric aerosols. When exposed to sunshine, not all aerosols have the same effect. Sunlight is reflected back into space by particles of sea salt. Particulate black carbon produced from the combustion of wood or fossil fuels reflects very little of the light that reaches it. Clouds have an effect on climate, and aerosols play a role in cloud formation. Each of the millions of water droplets that make up a cloud condenses onto a tiny particle resembling an aerosol. More aerosols can lead to a rise in cloud cover. When it comes to weather, not all clouds are created equal. Scientists are still learning more about this, but it's generally accepted that clouds can have a cooling influence on the climate by reflecting incoming solar radiation back into space.

Aerosols added to the atmosphere over the past century likely mitigated global warming. But now, because of advances in technology, companies, power plants, and vehicles are able to emit less air pollution, leading to a decrease in atmospheric aerosols. Although this is encouraging news for people's health (air pollution is a major issue), it also suggests that global warming will accelerate.

Due to their small size and light weight, these liquid droplets and solid particles can remain suspended in the air for extended periods before eventually falling to the ground. An aerosol is a suspension of solid or liquid particles in a gas (https://caice.ucsd.edu/introduction-to-aerosols/). The term "aerosol" probably brings to mind images of hairspray or spray paint. Hairspray and spray paint both use what is known as an aerosol spray when you press the trigger. However, because it is suspended in the air's gases, the particulate matter (PM) in the atmosphere is also called an aerosol. Aerosol particles are another name for particulate matter. The particles that make up particulate matter might be quite large or quite microscopic. Particles with a diameter of less than 2.5 micrometers (PM2.5) are considered fine, whereas those between 2.5 and 10 micrometers (PM10) are considered coarse (Hinds, 1999; Twomey, 1977).

1.2 THE SOURCES OF AEROSOL PARTICLES

The particles that make up aerosols (Li et al., 2023b) can originate either naturally or as a result of human activities (Figure 1.1). Figure 1.1 depicts a few of the many potential origins of aerosol particles.

Mineral dust from the desert is a common cause of PM2.5. Heated air is less dense than cold air; hence, it rises as it is heated. Convection is the method by which wind is created. Particulate matter from the ground can be lifted into the air by the wind and carried great distances, even between continents. Scientists have demonstrated that dust from China's Taklimakan desert can circle the globe in just 13 days (https://caice.ucsd.edu/introduction-to-aerosols/#what-is-an-aerosol).

Sea spray from the ocean is another natural source of particles. Consider how the wind feels while you're going down the shore; is it the same as the wind you'd experience if you were walking inland? Sea spray is the fine mist of water and other particles that you may feel when you're near the water. Algae, bacteria, and other marine organisms secrete salt and other organic chemicals into the ocean water. Sea water carrying these minerals and organic materials is released into the air as waves break. These microdroplets include water, which can evaporate, leaving behind a solid particle of sea salt and organic substances. Sea spray aerosols are a major contributor to atmospheric aerosol levels because the oceans cover around two-thirds of the Earth's surface. Smoke from fires, car exhaust, and industrial processes are all human-made sources of aerosol particles. What occurs when hydrocarbon fuels (like the gasoline we put in our cars) are burned? On a cold day, your car's exhaust pipe may show signs of smoke and even water condensation. Combustion is the process that occurs when fuel is burned. When fuel burns, it undergoes a chemical reaction with the oxygen (O_2) in the air, producing carbon dioxide (CO_2) and water (H_2O) as byproducts. However, carbon monoxide (CO) is also produced in the reaction of these fuels through incomplete combustion. Soot is a particle byproduct created from incomplete combustion; it contains the unburned carbon that would have otherwise been transformed to gaseous carbon monoxide (https://caice.ucsd.edu/

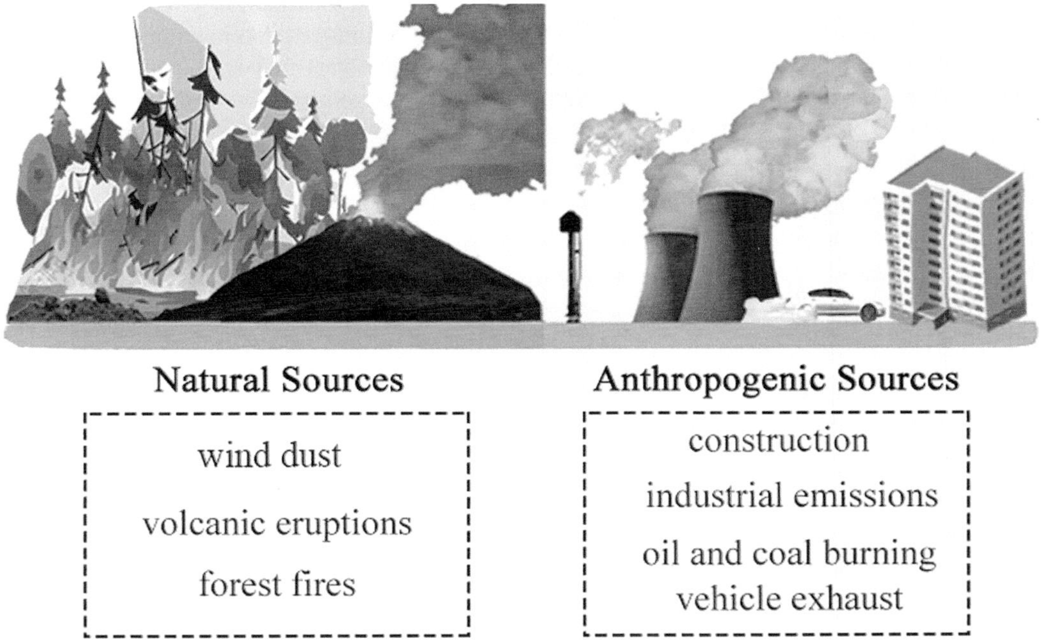

FIGURE 1.1 Natural and anthropogenic aerosol sources. (Li et al., 2023b. https://doi.org/10.1080/26395940.2023.2173657. http://creativecommons.org/licenses/by/4.0/.)

introduction-to-aerosols/#what-is-an-aerosol). Secondary sources of aerosol particles emerge from chemical reactions in our atmosphere and complement these primary sources, which include both natural and human activities. Aerosol particles are the result of chemical reactions between gases in the air, such as ozone and organic gases. The health of our community members and the planet's climate are both affected by these aerosol particles.

1.3 AEROSOLS' EFFECT AND CLIMATE

Sunlight, atmospheric gases, and aerosol particles are only a few of the many elements that affect Earth's climate. The impact of aerosol particles, which have both direct and indirect effects on our climate, is complex. The aerosol particles themselves are responsible for the direct effects, while the aerosol particles' role in cloud formation is responsible for the indirect effects. Radiative effects of absorbing and scattering aerosols are shown in Figure 1.2. The capacity of the particles to either absorb or scatter sunlight directly accounts for the direct aerosol impact.

Radiative effects of absorbing and scattering aerosols are shown in Figure 1.2. The capacity of the particles to either absorb or scatter sunlight directly accounts for the direct aerosol impact. The sun's light radiation reaches Earth and warms the globe, which in turn affects the global temperature (Li et al., 2022). However, the Earth does not directly receive all of the sun's rays. This is due to the scattering of around 25% of the sun's light back into space by aerosol particles (and the clouds they assist in generating). This has a cooling impact since less heat from the sun is absorbed by the ground (Figure 1.3).

The meaning of "light dispersion". Light can create new light sources when its energy collides with matter, such as a particle. Because this new light can go in any direction, we say that it is "scattered" (Figure 1.4).

The fact that light is an electromagnetic wave is crucial to grasping the phenomenon of light dispersion. It's an electromagnetic wave, which implies it consists of electric and magnetic forces. Charged species, including protons, electrons, and molecular dipoles, are susceptible to the effects of electric and magnetic fields. A molecule with an unbalanced distribution of charge, known as a "dipole moment," can feel the pull of one of these fields. Charged species migrate because of electromagnetic forces (Cohan et al., 2002). When a charged object moves, it creates a new electric and magnetic field, which travels through space as an electromagnetic wave (light). Scattering

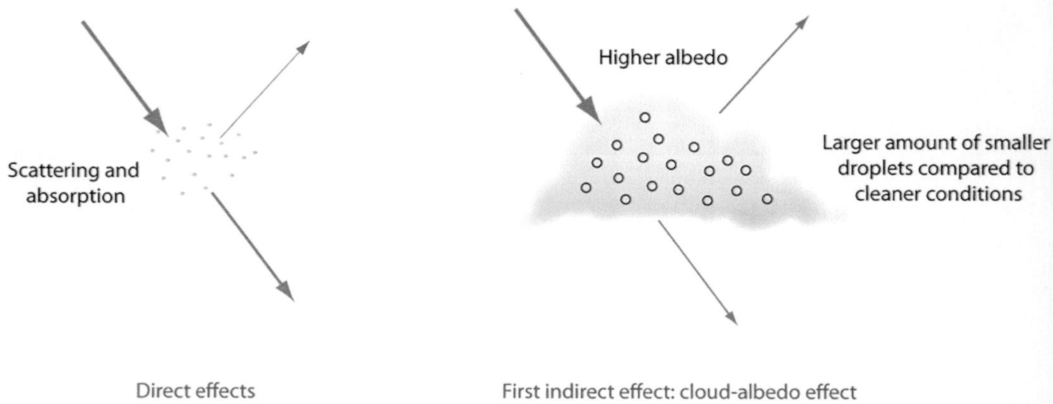

FIGURE 1.2 This diagram illustrates the direct and one of the indirect mechanisms by which aerosols affect the Earth's radiative budget. (https://en.m.wikipedia.org/wiki/File:Representation_of_the_direct_and_first_indirect_effect_of_aerosols_on_the_albedo_of_clouds.png . This file is licensed under the Creative Commons Attribution 3.0 Unparted license.)

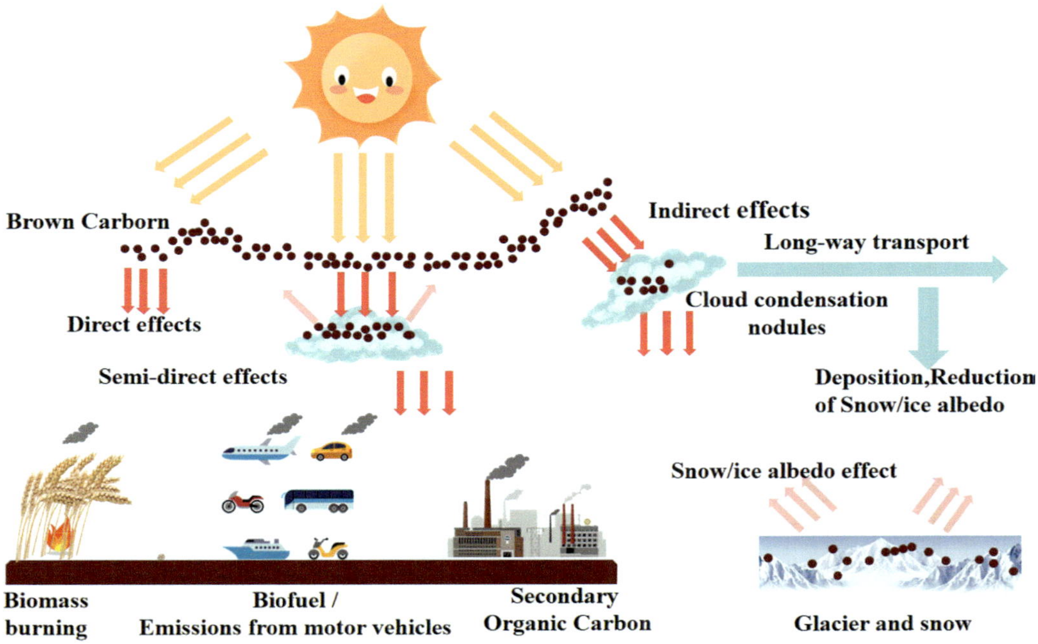

FIGURE 1.3 The major sources of aerosols and their effect on climate (Li et al., 2023a; https://creativecommons.org/licenses/by/4.0/ .)

occurs when a beam of light is dispersed in all directions (Trichili et al., 2021). The light scatters in different directions (Fahey et al., 2021) depending on the particle's size (Figure 1.5).

Very tiny particles disperse light roughly equally in all directions (Rayleigh scattering). More of the scattered light travels in the original direction as particle size rises (Mie scattering). While light can be scattered in any direction (https://physics.stackexchange.com/questions/128731/in-scattering-how-does-a-particle-know-which-direction-it-is-being-illuminate), the direct aerosol effect is caused by sunlight being deflected away from Earth's surface by aerosol particles in our atmosphere (Friedlander, 2000; Kokhanovsky, 2016).

Aerosols have an indirect effect on the environment because of their interaction with clouds. Water vapor and aerosol particles are essential for the formation of cloud droplets in our environment. A cloud droplet is formed when water vapor in the atmosphere is cooled enough to undergo condensation into the liquid phase. Cloud droplets arise when water vapor condenses onto aerosol particles, which serve as this surface (Figure 1.6).

There are distinct aerosol sizes (Seinfeld and Pandis, 1998). The size range of Aitken, accumulation, and coarse mode particles is: Aitken mode: 10–100 nanometers (nm), Accumulation mode: 100–1,000 nm, Coarse mode: Larger than 1,000 nm. Here,1µm = 1,000 nm and 1 nm = 0.001 µm. Nucleation is the smallest, with sizes between 1 and 10 nm. In Figure 1.7, modes are represented (Kaiser et al., 2014) in µm. For sea spray aerosol in particular, a size increase is typically seen during the transition from nascent SSA (which arises directly from breaking waves within 30s of production) to secondary SSA (aerosol formed after the process of chemical and atmospheric reaction).

1.4 AEROSOLS AND HUMAN HEALTH

Particles in the air called "aerosols" have been linked to numerous health issues in humans. Particulate matter is the most pervasive contaminant in terms of human health impacts. World

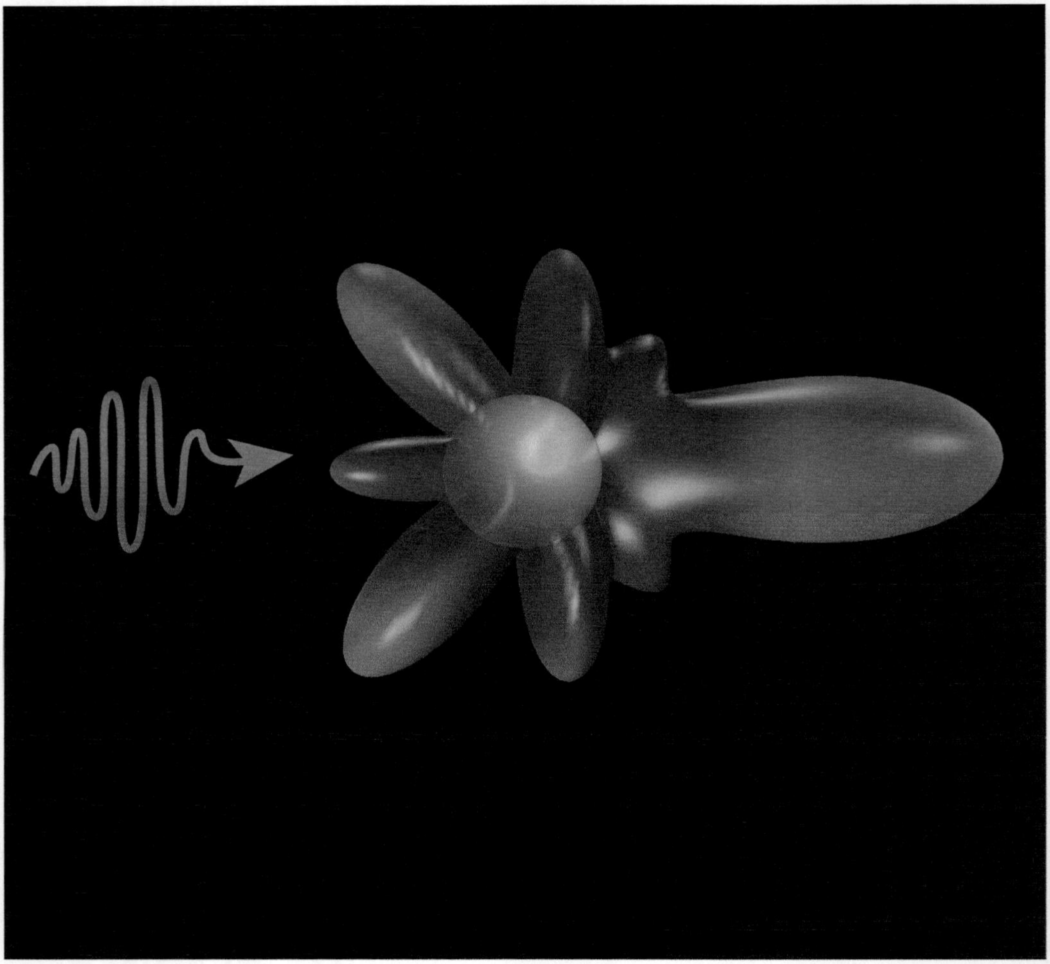

FIGURE 1.4 Mie scattering, artistic view (https://en.wikipedia.org/wiki/Mie_scattering#/media/File:Mie wik.svghttps://creativecommons.org/publicdomain/zero/1.0/.)

Health Organization (WHO) estimates place particulate pollution as a main cause of death globally, accounting for 7 million deaths annually (https://caice.ucsd.edu/introduction-to-aerosols/#what-is-an-aerosol).

Particles of varying sizes can reach various regions of the respiratory system. The "deposition fraction" refers to the amount of a given particle size that is retained by a given region of the respiratory system. Large particles rarely enter the deep regions of our lungs because they are deposited in the nasal cavity, the throat, and the voice box. However, particles that are smaller in size have a greater chance of entering deeper regions of our lungs, such as the trachea, bronchi, and alveoli.

The particles' size is proportional to the harm they do to human health. Larger than 100 nm particles cannot be safely breathed by humans. Airborne particles between 10 and 100 nm in size can be breathed in, but are often filtered out by the respiratory mucous membranes (Zuhara and Isaifan, 2018; Sagona et al., 2020). Particles smaller than 10 μm in diameter are considered inhalable since they can penetrate the lung tissue unscathed. The most dangerous to human health are these tiny particles (Sierra-Vargas and Teran, 2012). Air pollution from vehicles and factories often contains

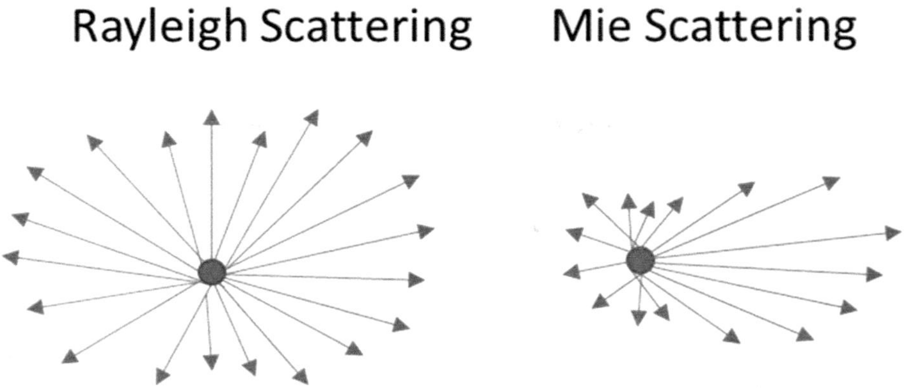

FIGURE 1.5 An illustration of Rayleigh and Mie scattering types (Fahey et al., 2021, https://creativecommons.org/licenses/by/4.0/ .)

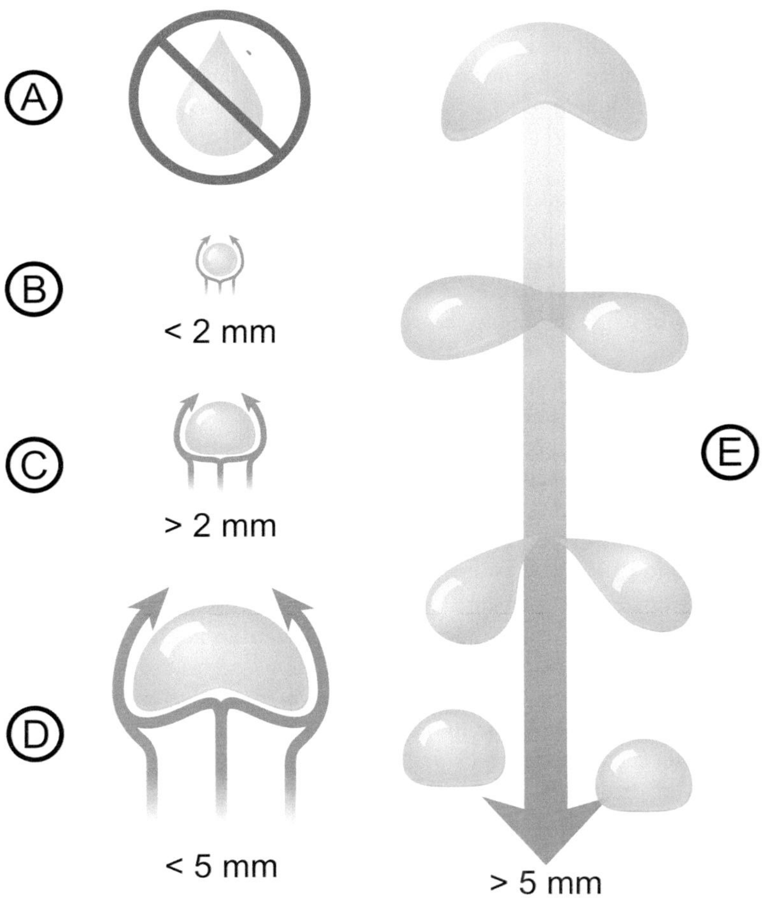

FIGURE 1.6 Formation of Cloud Droplets. The shape of raindrops depends on their size (https://en.wikipedia.org/wiki/Rain#/media/File:Raindrops_sizes.svg CC BY 3.0. Wikimedia Commons.)

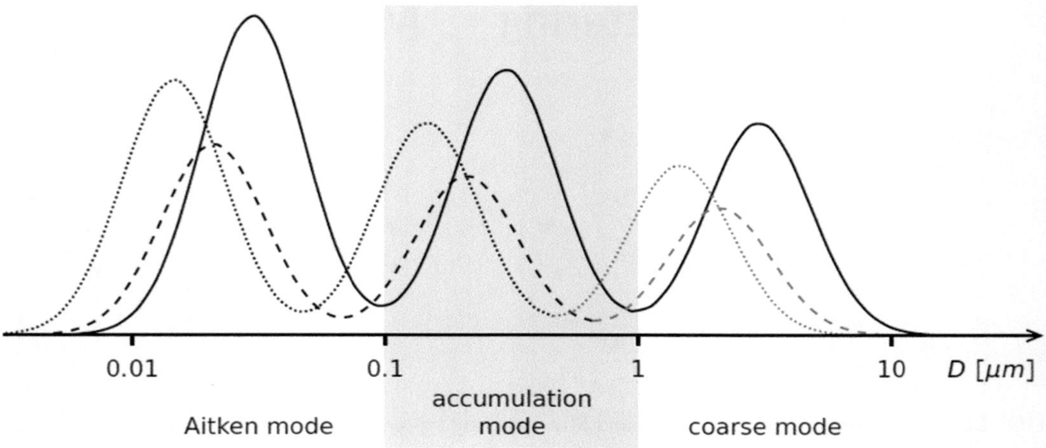

FIGURE 1.7 Schematic illustration of the MADE3 modes and aerosol composition. Each size range (Aitken, accumulation, and coarse mode size range) (Kaiser et al., 2014, https://creativecommons.org/licenses/by/3.0/.)

inhalable coarse particles with a diameter between 2.5 and 10 μm (PM10). Figure 1.8 illustrates how these particles typically lodge in the nasal cavity, throat, and voice box. Smoke and haze both contain fine particles with a diameter of less than 2.5 μm (PM2.5). These tiny particles penetrate deeper into the lungs, with many ending up in the trachea, bronchi, and alveoli.

Pathways linking PM lung exposure to cardiovascular and cerebrovascular diseases that cause morbidity and mortality (Lim and Kim, 2024). Major organs affected by PM exposure and related diseases and symptoms (Figure 1.9). The lung is the first target organ for air pollution and PM exposure is associated with reduced lung function, increased lung inflammation, asthma, respiratory infections, lung cancer, and exacerbation of COPD, which leads to systemic inflammation and oxidative stress affecting blood, vasculature, heart, and brain, ultimately contributing to the premature mortality.

The small fraction of particles that could pass through the upper airways into the alveolar region, such as PM2.5 fine particles (Santacroce et al., 2020), could be easily endocytosed and removed by alveolar macrophages. Bacterial genera and microbiome species of an adult in the upper respiratory tract are shown in Figure 1.10. Because of its position, the oropharynx is in constant contact with numerous microbes, both those that originate within and those that come from outside the body. The oral cavity, rhino-pharynx, larynx, lower respiratory system, and gastrointestinal tract are all geographically connected to it.

Even in otherwise healthy people, inhaled bacteria and viruses from the mouth and throat can spread to the lower respiratory system (Figure 1.10). When an adult's rhino-pharynx is in good health, it has been found to colonize many genera, including Gram-negative anaerobic species such as *Veillonella, Prevotella, Leptotrichia*, and *Fusobacterium*, as well as *Streptococcus, Haemophilus*, and *Neisseria* spp. The pharynx is home to many *Streptococcus* bacteria, including *S. pneumoniae*, *S. pyogenes* (which can cause septic shock and other serious illnesses), and *S. agalactiae*.

Particle pollution has been related to a variety of health issues, according to scientific research. These include 1. Exacerbation of asthma, reduced lung capacity, 2. Symptoms include a sore throat, a cough, and shortness of breath, 3. Cardiopulmonary causes of premature death, 4. Milder forms of heart attacks, and 5. Abnormal cardiac rhythm.

The National Ambient Air Quality Standards established by the United States Environmental Protection Agency intend to reduce exposure to harmful levels of air pollutants like fine and coarse

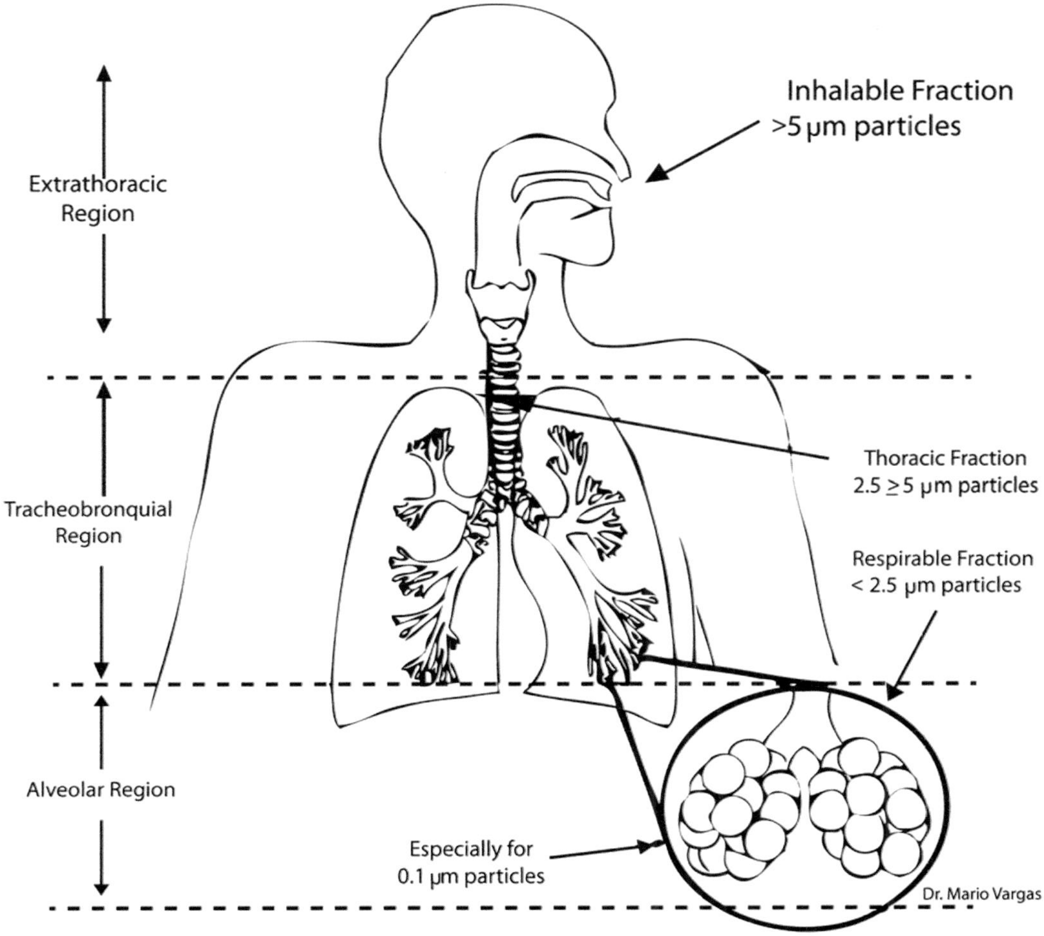

FIGURE 1.8 Various particles (the actual sizes of penetrating particles) to deposit in the respiratory pathway. (Sagona et al., 2020, https://doi.org/10.3390/atmos11060561.)

particles due to their known adverse effects on human health. The Air Quality Index (AQI) is a figure between zero and five hundred that the Environmental Protection Agency uses to report daily air quality (https://caice.ucsd.edu/introduction-to-aerosols/#what-is-an-aerosol). There is a graduated scale of health risks connected with each AQI value. Ozone (O_3), particle pollution (PM2.5 and PM10), carbon monoxide (CO), sulphur dioxide (SO_2), and nitrogen dioxide (NO_2) are the five principal air pollutants used to calculate the AQI (NO_2). The AQI index (Table 1.1) assigns a numerical value to the concentration of each of these contaminants in the air. Table 1.1 displays the PM2.5 and PM10 concentration levels together with the associated AQI index numbers. Each pollutant's AQI is determined daily, and the contaminant with the highest AQI is highlighted. The AQI is typically evaluated for particulate matter and ground-level ozone since they pose the greatest health risk in the United States. The PM2.5 AQI, for instance, was 51 on December 13, 2016, in Des Moines, IA. Health advisories were issued, downgrading the AQI from "good" to "moderate," and reading: "Unusually sensitive people may consider reducing prolonged or heavy exertion." One may check the air quality in the United States right now by visiting www.airnow.gov.

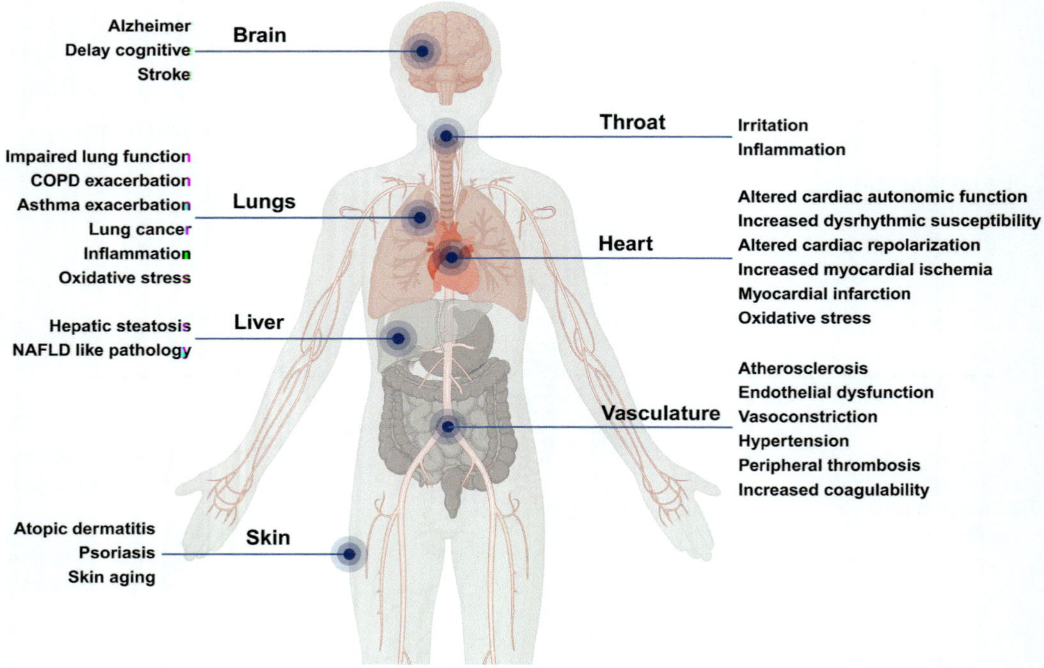

FIGURE 1.9 Major organs affected by PM exposure and related diseases and symptoms. (Lim and Kim 2024, https://doi.org/10.3390/antiox13101256.)

1.5 STUDIES ON ATMOSPHERIC AEROSOLS, IMPACTS, AND SOME HIGHLIGHTS

According to Mushtaq et al. (2022), people's varied interests and worries are a direct result of the severe negative impact of atmospheric aerosols on the ecosystem. Studying nucleation and growth mechanisms, formalizing particle dynamics, characterizing mechanisms for particle-size dispensation, and detecting chemical processes for atmospheric particle sources are all examples of the steps taken by researchers and scientists to gain a firm grasp on the fundamentals. Because of population development and other human activities, the environment has changed in ways that pollute various regions. The energy balance in our climatic circumstances and the radiation budget of Earth's atmosphere are both affected by environmental changes, such as changes in land use patterns, increased concentration of various greenhouse gases, and industrial pollution. The effects of climate change and environmental pollution on human health are widespread. Mushtaq et al. (2022) provide an overview of the present issues that must be addressed before full incorporation of new research perspectives on atmospheric aerosols, their quantification via multiple mechanisms, and their consequences.

According to Mushtaq et al. (2022), atmospheric aerosols play a crucial role in keeping Earth's atmosphere stable. Aerosols range in chemical complexity, particle size, and phase from solids to liquids to gases. To put it simply, they are a gaseous medium containing a suspension of solid or liquid particles. Particulate matter, sulphates, nitrates, mineral dust, sea salt, black carbon, organic carbon, etc., are all examples of aerosols. Coarse (10 μm), fine (2.5 μm), and ultrafine (0.1 μm) refer to the aerodynamic diameters of three distinct particle sizes. Both naturally occurring and artificially produced aerosols in the atmosphere have the potential to disrupt the delicate climate equilibrium of Earth (Gollakota et al., 2021). The chief sources of natural emissions include dirt,

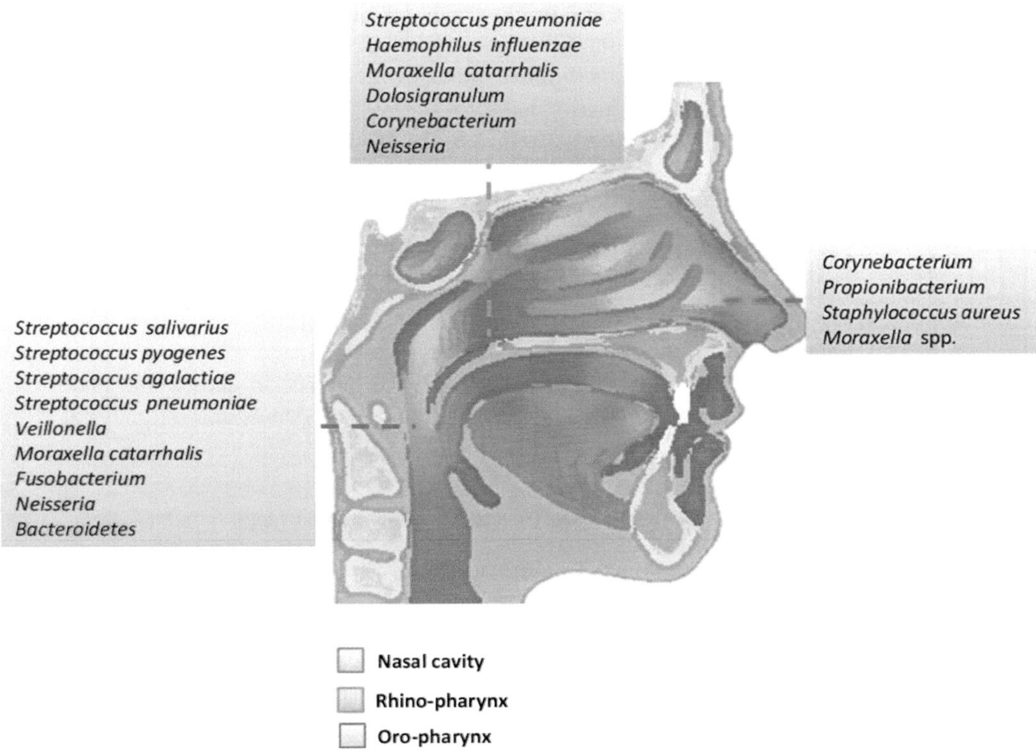

FIGURE 1.10 Main bacterial genera and microbiome species of an adult in the upper respiratory tract (Santacroce et al., 2020, https://doi.org/10.3390/biology9100318http://creativecommons.org/licenses/by/4.0/.)

TABLE 1.1
European air quality Index (PM 10 and PM 2.5 in μg/m³)

Air Quality	PM 2.5	PM 10
Very Good Air Quality	01-15	0-25
Good Air Quality	15-30	25-50
Fair Air Quality	30-55	50-90
Poor Air Quality	55-110	90-180
Very Poor Air Quality	>110	>180

Source: Gryech et al., 2020, http://creativecommons.org/licenses/by/4.0/

forest fires, volcanic eruptions, and so on; the primary sources of air pollution are human-made and include emissions from vehicles, the combustion of biomass and fuels, industrial emissions, petroleum refineries, and so on (Bisht et al., 2022).

Particles of sea salt produced from the ocean's surface, aerosols of mineral dust produced by the effect of wind erosion on arid ground, and aerosols of sulphate formed by the emission of sulphur dioxide from volcanic eruptions or by the formation of dimethyl sulphide from phytoplankton in the oceans are all examples of natural aerosols. Man-made activities like industrial production, mining, transportation, and biomass burning are the primary sources of anthropogenic aerosols (Gautam et al., 2021). Numerous studies have found that atmospheric aerosols' scattering and absorption properties are mostly to blame for winter's diminished visibility. Aerosols have a significant role in the radiation budget and atmosphere of Earth, and their increased emission has a direct impact on climate circumstances, especially the increase in smoke, fog, and cloudiness, which impair visibility largely during the winter season.

It has been found that particulate matter (PM) air pollution lowers crop production and reduces food productivity. The capacity for life and fish to thrive in aquatic environments is directly tied to their effects.

Airborne particles can travel great distances before settling to the ground or water, posing health and environmental risks (www.epa.gov/pm-pollution/health-and-environmental-effects-particulate-matter-pm). The results of this settling can vary greatly from substance to substance: 1. Acidifying water sources like lakes and rivers. 2. Altering the nutritional balance in oceans and major river systems, thirdly, soil nutrient depletion. 3. Threatening fragile ecosystems and agricultural crops. 4. Diminishing ecosystem diversity (Zuhara and Isaifan, 2018).

1.6 EFFECT OF AIR POLLUTION ON PLANTS AND ANIMALS

Pollutants in the air have a negative impact on the health of the environment and all living things in it. Air pollution can refer to a variety of atmospheric disturbances, including those caused by chemicals, organisms, and other physical factors. The effects of air pollution on humans are the first thing that comes to mind. It's easy to lose sight of how detrimental air pollution is to plant and animal life (https://greentumble.com/effect-of-pollution-on-plants/). An increasing body of research links exposure to pollution to abnormalities in animal behavior. Social and mating behavior in animals are directly affected by endocrine disruptors, heavy metals, and PCBs.

1.6.1 Effects of Aerosols on Photosynthesis

Air pollution has both direct and indirect effects on plants, causing a wide range of issues for farmers, as noted by Gerstl and Zardecki (1982). When compared to the impacts of fertilizers or chemical disease control techniques, the net effects of air pollution on agricultural crop yields are difficult to estimate numerically. However, rising levels of air pollution could pose a long-term, difficult-to-solve problem for farming. In this article, Gerstl and Zardecki (1982) focused on a single indirect aerosol effect on agricultural plant life: the diminished availability of photosynthetically active radiation (PAR) at ground level when pollution levels are high. Results from thorough model computations of solar radiation transport through the atmosphere with varied amounts of air pollution allow for confident quantification of this effect. In both rural and urban pollution scenarios, decreases in photosynthetically active radiation (PAR) owing to rising tropospheric aerosol loading have been calculated for the physiologically effective wavelength range between 0.35 and 0.72 m. For the northeastern United States in the summer, for instance, reducing the optical range from clear air conditions to 10 km due to rural and urban aerosols, respectively, resulted in a drop in PAR of roughly 20% and 33%.

1.6.2 The Ozone Layer's Declining

Ozone depletion is blamed for the "indirect" deaths of plants and animals. Studies show that the ozone hole has killed off over 30% of plant species. Ozone (O_3) damages most plants by inducing foliar ulcers. In addition to being able to directly disintegrate plant cells, it can also enter the stomata (the plant's breathing openings). O_3 has the potential to cause stunted development and, in the worst case, death. Brown patches, indicative of moderate ozone damage, are visible on potato leaf surfaces. The plants' DNA was altered, and photosynthesis was stifled by the sun's ultraviolet rays. Ozone molecules close to the Earth reduce plant respiration. They block the holes in leaves where photosynthesis and respiration occur. Without oxygen, a plant cannot perform photosynthesis and so cannot grow. According to a report from 2014, agricultural production in the worst-affected districts was practically halved from 1980 levels due to air pollution from black carbon and ground-level ozone in 2011.

1.6.3 An Abundance of Nitrogen

Plants can't thrive without the nutrient nitrogen. Increases in atmospheric nitrogen deposition, however, can be stressful for ecosystems. Fertilizer use is another benefit of nitrogen surplus. Some plants will benefit from this while others will suffer. The result is a broad list of negative effects on natural ecosystems, including shifts in plant and animal species, a rise in pest and disease outbreaks, and disturbance of ecosystem processes such as nutrient cycle and fire frequency.

1.6.4 Soil Pollution

Deposition of heavy metals and other air contaminants on soil has an initial effect on root performance. They block the plant from getting the nutrients it needs from the ground. The growth of plants will slow because of fewer resources being acquired. This is because of recent shifts in emphasis placed on plant organs. Pollution has the potential to shift the biological community's precarious trophic equilibrium. As a result, the diversity of plant life shifts. These alterations to the agroecosystems may reduce economic output.

1.6.5 Impact of Aerosols and Atmospheric Particles on Plant Leaf Proteins

Plant photosynthesis and the expression of proteins involved in the process can be directly impacted by the presence of air aerosols and particles. Yan et al. (2014a) used a novel combination of Geographic Information System (GIS) and proteomic techniques to examine the impact of aerosols and air particles on plant leaf proteins. Over six years (2007–2012), information on Beijing's aerosol optical depth (AOD) and PM1 data were gathered from various sites across the city. By analyzing the heavy metal concentration of the soil, Yan et al. (2014b) were able to isolate the impact of pollution changes in plant protein expression, examined in two locations chosen on the basis of AOD and PM1 that correspond to strong and weak diffuse solar radiations. Their data showed that the expression levels of plant ribulose bisphosphate carboxylase were greater in regions with significant diffuse solar radiation, while those of oxygen evolved in the enhancer protein and light-harvesting complex II protein were lower. Both regions had the same levels of expression for the beta component of ATP synthase and the chlorophyll a-b binding protein. Changes in the expression and function of these leaf proteins led them to the conclusion that plant photosynthesis, aided by diffuse solar radiation, is boosted by aerosols and atmospheric particles.

1.6.6 Impacts and Responses of Particulate Matter (Aerosols) Pollution on Vegetation

Due to increased industrialization and urbanization, atmospheric pollution has emerged as one of the most pressing problems of the previous decade. The climate, visibility, biogeochemical cycle, atmospheric reactivity, and human health are all profoundly influenced by atmospheric aerosol, often known as particulate matter (PM). Particulate matter (PM) is a type of air pollution that poses a significant risk to human and plant health. The morphological, biochemical, and physiological aspects of plants and their responses are profoundly affected by the notable change in the ambient environment observed due to particulate matter load in metropolitan regions. The significant effects of PM pollution on vegetation and other plant species were reviewed by Sharma and Saxena (2022), who focused on aspects such as morphological characteristics (leaf area, leaf number, stomata structure, flowering, growth, reproduction) and biochemical parameters (pigment content, enzymes, ascorbic acid, protein, sugar). Landscape architects and greenbelt designers can benefit greatly from biomonitoring studies that examine plant species' tolerance and sensitivity.

Aerosols, or atmospheric particles, have a significant impact on a wide range of environmental factors. Particles originate from sea spray and wildfires in a less polluted environment, while human activities like burning fossil fuels are the primary cause in ours. By reflecting the sun's rays away from Earth, aerosols act to chill the globe and have a similar impact on the climate as greenhouse gases. Industrial aerosols, according to the Intergovernmental Panel on Climate Change (IPCC), have significantly slowed the rise in global temperatures over the past 30 years. Premature deaths from aerosols are estimated at tens of thousands annually in the United Kingdom. We still don't know how aerosol pollution will respond to pricey mitigation methods, despite estimates that it costs the UK more than obesity.

1.6.7 Aerosol Sulfate Trends and Their Effect on Photosynthesis

Reducing air pollution is a complicated process involving many aspects. Acid rain, ozone, and nitrate deposition are just a few of the ways in which power plant SO_2 and NOx emissions were reduced thanks to the Clean Air Act. Aerosol pollution's indirect effects on ecosystems are better understood thanks to in-depth studies of scattered light and photosynthesis.

Scientists from CIRES at CSD, including Rebecca Ashenfelter, and the University of Michigan's Gretchen Keppel-Aleks, explored what might happen if sulphate emissions were reduced in the East (https://clasp.engin.umich.edu/people/keppel-aleks-gretchen/). In a study published in Geophysical Research Letters, a journal of the American Geophysical Union, researchers examined the impact of sulphate decreases on diffuse radiation and photosynthesis in the United States from 1995 to 2013 (https://en.wikipedia.org/wiki/Gretchen_Keppel-Aleks).

Both natural and human-caused processes can result in the emission of aerosols. Coal-fired power stations and other industrial operations are major contributors of sulphur dioxide (SO_2), one of the principal anthropogenic aerosols. Sulfate aerosols are produced when this gas is oxidized in the atmosphere. As a result of the Clean Air Act Amendments of 1990, SO_2 emissions were drastically cut, providing researchers with an ideal opportunity to study the impacts of decreased sulphate aerosol mass.

The effect of SO_2 emissions on light scattering and, by extension, photosynthetic rate (measured as gross primary production) was investigated. Since reliable emission data covering the period 1995–2013 is readily available for the eastern United States, this is an excellent area to investigate. During that time, SO_2 emissions in the United States dropped by 70% due to the implementation of emission controls. Keppel-Aleks and Washenfelder employed two methods to investigate the impact of this significant drop in SO_2 on ecological systems. For the eastern United States, one method combined ecological maps with previously documented connections between dispersed light and photosynthesis for various ecosystems. The second method involved simulating the impacts of reduced SO_2 emissions on aerosol optical depth, surface radiation, and photosynthesis using the Community Earth System Model. During the research period, both the satellite data and the model indicated declining aerosol optical depth in the eastern United States. Researchers were able to keep using this model as an accurate simulation of direct and dispersed light patterns after finding that it coincided with genuine satellite observations.

1.6.8 The Relationship between SO_2 Exposure and Plant Physiology

As Lee et al. (2017) emphasized, multiple industries and households contribute to air pollution. Thermal Power stations generate sulphur dioxide (SO_2), which is harmful to plant growth and human health, when fuel is burned in heating systems. The full extent of SO_2's negative effects on crops is, however, little understood. This review set out to synthesize the impact of SO_2 produced by heating systems on plant physiology, morphology, and biochemistry. In general, these processes are hindered when plants are exposed to SO_2. Several days after being exposed to SO_2, a reduction in physiological activities is first noticeable. Extensive exposure to SO_2 has deleterious effects on morphological and metabolic activity. However, low-level SO_2 exposure has a favorable effect on plants since sulphur is a necessary component for plant growth. Plants absorb SO_2 from the air through their stomata and utilize it to make sulfur-containing amino acids like methionine and cysteine. Plant development and many biochemical and physiological processes require sulfur-containing molecules. Specifically for agricultural purposes, Lee et al. (2017) found that SO_2 is a major greenhouse pollutant.

1.6.9 Phytotoxic Effects of SO_2 On Crop Plants: Total Chlorophyll Content

Sulfur dioxide has been known for a long time to be one of the most powerful phytotoxicants, as indicated by Chauhan (2015). Even though global SO_2 emissions have decreased over the past decade, the average SO_2 concentration over India has grown by roughly 71% between 2005 and 2012, as measured by NASA's Aura satellite. In other words, SO_2 poses an ever-greater risk to plants in India's most industrialized areas. The majority of SO_2 emissions come from coal-fired thermal power facilities. The impacts of SO_2 on plants are twofold. Foliar necrosis occurs after even brief contact, but the real damage to crops, woods, and grasses comes from persistent injury brought on by exposure to low levels of SO_2 (Chauhan, 2015). Possible consequences include stunted plant development and reduced harvests. According to Darall (1989), a better understanding of the physiological and biochemical events occurring in affected plants before the appearance of visible injury, symptoms can be gained by using dose-response relationships based on controlled fumigations under quasi-field conditions or well-defined environmental conditions.

The total chlorophyll content in the leaves of three economically significant plant species was investigated by Chauhan (2015). These species were *Vigna radiata* (mung bean), *Lycopersicon esculentum* [Tomato], and *Zea mays* (maize). Fumigation tests were conducted under controlled conditions for a total of 60 days with three distinct SO_2 treatments: T-1=0.05 ppm (134.0g m-3 SO_2) [x 4h], T-2=0.1 ppm (268.0 g m-3 SO_2) [x 2h], and T-3=0.2 ppm (536.0 g m-3 SO_2) [x 1h]. The length of time maize was exposed to was increased to 75 days. There was a consistent pattern in the reactions of *V. radiata* and *S. esculentum* plants to elevated levels of sulfuric acid. The order of chlorophyll loss was T-1 > T-2 > T-3. Surprisingly, the overall chlorophyll content was not drastically altered by SO_2-treatment alone. What mattered was how the two factors (treatment with SO_2 and length of fumigation) interacted with one another. After 15 days of fumigation, there was a noticeable effect on *Z. mays*. The interaction between the two factors was likewise statistically significant across all age groups.

1.6.10 Aerosol Pollution Has a Huge Impact on India's Infant Mortality, Crop Productivity

The regional weather patterns and terrain of India render the country very sensitive to aerosol pollution, as indicated by Rohini Krishnamurthy (www.downtoearth.org.in/news/environment/aerosol-pollution-has-huge-impact-on-india-s-infant-mortality-crop-productivity-study-85112). India had the largest estimated number of newborn mortalities from aerosols (three million) out of the eight regions studied. According to the study published in Science Advances, crop productivity dropped by 12.5 metric tons due to aerosol pollution. National Aeronautics and Space Administration (NASA) noted in a blog that aerosols are known to harm the climate and human health, specifically black carbon, organic carbon, carbon monoxide, nitrogen oxides, volatile organic compounds, and sulphur dioxide. They linger in the air for weeks.

1.7 TOPOGRAPHICAL DISADVANTAGE

The degree of damage caused by these transient pollutants varies according to their point of origin. According to Geeta Persad, a professor at the University of Texas at Austin's Jackson School of Geosciences and the study's lead author, "aerosols that are emitted in the country are also much more likely to stay local because of regional weather patterns and topography." This has a disproportionately negative effect on infant mortality and crop yields in India. The aerosols' potential range is affected by regional weather patterns. In addition, the Himalayas are mentioned in the report as a geographical barrier to ventilation. Aerosol emissions from throughout the world also threaten the country. "Because of India's position in global atmospheric circulation, emissions from other regions can more easily alter India's temperature and rainfall patterns," Persad said.

The researchers also cited India's vast agricultural acreage and the country's big and fragile newborn population as contributing factors. They compound one another to greater social detriment. According to the study, China and India account for the majority of global aerosol emissions. Aerosol emissions are expected to rise, according to the paper's experts, particularly in Indonesia, East Africa, South Africa, and Brazil. For eight regions (including India, the United States, Europe, China, Brazil, East Africa, Indonesia, and South Africa), researchers utilized computer modelling to map the effects on infant mortality, staple crop production, and gross domestic product (GDP). The researchers concluded that infant mortality, staple crop production, and GDP are all important indicators of prosperity. According to the data, aerosols caused the most mortality among infants in India and East Africa. The number of newborn fatalities that were caused by Europe was four times higher than the regional consequences. Due to atmospheric circulation patterns, "European aerosol emissions get widely distributed over many parts of the African subcontinent," Persad said, adding that this is especially concerning because many children in these areas are particularly susceptible to disease as a result of poor air quality. She went on to say that the effects wouldn't be as severe in Europe because there aren't as many babies, and those who are born are healthier.

Northern Africa, West Asia, and South Asia all saw decreases in crop productivity. The United States and China took the biggest hits to their GDP from aerosol emissions, the study found. The study found economic benefits for India. NASA reports that this type of pollution can have a cooling effect since it reflects sunlight into space. Maximum economic output appears to be related to temperature, according to research. Some relief from the heat would be welcome in places like India. Persad said that crops like wheat that are sensitive to temperature changes can benefit from a mild cooling. The United States, Persad said, "may be at the cooler end of the best growth conditions for a certain crop," so any additional cooling would be detrimental.

The study found that the societal cost, an assessment of the economic expenses of greenhouse gases, can be increased by as much as 66% when aerosols are produced alongside carbon dioxide. Co-lead author Jennifer, who holds the Marshall Saunders Chancellor's Endowed Chair in Global Climate Policy and Research at the UC San Diego School of Global Policy and Strategy, said that the study "highlights how the detrimental effects of our emissions are typically underestimated." Persad pointed out that aerosols, which chill the climate, are typically seen as having a silver lining in that they counterbalance carbon dioxide-driven warming. She went on to say that when all the effects were considered, it became clear that no location gained locally or globally from aerosol emissions.

1.8 AEROSOLS AND CLOUDS

Clouds aren't the only thing that weather can be affected by aerosols (https://ethicalunicorn.com/2019/04/29/everything-you-need-to-know-about-aerosols-air-pollution/). Aerosol effects can be more variable than those of greenhouse gases because of how they influence clouds, which spread and have a relatively uniform impact across the Earth. Scientists know that clouds develop when water evaporates and condenses at a high enough temperature, but they also know that aerosols play a role in cloud formation by acting as small "seeds," or condensation nuclei. In non-polluted regions, the most common condensation nuclei are natural aerosols such as sulphates, sea salt, or ammonium salts. However, clouds formed from polluted air typically feature more droplets of a smaller size. By scattering more light and becoming more reflective while shading the ground, these droplets make clouds appear brighter than usual, resulting in net cooling. Scientists have just recently begun to quantify this effect, which is called the cloud albedo effect.

Aerosols and clouds are two fundamental components of the Earth's atmosphere (Cairo et al., 2024), exerting significant influences on climate, weather, and atmospheric chemistry. An aerosol is a suspension of tiny solid or liquid particles in air. Aerosol particles originate from natural sources such as dust storms, volcanic eruptions, and sea spray, as well as anthropogenic activities like industrial emissions and vehicle exhaust. Clouds, on the other hand, consist of water droplets or ice crystals suspended in the atmosphere, forming in response to changes in temperature, humidity, and atmospheric dynamics (Fan et al., 2013). Cloud formation due to clean and Polluted atmospheres is shown in Figure 1.11. Clouds play a key role in Earth's radiation budget, and aerosols serve as the

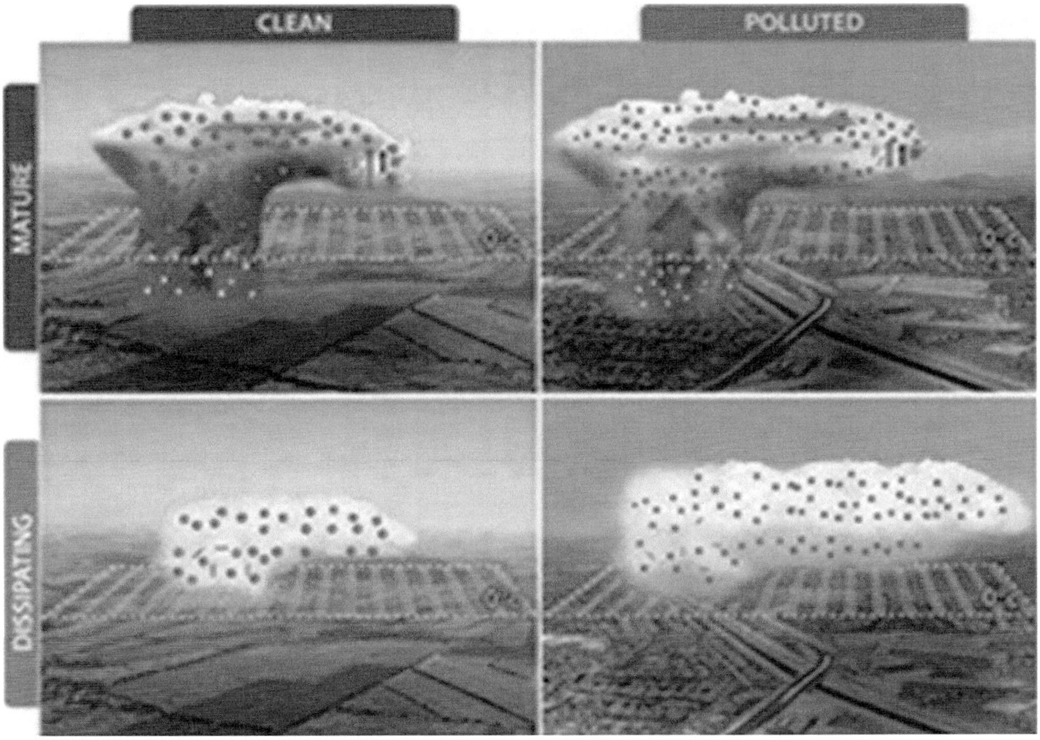

FIGURE 1.11 Schematic illustration of the differences in cloud-top height (CTH), cloud fractions, and cloud thickness for storms in clean and polluted environments. (Cairo et al., 2024. https://doi.org/10.3390/rs16152788. Open access.)

seeds upon which cloud droplets form. Anthropogenic activity has led to an increase in aerosol particle concentrations globally and an increase in those particles that act as cloud condensation nuclei (CCN). Tiny particles in the atmosphere affect the formation of clouds and precipitation.

It is generally accepted that clouds help keep Earth's surface temperature down by increasing the planet's reflectivity and shading 60 percent of the world at any given moment. Therefore, a long-term decline in cloudiness could have significant negative effects, while a 5% increase in cloud reflectivity could offset the whole rise in greenhouse gas emissions from the modern industrial age. When averaged throughout the globe, the cooling caused is less than half of the warming generated by greenhouse gases; unfortunately, the effects don't simply cancel each other out because aerosols aren't dispersed around the earth in the same consistent fashion as greenhouse gases. However, the consequences are much more pronounced at the regional scale, which is why climatologists are putting a premium on learning more about clouds and the feedback cycles that involve them and how they affect regional climate systems. Aerosol types are also crucial to comprehending the relationship between aerosols and clouds. Black carbon can either prevent cloud formation (with potentially catastrophic results) or warm the surrounding atmosphere and cause droplets to evaporate, turning clouds into a haze that suppresses precipitation. Reflective aerosols tend to brighten clouds and make them persist longer.

The role of clouds is the "largest single uncertainty in climate prediction," according to climatologists (https://ethicalunicorn.com/2019/04/29/everything-you-need-to-know-about-aerosols-air-pollution/), but the specifics of aerosol effects are still not well understood because most instruments can't yet measure aerosols within clouds. Only around a third of the models used by the Fourth IPCC examined aerosol effects, and of those, only a small fraction considered sulphate aerosols. As we understand more about aerosols, we can perhaps eliminate the more dangerous ones, like black carbon, from the air.

Clouds play a crucial role in maintaining Earth's weather. In addition to reducing surface temperatures, clouds also reduce atmospheric temperatures by reflecting and refracting sunlight (https://scitechdaily.com/science-made-simple-what-are-clouds-and-aerosols/). They also contribute to global warming by retaining heat and moisture in the atmosphere. Water is carried by clouds, which in turn affect climate, vegetation, erosion, and other aspects of our planet. The complexity of clouds makes them a major source of uncertainty in climate and Earth system models. Understanding the formation of clouds and their effects on the weather and climate is crucial for climate scientists in order to make accurate predictions about the future of our planet. The aerosols in the air are usually the catalyst for cloud formation. Clouds can be composed of ice crystals, water droplets, or both. When the sun heats the Earth, water in the form of liquid evaporates from the oceans, lakes, soil, and plants. Evaporation is the term for this phenomenon. Rising air cools and loses some of its capacity to carry moisture as it travels higher into the atmosphere. Water vapor in the air can condense onto aerosol particles as cloud droplets or ice crystals (https://scitechdaily.com/science-made-simple-what-are-clouds-and-aerosols/) if the air is cold enough and the aerosol particles have the correct shape and size. The temperature of the air determines whether water droplets or ice crystals form in clouds. And not all aerosols are created equal when it comes to seeding clouds. A cloud forms when a large number of cloud droplets or ice crystals form and grow together.

Sea spray, volcanic ash, and dirt and rock dust are all examples of natural aerosols. These atoms come from nature, yet they aren't animals or plants. Other sources of aerosols include the breakdown of organic matter and the combustion of carbon-based compounds. Water droplets can form from sea salt particles, and ice crystals can form from dust particles. Aerosols can also be created in a lab. Black carbon is a form of aerosol that can be found in the atmosphere. Burning fossil fuels like oil, gasoline, and coal, as well as burning wood and other plant material, are the primary sources of black carbon. Researchers have concluded that black carbon is a major factor in our planet's rising climate.

1.8.1 Marine Gases Affect Cloud Formation

The term "radiative force" (RF) refers to the overall shift in the Earth system's energy balance that results from external perturbations. The "direct effects" of aerosols on the global RF occur when they absorb and reflect incoming solar radiation, whereas the "indirect effects" occur when they serve as cloud condensation nuclei (CCN) or ice nuclei. It has been pointed out that one of the major sources of uncertainty in the indirect effects is the unpredictability of natural background aerosol production (Hoose et al., 2009).

Since there are essentially no observations to constrain which model is correct, the "substantial uncertainty about what sources are most important for CCN" is reflected in the simulations of both global climate models (GCMs) and global aerosol transport models that parameterize natural aerosol production and the corresponding budget of CCN. Even the sign of the radiative forcing that occurs on Earth as a result of aerosol indirect effects involving natural marine particles is unknown. Therefore, it is crucial to comprehend and predict how marine aerosols' shifting circumstances over the oceans may modify marine boundary layer (MBL) cloud structures (Wood et al., 2015).

There are countless minute processes, chemical mechanisms, local weather occurrences, and other factors to consider while studying the global climate and its changes. Aerosols, tiny particles dispersed in the air, have had a significant influence on our climate change since the Industrial Revolution (Miller et al., 2021). Most aerosols produced by humans are produced when fossil fuels are burned. However, plants, volcanic eruptions, and chemical processes in the ocean are all natural sources of aerosols.

Aerosols can either warm or cool the atmosphere. Either the Sun's heat is reflected back into space, or it is trapped near the Earth's surface, both of which have direct effects on the climate. To a lesser extent, aerosols can have an indirect impact on climate by altering cloud formation; clouds have a large impact on climate, but the details aren't fully known. Miller et al. (2021) set out to study the effects of different marine aerosols on cloud development and cloud movement at sea. Water vapor in the air condenses into liquid, leading to the formation of clouds. Particles in the air, such as dust or aerosols, become encrusted with water droplets. Water vapor condenses onto tiny particles called cloud condensation nuclei (CCN), which can then alter cloud dynamics like droplet size and concentration. Understanding cloud dynamics is crucial to comprehending climate change since clouds may both cool and warm the atmosphere.

According to Miller et al. (2021), sea salt and sulfur-containing chemicals are the most frequent aerosols in the marine boundary layer, the region of the atmosphere directly above the ocean's surface. Dimethyl sulphide (DMS), a molecule generated by sea algae and phytoplankton, is the sulphur source for these compounds. In particular, the researchers looked into methanesulfonic acid (MSA), a prevalent but poorly known intermediate in the pathway from DMS to sulphate CCN. Researchers focused on how the presence of DMS (and consequently its CCN by-products) altered cloud droplet size and droplet concentration. Twenty research flights were flown over the North Atlantic, not far from the Azores, and their data was utilized. The data showed no association between DMS and the total quantity of cloud particles; however, there was a modest but statistically significant positive correlation between the presence of DMS and the size of cloud droplets.

More observations of marine biogenic gases are required to properly understand their effect on cloud formation, according to the findings of Miller et al. (2021). They point out that DMS and MSA data collection was lacking a complete picture of the complex cloud dynamics that occur over the ocean because of its infrequent study flights. Biogenic gas measurements are valuable in the context of cloud microphysics only when cloud structure is being observed simultaneously; therefore, future investigations should incorporate more consistent monitoring of both biogenic gases at the ocean's surface and cloud microphysical structure.

Cloud formation relies on aerosol particles, and the varied types of aerosols influence how clouds act. To quote Philip Stier, professor of atmospheric physics at Oxford University in the UK (https://ec.europa.eu/research-and-innovation/en/horizon-magazine/black-carbon-and-other-pollut ion-seeds-clouds-were-just-beginning-understand-climate-implications), "every single cloud droplet normally forms on an aerosol particle," because clouds cannot form from pure water under atmospheric conditions. Molecules released by plants, sulphate compounds from volcanoes, soot from automobile tailpipes, and other sources can serve as nuclei for developing cloud droplets. Cloud formation and aerosol science are complicated, yet aerosol features are crucial. One must understand their magnitude, their make-up, and the dynamics at play between them. Floating sea salt, for instance, easily absorbs moisture, while pure black carbon has a tendency to reject it. Once clouds have been seeded, the clouds themselves will exhibit unique characteristics. Professor Stier has stated that "a cloud in a polluted location will generally start from more aerosols and therefore create more droplets" (https://ec.europa.eu/research-and-innovation/en/horizon-magazine/black-carbon-and-other-pollution-seeds-clouds-were-just-beginning-understand-climate-implications). Water and smaller droplets tend to accumulate near the end of a cloud that has formed around tiny man-made particles. The droplets in clouds that form over salt or desert dust are larger than average because they are attracted to the larger particles. "If the air is very clean, then cloud droplets start considerably bigger, and these clouds can rain out quite easily," Prof. Stier elaborated. The question that needs answering, however, is the effect that aerosols have on rainfall at a regional scale. Aerosol pollution causes clouds to seem brighter and fuller with water droplets. Light is reflected, and the air is cooled. It's possible that such clouds have a longer lifespan as well. Clouds and tiny particles of aerosol make it difficult to predict the weather. Scientists continue having trouble understanding clouds on both the microscopic and the macroscopic scales due to their complexity and the difficulty of doing the necessary computations.

1.9 IMPACT OF AIR POLLUTION ON TERRESTRIAL ECOSYSTEMS

The marine and terrestrial ecosystems in the Mediterranean Basin are among the most diverse in the world. There are between 15,000 and 25,000 plant species in the Mediterranean, with 60 percent of those being local or endemic to the area. Worldwide, terrestrial vegetation produces over half of the oxygen we breathe. Plants' photosynthesis, respiration, and transpiration all contribute to the regional atmospheric water cycle. Vegetation can ameliorate the microclimate of metropolitan areas by decreasing the volume of ambient noise and the strength of atmospheric ionization, therefore mitigating the heat island effect. In addition to providing us with food and shelter, plants also serve as a valuable source of renewable energy through recycling. Agriculture has always been an essential element of human life and a significant economic factor in Mediterranean countries due to the region's generally moderate climate and the fertile nature of most soils. Land use shifts, environmental stressors, and especially air pollution and climate change pose threats to the region's vegetation. Although the mechanisms linked with ecosystem resilience are not fully understood, it is known that Mediterranean vegetation is very resilient to perturbations. This is likely attributed to its past evolution under varied pressures and to its enhanced biodiversity.

Multiple pathways are used by air pollutants to influence plant physiology: a) absorption through the leaf stomata or epidermis (Heath, 2008), b) through the root system after air pollutant deposition onto soils, c) adsorption of aerosols onto the leaves, reducing the penetrating light and blocking the opening of stomata, and d) modification of the Photosynthetically Active Radiation that reaches the plant. Vegetation exposed to air pollutants can sustain either short-term, acute damage when exposed to high pollutant concentrations or long-term, chronic damage when exposed to sub-lethal concentrations. Mills et al. (2018) stated that wheat yields globally, owing to O_3 exposure, were expected to drop by 6% to 9% in the south and north hemispheres, respectively, leading to actual total grain losses of around 85 Tg annually.

How air pollution affects plants on land, especially in the Mediterranean, was given by Dulac et al. (2022). Direct effects include phytotoxicity due to stomatal and nonstomatal uptake of ozone, sulphur, nitrogen, and other pollutants by plants, nutrient balance modification by atmospheric deposition, transfer of plant diseases by aerosols, and acid rain. Indirect effects include temperature increase due to greenhouse gases and diffuse radiation due to aerosols, both of which affect vegetation. Kanakidou et al. (2022) investigated these effects and presented data on plant physiological responses and the method of action of deposition when nutrient levels are exceeded.

1.9.1 The Impact of Air Pollution on Terrestrial Managed and Natural Vegetation

The impact of air pollution on terrestrial managed and natural vegetation was investigated by Stevens et al. (2020). Causes and methods of damage date at least to the 1600s, but were not examined systematically until the late 20th century. After the Industrial Revolution, air quality in metropolitan areas in the United Kingdom deteriorated significantly, with highly phytotoxic SO_2 and NO_2 concentrations persisting until the middle of the 20th century. There has been a dramatic improvement in air quality and knowledge of the dangers posed by pollution since then. Pollutants in the air continue to be a problem for both wild and cultivated environments. The effects of air pollution on vegetation are examined using four different types of threats. The consequences of primary emissions of SO_2 and NO_2 in the gas phase, and how they affect lichens in urban settings, are explored. Acidity from deposited sulphur and nitrogen compounds is studied, as is its impact on forest decline due to both wet and dry deposition. The decrease of heathlands in the Netherlands is used as an example of ecosystem eutrophication caused by nitrogen deposition, and the effects of ground-level ozone at phytotoxic amounts on seminatural vegetation are explored. While air quality is improving overall, Stevens et al. (2020) indicate that much more may be done to reduce eutrophication's negative impacts on both cultivated and wild ecosystems.

Polluted air has a deleterious effect on both land and sea ecosystems, leading to a general decline in biodiversity. Vegetation is harmed and biodiversity is lost as a result of ground-level ozone. As of 2020, 59% of the combined forest area in the 32 EEA member nations has already reached levels above which they needed to be protected. Damage from ozone to crops leads to lower harvests. As reported on the EEA's website (www.eea.europa.eu/publications/air-quality-in-europe-2022/impacts-of-air-pollution-on-ecosystems), the percentage of agricultural land exposed to levels above the target value for the protection of vegetation in EEA member countries decreased from 2000 to 2020. However, in 2019, 35 European nations lost a combined EUR 1,418 million due to the effects of ground-level ozone on wheat yields. 75% of the 27 EU Member States' overall ecosystems had nitrogen deposition levels that were harmful in 2020. Since 2005, this decline has been 12 percent. The zero pollution action plan has established a goal of a 25 percent decrease in levels from 2005 by the year 2030. Successful attempts to limit sulphur dioxide emissions in recent decades have resulted in only 4 percent of the European ecosystem in 2020 being acidified at levels above which the deposition of sulphur dioxide, nitrogen oxides, and ammonia drives the acidification of ecosystems.

1.10 CONCLUSION

The particles that make up aerosols can originate either naturally or as a result of human activities. Aerosols and atmospheric particles can diffuse and absorb solar radiation, and directly affect plant photosynthesis and related protein expression. It is highly important to understand the expression profile of plant leaf proteins related to plant photosynthesis, productivity, and health status under different aerosol and atmospheric particle pollution conditions over a long-term period. Plant photosynthesis and the expression of associated proteins are directly impacted by aerosols and air particles, which have the ability to scatter and absorb solar light. The expression profile of plant leaf proteins linked to photosynthesis, productivity, and health status under various aerosol and air particle pollution situations over an extended period of time is crucial. Fine particles that float in the atmosphere are called aerosols. Although many are natural, their numbers haven't changed significantly throughout the ages. However, human-caused aerosols, which are released by automobile exhausts, smokestacks, wildfires, and even clothes dryers, have grown significantly in tandem with the greenhouse gases that drive climate change. In excess, aerosol particles can cause death or irreversible respiratory damage in addition to irritating the lungs. Pollutant deposition from the atmosphere can impact site and soil conditions, which in turn can impact forest tree health. The quantity of aerosol in the troposphere has grown due to human activities such as transportation, industry, agriculture, and power generation, although emissions are currently declining to improve air quality. Both local and upwind aerosol sources have an impact on a specific region because aerosols travel great distances during their brief time in the atmosphere. The ability of trees to absorb pollutants will increase if they are placed close to pollution sources. Although the protection of human health against the risks associated with this type of exposure is not only a legal requirement but also a logical consequence of today's state of knowledge, both the control of exposure to microorganisms and the resultant assurance of safe living and working conditions are not treated with due seriousness.

REFERENCES

Bisht, L., V. Gupta, A. Singh, A.S. Gautam, and Gautam, S. 2022. Heavy metal concentration and its distribution analysis in urban road dust: A case study from most populated city of Indian state of Uttarakhand. *Spatial Spatio-Temporal Epidemiol*. 40: 100470. https://doi.org/10.1016/j.sste.2021.100470

Cairo, F., L. Di Liberto, D. Dionisi., and M. Snels. 2024. Understanding aerosol–cloud interactions through lidar techniques: A review. *Remote Sens*. 16: 2788. https://doi.org/10.3390/rs16152788

Chauhan, A. 2015. Phytotoxic effects of SO_2 on crop plants total chlorophyll content. *Intern. J. Chem. Appl.* 7(1): 51–61.

Cohan, D.S., J. Xu, R. Greenwald, M.H. Bergin, and W.L. Chameides. 2002. Impact of atmospheric aerosol light scattering and absorption on terrestrial net primary productivity. *Global Biogeochem. Cycles*. 16(4): 1090. https://doi.org/10.1029/2001GB001441

Darall., N.M. 1989. The effect of air pollutants on physiological processes in plants. *Plant Cell Environ.* 12: 1–30.

Dulac, F., S. Sauvage, and E. Hamonou (eds.). 2022. *Atmospheric Chemistry in the Mediterranean Region: 2 – From Air Pollutant Sources to Impacts*. Springer.

Fahey, T., M. Maidul Islam, A. Gardi, and R. Sabatini. 2021. Laser beam atmospheric propagation modelling for aerospace LIDAR applications. *Atmosphere*. 12: 918. https:// doi.org/10.3390/atmos12070918

Fan, J., L.R. Leung, D. Rosenfeld, Q. Chen, Z. Li, J. Zhang, and H. Yan. 2013. Microphysical effects determine macrophysical response for aerosol impacts on deep convective clouds. *Proc. Natl. Acad. Sci. USA*. 110: E4581–E4590.

Friedlander, S.K. 2000. *Smoke, Dust, and Haze: Fundamentals of Aerosol Dynamics*. Oxford University Press.

Gautam, S., A.K. Patra, J. Brema, P.V. Raj, K. Raimond, S.S. Abraham, and K.R. Chudugudu,. 2021. Prediction of various sizes of particles in deep opencast copper mine using recurrent neural network: A machine learning approach. *J. Inst. Eng. India Ser. A*. 103: 283–294. https://doi.org/10.1007/s40030-021-00589-y

Gerstl, S., and A. Zardecki. 1982. Effects of aerosols on photosynthesis. *Nature* 300: 436–437. https://doi.org/10.1038/300436a0

Gollakota, A.R.K., S. Gautam, M. Santosh, H.A. Sudan, R. Gandhi, V.S. Jebadurai, and C.M. Shu. 2021. Bioaerosols: Characterization, pathways, sampling strategies, and challenges to geo-environment and health. *Gondwana Res*. 99: 178–203. https://doi.org/10.1016/j.gr.2021.07.003

Griffin, R.J. 2013. The sources and impacts of tropospheric particulate matter. *Nature Education Knowledge*. 4(5): 1. www.nature.com/scitable/knowledge/library/the-sources-and-impacts-of-tropospheric-particulate-102760478/

Gryech, I., Y. Ben-Aboud, B. Guermah, N. Sbihi, M. Ghogho, and A. Khobbane. 2020. More air: A low-cost urban air pollution monitoring system. *Sensors*. 20: 998. https://doi.org/10.3390/s20040998

Heath, R.L. 2008, Modification of the biochemical pathways of plants induced by ozone: What are varied routes to change? Environ. *Pollut*. 155: 453–463. https://doi.org/10.1016/j.envpol.2008.03.010

Hinds, W. 1999. *Aerosol Technology: Properties, Behavior, and Measurement of Airborne Particles*. Wiley.

Hoose, C., J.E. Kristjánsson, T.Iversen, A. Kirkevåg, Ø .Seland, and A. Gettelman . 2009. Constraining cloud droplet number concentration in GCMs suppresses the aerosol indirect effect. *Geophysical Research Letters*, 36: 1–5. https://doi.org/10.1029/2009gl038568

Kaiser, J.C., J. Hendricks, M. Righi, N. Riemer, R.A. Zaveri, S. Metzger, and V. Aquila. 2014. The MESSy aerosol submodel MADE3 (v2.0b): description and a box model test. *Geosci. Model Dev. Discuss.* 7: 691–739.

Kanakidou, M., M. Sfakianaki, and A. Probst. 2022, Impact of air pollution on terrestrial ecosystems. In: Dulac, F., S. Sauvage, and E. Hamonou (eds.) *Atmospheric Chemistry in the Mediterranean: From Air Pollutant Sources to Impacts*. Springer. https://hal.science/hal-03377002/document

Kokhanovsky, A.A. 2016. *Aerosol Optics – Light Absorbtion and Scattering by Particles in the Atmosphere.* John Wiley.

Lee, H.K., I. Khaine, M.J. Kwak, J.H. Jang, T.Y. Lee, J.K. Lee, I.R. Kim,W.I. Kim, K.S. Oh, and S.Y. Woo. 2017. The relationship between SO_2 exposure and plant physiology: A mini review. *Hortic. Environ. Biotechnol.* 58: 523–529. https://doi.org/10.1007/s13580-017-0053-0.

Li, J., B.E. Carlson, Y.L. Yung, D. Lv, J. Hansen, J.E. Penner, H. Liao, V. Ramaswamy, R.A. Kahn, P. Zhang, O. Dubovik, A.A. Lacis, L. Zhang and Y. Dong. 2022. Scattering and absorbing aerosols in the climate system. *Nat. Rev. Earth Environ.* 3: 363–379. https://doi.org/10.1038/s43017-022-00296-7

Li, S., H. Zhang, Z. Wang and Y. Chen. 2023a. Advances in the research on brown carbon aerosols: Its concentrations, radiative forcing, and effects on climate. *Aerosol Air Qual. Res.* 23: 220336.

Li, Y., X. Chen, C. Sonne, S.S. Lam, Y. Yang, N.L. Ma, and W. Peng. 2023b. Reduction and control of air pollution: Based on plant-microbe interactions. *Environ. Poll. Bioavailabil.* 35(1). https://doi.org/10.1080/26395940.2023.2173657.

Lim, E.Y and G.-D. Kim, 2024. Particulate matter-induced emerging health effects associated with oxidative stress and inflammation. *Antioxidants*. 13: 1256. https://doi.org/10.3390/antiox13101256

Mansour, K., M. Rinaldi, J. Preißler, S. Decesari, J. Ovadnevaite, D. cburnis, M. Paglione, M.C. Facchini, and C. O'Dowd, C. Phytoplankton impact on marine cloud microphysical properties over the north-east Atlantic Ocean. *J. Geophys. Res. Atmos.* 127(10): e2021JD036355. https://doi.org/10.1029/2021JD036355

Miller, M.A., M. Zackary, Q. Zheng, L. Trabachino, L.M. Russell, J.E. Shilling and M.A. Zawadowicz, 2021. Observed relationships between cloud droplet effective radius and biogenic gas concentrations in summertime marine stratocumulus over the eastern North Atlantic. *Earth Space Sci.* 9(2): e2021EA001929. https://doi.org/10.1029/2021EA001929

Mills, G., K. Sharps, D. Simpson, H. Pleijel, M. Frei, K. Burkey, L. Emberson, J. Uddling, M. Broberg, Z. Feng, K. Kobayashi and M. Agrawal . 2018. Closing the global ozone yield gap: Quantification and benefits for multitrees tolerance, *Global Chang. Biol.* 24: 48694893.

Mushtaq, Z., M. Sharma, P.Bangotra, A.S. Gautam, and S. Gautam. 2022. Atmospheric aerosols: Some highlights and highlighters, past to recent years. *Aerosol Sci. Eng.* 6: 135–145. https://doi.org/10.1007/s41810-022-00133-w

Sagona, J.A., Secondo, L.E., and Mainelis, G. 2020. Comparison of two models to estimate deposition of fungi and bacteria in the human respiratory tract. *Atmosphere*. 11: 561. https://doi.org/10.3390/atmos11060561

Santacroce, L., I.A. Charitos, A. Ballini, F. Inchingolo, P. Luperto, E. De Nitto, and S. Topi. 2020. The human respiratory system and its microbiome at a glimpse. *Biology*. 9: 318. https://doi.org/10.3390/biology9100318

Seinfeld, J.H., and S.N. Pandis. 1998. *Atmospheric Chemistry and Physics*. John Wiley.

Sharma, P., and P. Saxena. 2022. Impacts and responses of particulate matter pollution on vegetation. In: Sonwani, S., and A. Shukla. (eds.) *Airborne Particulate Matter.* Springer. https://link.springer.com/chapter/10.1007/978-981-16-5387-2_10

Sierra-Vargas, M.P., and L.M. Teran. 2012. Air pollution: Impact and prevention. *Respirology*. 17(7): 1031–1038. https://doi.org/10.1111/j.1440-1843.2012.02213

Stevens, C.J., J.N.B. Bell, P. Brimblecombe, C.M. Clark, N.B. Dise, D. Fowler, G.M. Lovett, and P.A. Wolseley. 2020. The impact of air pollution on terrestrial managed and natural vegetation. *Phil. Trans. R. Soc. A* 378: 20190317. http://dx.doi.org/10.1098/rsta.2019.0317

Trichili, A., A. Ragheb, D. Briantcev, M.A. Esmail, M. Altamimi, I. Ashry, B.S. Ooi, S. Alshebeili, and M.S. Alouini. 2021. Retrofitting FSO systems in existing RF Infrastructure: A non-zero sum game technology. *IEEE Open J. Commun. Soc.* 2: 2597–2615.

Twomey, S. 1987. Pollution and planetary albedo. *Atmospheric Environ.* 8: 1251–1256.

Wood, R., M. Wyant, C.S. Bretherton, J. Rémillard, P. Kollias, J. Fletcher, J. Stemmler, S. de Szoeke, S. Yuter, M. Miller, D. Mechem, G. Tselioudis, J. Chiu, J.A.L . Man, E.J. O'Conner, R.J. Hogan, X. Dong, M. Miller, V. Ghate, A. Jefferson, Q. Min, P. Minnis, R. Palikonda, B. Albrecht, E. Luke, C. Hannay, and Y. Lin. 2015. Clouds, aerosol, and precipitation in the marine boundary layer: An ARM Mobile Facility deployment. *Bull. Amer. Meteorol. Soc.* 96(3): 419–440. https://doi.org/10.1175/BAMS-D-13-00180.1

Yamaguchi, M., and T. Izuta. 2017. Effects of aerosol particles on plants. In: Izuta, T. (ed.) *Air Pollution Impacts on Plants in East Asia.* Springer. https://doi.org/10.1007/978-4-431-56438-6_19

Yan, X., W.Z. Shi, W.J. Zhao, and N.N. Luo. 2014a. Impact of aerosols and atmospheric particles on plant leaf proteins. *Atmospheric Environ.* 88: 115–122. https://doi.org/10.1016/j.atmosenv.2014.01.044

Yan, X., W. Shi, W. Zhao, and N. Luo. 2014b. Estimation of atmospheric dust deposition on plant leaves based on spectral features. *Spectrosc. Lett.* 47: 536–542.

Zuhara, S., and R. Isaifan. 2018. The impact of criteria air pollutants on soil and water: A review. *J. Environ. Sci. Pollution Res.* 4(2): 278–284. https://doi.org/10.30799/jespr.133.18040205

2 Bioaerosols and Human Health

2.1 INTRODUCTION

Humans are constantly surrounded by microorganisms, most of which we don't give much thought to unless they're linked to illness or deterioration (Feazel et al., 2009). Dust was initially discovered to be contagious by Charles Darwin, but airborne pathogens were first studied by Louis Pasteur. Several bioaerosols had already been cultured and identified before Pasteur's investigation. Humans take in between 12,000 and 14,000 liters of air every day while at rest. A person's breathing rate and the volume of air processed by the lungs both increase during physical activity. Thus, the importance of good air quality to human health cannot be overstated. Bacteria, actinomycetes, fungi, pollen, molds, and viruses are only some of the airborne microorganisms that fall under the umbrella term "bioaerosol," which is also known as "organic dust." Enzymes, endotoxins, mycotoxins, peptidoglycans, glucans, and volatile organic compounds (VOCs) are all examples of things that might be described as "organic." Typically, the particles that make up bioaerosols have a diameter of less than 20 microns and are thus invisible to the naked eye. Most molds and yeasts are significantly larger than bacteria, which are in turn much larger than viruses, with the size ranges being roughly: molds/yeasts (1-50 μm), bacteria (0.3-15 μm), and viruses (0.02-0.25 μm) respectively (Mack et al., 2019). Particulate matter (PM), bioaerosols, and reference particles are shown side-by-side for size comparison in Figure 2.1.

Bioaerosols can change in time, place, size, shape, mass, and composition, but they are always there. Because of this, certain birds can stay in the air for a long time while others quickly return to the ground. The concentration of particles in the air depends on a number of parameters, including the ambient temperature, humidity, and wind speed in open air. Exposure to biological agents has become more common in a number of newly developed industrial operations (Douwes et al., 2008).

Wind turbulence over a surface is the typical mechanism by which bioaerosols are dispersed into the atmosphere. Bioaerosols, once released into the atmosphere, can be carried either locally or globally, depending on the circumstances. Bioaerosols are spread locally by winds of a comparable direction and strength, whereas global transport is often facilitated by tropical storms and dust plumes. Bioaerosols are formed when sea spray and bubbles rise to the surface of the water. Bioaerosols have the potential to spread pathogens, endotoxins, and allergens that are harmful to humans. Bioaerosols are the biological constituents of air and include both living and dead microorganisms (such as bacteria, fungi, and viruses) as well as antigenic compounds of biological origin (such as animal and plant debris, endotoxins, mycotoxins, (13)—D-glucan, proteins, and any other microbial metabolites) (Després et al., 2012). Live and dead microbes, as well as their aggregates, fragments, and products

DOI: 10.1201/9781003618164-2

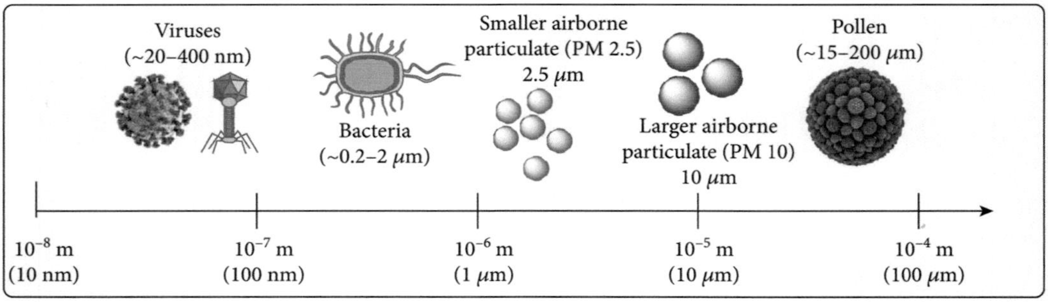

FIGURE 2.1 Relative size chart of common airborne contaminants and pathogens. (Chua et al. (2020), https://doi.org/10.34133/2020/7286735. Creative Commons Attribution License (CC BY 4.0).)

such as bacterial endotoxins, (1-3)-D-glucans, and mycotoxins, make up bioaerosols. All of these infectious organisms can be transported by a variety of different particles. The components and health effects of bioaerosols (Kim et al., 2018) are given in Figure 2.2. Airborne bioaerosols and their impact on human health are illustrated in Figure 2.3. Modification and coupled use of technologies are an essential envisioned need for bioaerosol study – an emerging public health concern (Ghimire et al., 2022).

Any activity, from nature to industry, can result in the aerosolization of microorganisms (or microbes) and their components (Hospodsky et al., 2012) (e.g., human, soil, water). Fungal spores, bacterial endotoxin, fungal mycotoxins, and allergens are all examples of microorganisms that can be found in bioaerosols. In addition to being a major means of disease transmission, bioaerosols can also cause a wide range of adverse health effects by inhalation and possible ingestion. Aerosol scientists place a premium on exposure monitoring because it is difficult to draw a direct line between bioaerosols and health issues (Han et al., 2020; Wei et al., 2017).

The pneumonic plague, smallpox, inhalation anthrax, tuberculosis, influenza, and even chicken pox were all airborne diseases that caused widespread death and destruction throughout human history. More people died from these diseases than from any wars or natural catastrophes combined. Each new epidemic has been more catastrophic than the last because of the exponential growth of cities and the globalization of the human population. Controlling them led to major breakthroughs in the fields of biology, medicine, public health, engineering, and even architecture.

On the flip side, in the last century, contamination control technologies have shifted from people to objects due to the rise and development of bio- and nanotechnology, along with analytical laboratories, the pharmaceutical and medical device industries, the space programme, and semiconductor manufacturing, to name a few. Safe and lucrative operations in these sectors have been hampered for decades by issues related to airborne contaminants and their management. The motto for the creation of clean rooms, improvements in asepsis, and the management of indoor air pollutants became the prevention of aerosol production and the capture of accidentally released viable and non-viable particles. The roles of biological safety officer, sanitarian, industrial hygienist, and contamination control technician were defined, and their expertise was on display. It's all about the air, whether the goal is to keep people or things safe.

The recent rash of food poisoning has made bioaerosols contribute to contamination in our manufacturing and retail facilities. Unresolved are the following questions: Is the potential for contamination raised by the fact that different stages of processing all provide their own unique set of droplets and droplet nuclei? How much does the danger increase if it does? Since we operate in a less-than-sterile setting, we deal mainly with wet processes. Dicing, chopping, pureeing, grinding, mixing, filling, and washing all produce droplets and droplet nuclei as byproducts of the processes themselves. We employ spraying and scrubbing as part of our standard operating procedure for cleaning,

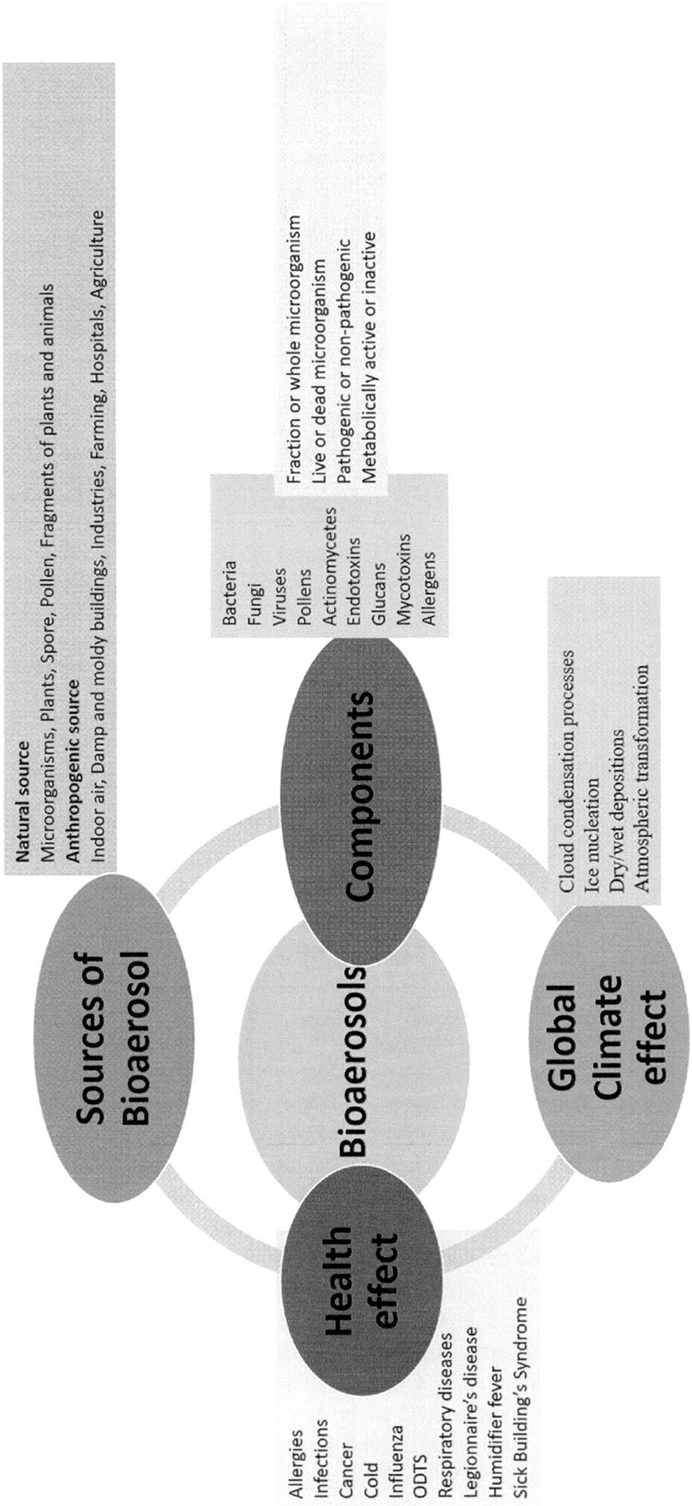

FIGURE 2.2 Bioaerosols and their impact on health and the environment (Ghimire et al. (2022), https://creativecommons.org/licenses/by/4.0/.)

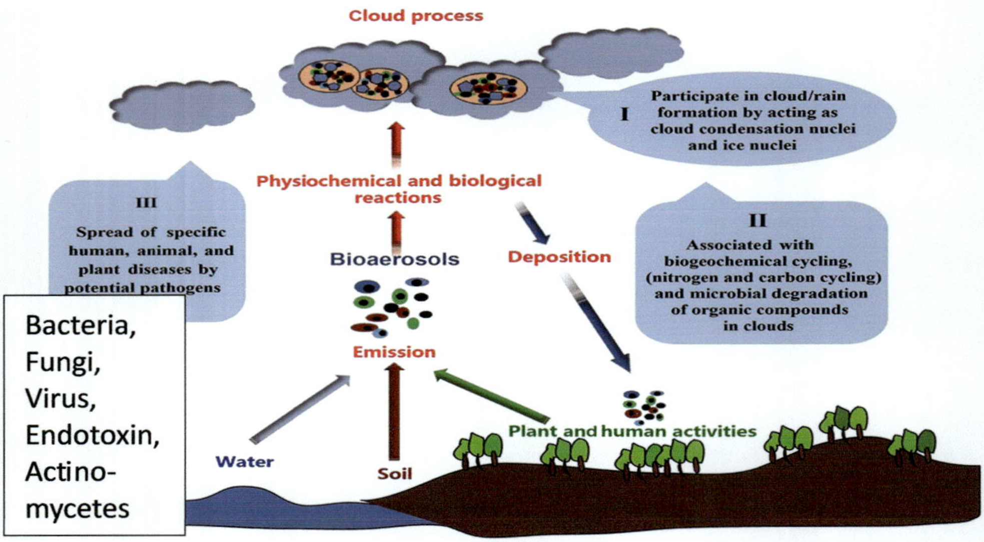

FIGURE 2.3 Bio aerosols are emitted from various terrestrial environments, e.g., Wind, soil, water, plants, animals, and or human beings (Wei et al. (2017), https://creativecommons.org/licenses/by/3.0/)

especially for clean-in-place equipment and terminals. This means a huge number of aerosols will be released.

2.2 HEALTH PROBLEMS CAUSED BY BIOAEROSOLS

Factors such as environmental circumstances that affect the survival of the microorganisms in the air and the behavior of the bioaerosol particles are important in determining whether or not bioaerosols pose a threat to human health. Noninfectious diseases and infectious diseases are the two most common types of illnesses linked to bioaerosol exposure.

2.2.1 Noninfectious Diseases

Hypersensitivity, allergies, and asthma are just a few of the noninfectious disorders linked to bioaerosols (Eduard et al., 2012). Workers who are constantly exposed to airborne microorganisms are at risk of becoming sensitized and eventually developing occupational illnesses. Hypersensitivity pneumonitis continues to be diagnosed in workers exposed to bioaerosols in a variety of settings (Tjalvin et al., 2018). Lipopolysaccharides (LPS) are a class of microbial allergens and endotoxins produced by gram-negative bacteria that pose a serious threat to human health. Inflammation of the airways and intestines, as well as other symptoms including diarrhea, weariness, and nose irritation, have been linked to endotoxins in a number of studies. Mold growth inside buildings is linked to the presence of mycotoxins, which are secondary metabolites produced by fungi and are known to cause cancer, hepatitis, and nephritis. Fungal β- 1,3-glucan, a component of fungal cell walls, has been linked to pulmonary inflammation, heightened vulnerability to endotoxins, and pulmonary embolisms (Mbareche et al., 2019). Changes in airway responsiveness and chest symptoms following indoor exposure have been linked to both β- 1,3 glucan and endotoxins. Secondary metabolites of molds and bacteria, known as microbial volatile organic compounds (MVOCs), have been shown to irritate the eyes and nose when inhaled (Schenkel et al., 2015). In addition, a recent study found that high humidity in the home may increase the risk of asthma and allergies in children are exposed to MVOCs (Choi et al., 2017). The health effects of aerosol exposure are complicated by the biological component of inhalable and respirable particulate matter (PM2.5 [aerodynamic diameter 2.5 micrometers] and PM10 [aerodynamic diameter 10 micrometers]).

2.2.2 Infectious Diseases

Exposure to biological agents (e.g., bacteria, viruses, fungi, and parasites) result in infectious diseases, which can be spread through licking, touching, biting, coughing, sneezing, breathing, and vector-borne transmission. Livestock farms, veterinary clinics, abattoirs, and animal shop employees are commonly blamed for the spread of zoonotic illnesses (such as Q-fever, brucellosis, anthrax, and avian and swine flu). In addition to being a major means of disease transmission, bioaerosols can cause a wide range of adverse health effects by inhalation and even ingestion. Aerosol scientists place a premium on exposure monitoring because it is difficult to draw a causal link between bioaerosols and health issues (Mbareches et al., 2019). Bioaerosol exposure studies are notoriously challenging due to the wide variety of factors that can influence bioaerosol concentration, the different foci of different investigations, and the lack of uniformity in experimental design and methodology. Because of this, the effects of bioaerosol exposure on human health remain largely unknown. Mbareche et al. (2019) briefly described a recent advancement in bioaerosol exposure investigations, which is backed up by research on various relevant topics. Their primary focus is on enhancing the understanding of bioaerosol data through the creation of new tools and study designs, and the suggestion of new considerations for bioaerosol exposure guidelines. The primary findings and observations concern the incompatibility of comparing results from bioaerosol investigations due to the variation in the methods used. In addition, the silo mentality contributes to widening the existing knowledge gap about bioaerosol exposure. Mbareche et al. (2019)'s contribution draws heavily on examples from occupational exposure, yet the issues discussed are universal. The research of Mbareche et al. (2019) provides detailed recommendations for planning a comprehensive bioaerosol study that accounts for all the essential factors in elucidating the true effects of exposure to bioaerosols in the short term. These authors argued that a public database on bioaerosols like the one they proposed would provide useful data for regulating public health.

Most human infections are caused by inhaling or coming into contact with *C. psittaci*-infected aerosols or tissues from infected birds. Inhalation of a mist or vapor carrying *Legionella* bacteria exposes people to the disease. This type of infection can cause anything from a slight fever (Pontiac fever) to a life-threatening lung infection (pneumonia) (Kim et al., 2018). Inhaling air infected with *Mycobacterium tuberculosis* (Mtb) can lead to the development of tuberculosis in the lungs (TB). It was discovered that when TB patients coughed, sneezed, talked, or expelled air or droplets containing *M. tuberculosis*, the bacteria were released from their lungs or larynx into the environment.

A bioaerosol is a group of biological particles that have accumulated in the air. Bioaerosols are microscopic particles or molecules released into the air from living or dead microbes, plants, or animals. These chemicals have the potential to produce a wide range of health problems when exposed, including infectious diseases, allergies, acute toxic effects, respiratory diseases, neurological impacts, and even cancer. Asthma, hay fever, organic dust toxic syndrome, hypersensitivity pneumonitis, and chronic bronchitis are only some of the respiratory symptoms and diseases that can be caused by exposure to noninfectious bioaerosols. Lung function impairment due to bioaerosol exposure has been documented, suggesting a possible link to sick building syndrome (Cai et al., 2019). Interestingly, it has been proposed that early life exposure to moderate amounts of bioaerosols can lessen the chance of developing allergies and allergic asthma.

Surface moisture films foster the growth of bioaerosols. Multiple studies have found that bioaerosols can cause discomfort for building occupants through a variety of pathways, including the emission of irritants, the release of allergen-causing pieces, spores, or living organisms, and the secretion of complex poisons. Several investigations have found an association between the presence of endotoxin in the air and the presence of symptoms in the home. Measurements of airborne bacteria or endotoxin are less reliable as predictors of discomfort than endotoxin concentrations on cooling coils themselves. This indicates that moisture and growth on the coils, rather than in the air or ductwork, have a role in symptom generation, and that a bacterial component, presumably endotoxin, may play a role in symptom generation in combination with other bioaerosols. Waste-related microbiological exposures can occur either indoors, where waste is stored, or outdoors, where it is collected, and may be affected by the sorting, transfer, or cleaning processes. *Aspergillus* spp. is one of the most common mold genera in these kinds of heavily polluted settings. *Aureobasidium pullulans, Epicoccum nigrum, Fusarium sambucinum, Beauveria bassiana, Rhizopus stolonifer, Phoma, Absidia, Mucor, Rhizopus, Penicillium,* and yeasts are just a few of the other genera/species that are also rather common. It's also worth noting that workers may be exposed to the same risks as the general public; however, their exposure and danger may be different due to workers' waste disposal and, consequently, nearer to pollution sources and a more contaminated interior environment.

Bacteria, fungi, and viruses all have a role in the spread of infectious diseases. Any of these, if they go airborne, can spread from person to person. Because they can spread from person to person even at low concentrations, microorganisms like legionellosis, TB, and anthrax pose serious threats to public health. Aerosolization of contaminated water is a potential source of *Legionella pneumophila*, the bacterium responsible for legionellosis (Choi et al., 2017). *Mycobacterium tuberculosis* can be spread from person to person through coughing, sneezing, and even casual conversation. Inhaling spores of the bacterium *Bacillus anthracis* causes the disease anthrax, which has been related to acts of bioterrorism. Aerosols can also spread pathogens like *Chlamydia psittaci* and *Pseudomonas aeruginosa* (Morawska, 2006). These include aspergillosis (*Aspergillus fumigatus*), candidiasis (*Candida albicans),* cryptococcosis (*Cryptococcus neoformans*), mucormycosis (*Rhizopus oryzae*), pneumocystis (*Pneumocystis jirovecii*), and coccidioidomycosis (*Coccidioides immitis*) (Brown et al., 2012). SARS (severe acute respiratory syndrome), respiratory syncytial virus, and enteric viruses are all examples of viruses that can spread quickly through bioaerosols. Viral infections such as rubella and influenza A/B (Bonifait et al., 2015), norovirus, respiratory syncytial virus, hantavirus, varicella-zoster virus, and mumps virus have all been hypothesized to enter humans' gastrointestinal systems via these routes.

2.3 INDOOR BIOAEROSOLS

Bioaerosols are ubiquitous in the environment, but when their concentrations and kinds are contained within acceptable parameters, they rarely cause any difficulties. When breathed in, however, certain bioaerosols can lead to illnesses such as pneumonia, asthma, rhinitis (such as the common cold or hay fever), and respiratory infections (Nazaroff, 2016). Indoor bioaerosols are produced when microorganisms enter a building, colonize an indoor surface, reproduce, and then escape into the surrounding air. The HVAC system (Heating, Ventilation, and Air Conditioning), the doors, windows, and cracks in the walls, the potable drinking water system, and the shoes and clothing of persons working or visiting in the building are all potential entry points for microorganisms. Whether or not microorganisms develop inside depends on factors such as water, humidity, temperature, nutrition sources (such as sheetrock, wood paneling, cellulose ceiling tiles, carpets, upholstered furniture, and fiberglass-lined air ducts), and oxygen. Fungi and bacteria are the most frequent types of microorganisms found in buildings (National Academies of Sciences, Engineering, and Medicine, 2005). Some fungi also create mycotoxins or volatile chemical compounds, in addition to the spores that are dispersed into the air. *Penicillium, Aspergillus*, and *Stachybotrys chartorum* are examples of fungi commonly found indoors that can be harmful to human health. Toxins called endotoxins are made by some bacteria, and volatile organic chemicals are emitted by others. Similar pathways are used by nonmicrobial bioaerosols (such as pollen and cat dander) to enter buildings. These don't breed, but too many of them could cause trouble. There is a wide variety of natural and manmade processes that can produce bioaerosols in enclosed spaces. Food, wood, paper, textiles, and moist building materials are all prime environments for the growth of microbes, which can then spread through the air as either liquid or dry particles. Their little weight and compact dimensions make them easy to transport and allow them to remain airborne for extended periods.

Humans are a major contributor to the spread of airborne germs in non-industrial indoor settings. Specifically, the release of biological particulate matter into the air can be caused by human speech, sneezing, coughing, strolling, washing, and flushing the toilet. Fungal spores can be released into the air from a wide variety of sources, including food, houseplants and flowerpots, dust, dogs and their beddings, textiles, carpets, wood material, and furniture stuffing. Several studies have found that the levels of indoor bioaerosols are greatly affected by factors such as the moisture content of building material, relative humidity and temperature, outside concentrations, air exchange rates, and the number of people and pets in the home. Warmer climates have been found to have higher bioaerosol concentrations than cooler ones. Particles of biological and non-biological origin are constantly being distributed and displaced in the interior air, making it a highly dynamic system (Sadigh et al., 2021). Indoor environmental influences, to which people are exposed for more than 90 percent of their lives (in places like homes, workplaces, and schools), have been linked to a variety of unfavorable health outcomes. Bioaerosols, whether living or nonliving, can float through the air for extended periods. As a result, bioaerosols contribute to 5-34 percent of interior air pollution due to the increased exposure they provide in closed spaces. Some types of microorganisms found in indoor air, such as bacteria, fungi, and viruses, can be harmful to human health. Depending on how they enter the body, bioaerosols can have a wide range of adverse health impacts. In susceptible people, inhaling bacterial bioaerosols can trigger a variety of adverse health effects, including upper and lower respiratory tract inflammation, irritation, and allergy reactions. The relevance of indoor air quality and its effect on students' health is widely recognized in the field of public health.

Bioaerosols can cause a wide range of adverse health effects in humans, from infectious infections to acute toxic effects, allergies, and cancer, depending on how they enter the body (via inhalation, ingestion, or skin absorption). The air we breathe is a major vector for the spread of these germs. Exposure to bioaerosols has a number of negative health effects, including respiratory illnesses and diminished lung function (Sadigh et al., 2021; Ghosh et al., 2015). The danger of exposure to bioaerosols is significantly higher for youngsters and the elderly, since they are given a proportionately larger amount of air relative to their body weight. On average, a person takes in between 12,000 and 14,000 L of air per 24 hours. Because of this, the respiratory system can be protected from more than 99 percent of airborne pathogens. The amount and variety of microorganisms in the air are major factors in respiratory tract infections.

Physical qualities and environmental conditions have been found to be two of the most important determinants of bioaerosols in a number of studies. Environmental elements include things like the moisture content of building materials, the density of the air, relative humidity and temperature, air exchange rates, and human activities, whereas physical features include things like the size, density, and form of droplets or particles. Microbes grow best at optimal conditions of high average temperature and high relative humidity, which makes them an ideal bioaerosol source. Kallawicha and Chao (2022) note that factors such as the strength of sunlight, the size of air currents, the direction of winds, and the speed of winds all play significant roles in the concentration of bioaerosols and their movement from one habitat to another.

2.4 HUMAN HEALTH RISKS FROM BIOAEROSOLS

A multidisciplinary perspective defines bioaerosols as airborne particles that are either alive or derived from life (bacteria, viruses, and fungi). Complex and very varied, bioaerosols might have either a natural or an artificial origin. Mold contamination in indoor environments and the incidence of associated health problems have brought increased focus in recent years on the sampling and monitoring of airborne microorganisms (Kandle et al., 2015). About 5-34 percent of indoor air pollution is caused by bioaerosols. Bioaerosols can contain bacterial cells and cellular fragments, fungal spores, and metabolic byproducts in the form of particulate matter, liquids, or volatile chemical compounds. There are many different types of microorganisms in the air, and they serve as a medium for their spread. Humans are exposed to airborne microbes through three main channels: inhalation, ingestion, and skin contact. Particles in bioaerosols typically range in size from 0.3 to 100 μm; larger particles are less likely to be dispersed in the air and more likely to settle out quickly. Biological aggregates shield their inhabitants from desiccation, UV radiation from the sun, ozone, and other airborne pollutants. Skin cells, dust, and other organic and inorganic particles frequently harbor bacterial cells and virus particles. Fungus spores are discharged from plant surfaces and the soil and float on other particulate matter during agricultural processes like harvesting and tilling. This "rafting" alters the bioaerosol's aerodynamic properties, which in turn impacts the cells' ability to survive. Biological matter is typically enveloped by a thin layer of water when it is dispersed by water sources (such as splash, rainfall, or cooling towers and fountains). The aerodynamic properties are altered, and the microbes' chances of survival are increased because of this moisture layer.

Bioaerosols are defined as airborne particles (Figure 2.4) that are either alive or derived from life (bacteria, viruses, and fungi). They can have either natural or anthropogenic roots and can be found just about anywhere. Mold contamination in interior spaces, the possibility of bioterrorism, and the occurrence of associated health impacts, such as infectious diseases, acute toxic effects, allergies, and cancer, have all drawn attention to the collection and study of airborne microorganisms in recent years. From 5-34% of polluted indoor air is caused by bioaerosols (Stetzenbach, 2007). The bioaerosols themselves are as diverse as the surfaces from which they emerge, whether they are natural or manmade (Smets et al., 2016; Fröhlich-Nowoisky et al., 2016). The furniture and construction materials, as well as the fungal contamination within the wall, ceiling, and floor cavities can all contribute to the spread of mold. Figure 2.4 depicts natural bioaerosol sources and their subsequent transport.

A wide variety of viral and bacterial diseases, including measles and tuberculosis, can infect humans and spread from an infected person to a healthy person via the air. Particles in a bioaerosol can range in size from 0.3 to 100 μm, but only the percentage of particles between 1 and 10 μm that can be inhaled safely is of real significance. Particles larger than 1.0 μm are deposited on surfaces, while bio aerosols in the 1.0–5.0 μm range can float in the air for extended periods. Therefore, the risk of airborne infection is particularly significant in high-risk settings like hospitals and clinics, as well as in confined places like trains and airplanes (Xie et al., 2021). An estimated 1 in 10 hospitalized patients will contract a nosocomial infection (i.e., an infection acquired during hospitalization). Many of these illnesses are spread through direct human contact, but there is growing evidence that others can be spread through the air. It has been estimated that 10% to 20% of all endemic nosocomial illnesses are spread through the air. While the airborne route is known to play a significant role in the transmission of illnesses like tuberculosis (TB), measles, and pulmonary aspergillosis, its overall contribution to the burden of nosocomial infection remains unknown. Since the function of aerosolized microorganisms is not well understood, the significance of airborne microflora in hospitals has been generally overlooked for a considerable amount of time.

2.5 SOURCES OF BIOAEROSOLS: NATURAL AND ANTHROPOGENIC

Both natural and human-made processes contribute. Bioaerosols are airborne particles that originate from living organisms, such as plants, soil, and, most significantly, animals. Cloud droplets are a common location for bioaerosols that have been released into the atmosphere and can travel thousands of kilometers before being deposited. As such, they play a crucial role in the global transfer of living organisms. Bioaerosols, which have an aerodynamic diameter of up to 100

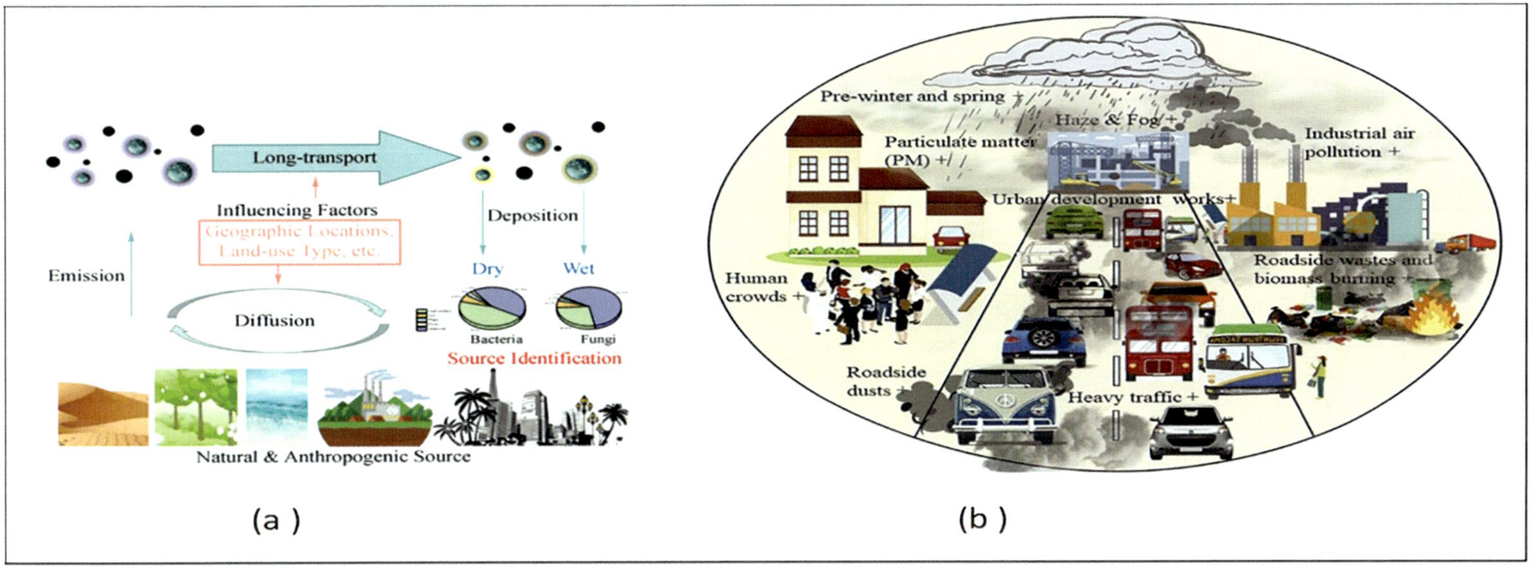

FIGURE 2.4 (a) The source and transport of bioaerosols in the air. Xie et al. (2021): https://doi.org/10.1007/s11783-020-1336-8 Open Access and (b) Shammi et al. (2021), License CC BY 4.0. https://doi.org/10.3389/fenvs.2021.698215.

μm, are solid airborne particles of biological origin suspended in the gaseous medium and may be found all throughout the Earth's atmosphere. There is a wide variety of natural sources that emit bioaerosols into the atmosphere, and these emissions can be roughly divided into two categories: liquid (droplets) and dry (particles). Both natural and artificial processes, such as the built environment, contribute to their formation (Alsved et al., 2020). Humans interact with the built environment, which consists of structures like homes and parks as well as transportation networks, power grids, and other essential services. When microorganisms are released into the environment, they often travel with other substances, such as water, solutes, or inorganic solids, from which they finally settle. After being deposited on a surface, bioaerosols can be resuspended by wind or by opening windows and doors inside a building. Aerosol concentration peaks, particle size distribution changes, and seasonal shifts in biodiversity are all phenomena that can be observed in their natural environments. Microorganisms (bacteria and fungi) that cause respiratory disorders and are important opportunistic pathogens that affect organs like the eyes, skin, and urinary system have been found in studies on natural and anthropogenic bioaerosols.

Natural microorganisms are typically dispersed via vectors such as the ground, water, vegetation, and animals. The majority of bioaerosols on Earth come from soil. Soil's high concentration of organic matter makes it an ideal environment for the proliferation of microbial communities. Compared to forest soils, which are estimated to have 4 × 10^7 cells/g of prokaryotes (bacteria and archaea together), desert and cultivated soils are estimated to contain 2,109 cells/g. It was calculated that between 1.5 × 10^23 and 3.5 × 10^24 cells/year are released from the land and become bacteria. Bio aerosols are released into the air in significant quantities when the wind blows (Xie et al., 2021). Bioaerosols can be released from the soil into the air when it rains. Droplets containing soil microorganisms are spread as rain falls on a soil surface. It is reported that 0.01 of bacteria in soil could enter the atmosphere when a raindrop falls, and the total number of bacteria dispersed by raindrops can range from 1.2 x 10^22 to 8.5 x 10^23 cells per year (Joung et al., 2017).

Outdoor bioaerosols are a heterogeneous mixture of biological particles that can originate from either natural or manmade processes. Soil fragments, microorganisms, and plant particles like pollen come from plants, while animal byproducts and bioaerosols originate in water. Both natural processes and anthropogenic (human-induced) activities can lead to the release of bioaerosols into the atmosphere. They are notorious for their infectious and bio-precipitative properties. Depending on their size and the conditions in which they form, bioaerosols can account for anywhere from 25 to 50 percent of the overall PM10 (particulate matter with aerodynamic diameter 10) mass burden, and they are known to affect things like visibility, climate, and ecosystem health (Krishnamoorthy et al., 2020). It's also worth noting that PM10, a pollutant of non-biological origin, is held to a very severe standard in terms of the "threshold limit" of mass before it's deemed to be dangerous to humans. Rather, the mass burden of bioaerosols is unrelated to the unfavorable health effects associated with human exposure to bioaerosols. As a result, bioaerosols in the air pose the same risks to human health in both dirty and clean areas. Microorganisms (bacteria and fungi) that cause respiratory disorders and are important opportunistic pathogens that affect organs like the eyes, skin, and urinary system have been found in studies on natural and anthropogenic bioaerosols. Studies of workers' exposure have shown that wastewater treatment facilities, animal breeding activities, and slaughterhouses all contain relatively high levels of bacterial bioaerosols. In addition, numerous scientists throughout the world have confirmed the presence of microbial toxins in bioaerosols, which can result in fatal illnesses for humans and animals. Anthropogenic bioaerosols with enhanced allergenic characteristics can induce more severe respiratory diseases through inhalation, and consumption of food contaminated by these aerosols leads to deadly enteric infections through ingestion. Therefore, we need a better knowledge of the bioaerosols and how they contribute to climate change and health problems.

Outside of these operations, studies have demonstrated large quantities of bacterial bioaerosols in the air, with concentrations decreasing with distance. The total number of microorganisms in the air is affected by several factors, including land use changes that include facilities that can emit bioaerosols, like wastewater and sewage treatment plants. Bioaerosols are another name for airborne microbes. To be transported and to remain viable in the environment, pathogenic bioaerosols are dependent on the prevailing physicochemical qualities of the atmosphere. For example, a high temperature is detrimental to the life of airborne bacteria and viruses, especially the latter. Due to increasing global air circulation brought on by climate change, bioaerosols may be transported from one region to another. In turn, airborne microbes can serve as cloud condensation nuclei, which can have far-reaching climate effects. It's concerning that climate change may disrupt the metabolic processes of microbes, leading to new and reemerging diseases. Composting, other agricultural processes, and industrial activities have also been singled out as human-caused sources of bioaerosols (Kim et al., 2018). This explains why bioaerosols can be found anywhere where there are people, whether indoors or outdoors. Visibility is just one manner in which bioaerosols can severely impact human and environmental health.

Bioaerosols are a class of biogenic organic aerosols that consist of very diverse airborne particles or large molecules with sizes ranging from 10 nm to 100 nm. Metabolites, pollen, cell debris, and biofilms are all byproducts of living creatures, although they can also be dead or dormant (like bacteria, viruses, and fungi). A consistent classification of species is not possible, but has to be related to its source, as many potential bioaerosol species, particularly among the metabolites, could also originate from other primary or secondary biogenic sources (for example, mannitol from the *Tamarix gallica* plant, but also common in fungi) (Ariya et al., 2009). Thus, bioaerosol species will add to the total organic aerosol budget, while it is unclear how much of an impact they will have.

2.6 ANTHROPOGENIC SOURCES OF PHTES (POTENTIALLY HARMFUL TRACE ELEMENTS)

Human-made PHTEs (Potentially harmful trace elements) come mostly from high-temperature combustion operations, which cause trace elements to volatilize or be released as very fine aerosols. In the absence of a high-temperature activity, such as in the case of erosion or dust emission, emissions are typically confined to a smaller area (i.e., mining activities). The following sources and activities can be used as broad categories for the events: Combustion, which includes the burning of wood, coal, and oil, is the primary anthropogenic cause of PHTEs emission (As, Cd, Cu, Hg, Ni, Pb, Sb, Se, V, and Zn). Dust is released close to the extraction and exploitation sites in the metallurgical industry, and aerosols rich in trace elements are released during the high-temperature processing of ores (Cd, Cu, Ni, Pb, V, and Zn). What percentage of specific elements is released in the aerosols is determined by the treated ore. High-temperature processing and production are examples of third-tier industrial processes (As, Cr, Cu, Ni, Pb, and Zn). Cd, Cu, Fe, Ni, Pb, and Zn are all found in road traffic; Cu and Sb are lost through eroding brake pads and train rails (Cu). Refuse management: Residential garbage burning (As, Cr, Cu, Ni, Pb, Sb, Se, V, and Zn). Humans are undeniably to blame for the spread of aerosols and PHTEs through anthropogenic activities, but they also bear considerable responsibility for the acceleration of naturally occurring dispersal mechanisms. For example, forest fires, which have been a natural aspect of plant evolution and ecological renewal for more than 400 million years, are today mostly caused and directed by humans. The natural suspension of dust and the spread of PHTEs are both aided by the degradation of semiarid zones and the expansion of agriculture on bare soil. Anthropogenic sources of PHTE contamination mostly come from metal mining and smelting, fossil fuel combustion, and the use of agrochemicals. Atmospheric deposition is also a major source of PHTE contamination in otherwise pristine lakes in remote areas. However, because the spatial heterogeneity of PHTE pollution may have varying effects on the contamination in these lakes, it is necessary to analyze each lake separately.

2.7 SOURCES OF BIOAEROSOLS INSIDE THE HOSPITAL

Hospital-acquired infections and the spread of germs are serious problems that need to be addressed. Patients with infectious diseases are the hospital's most likely source of airborne microorganisms (Kavita et al., 2014). When pathogenic microbes are dispersed into the air from an infected person and land on a vulnerable person, we have airborne transmission. Patients, visitors, and hospital staff all contribute to the spread of germs in the hospital environment. It is known that contamination can come from a variety of outside sources, including personal items, food, and fruit brought in from the outside. Gram-positive cocci bacteria are the most common type of bacteria found in occupied indoor environments. High viable counts are indicators of overcrowding and inadequate ventilation, despite the fact that they pose no direct threat to human health. Kaur and Hans (2007) hypothesized a positive association between bacterial counts and occupant density among hospitals. Airborne germs can also originate from personal items such as clothing and bedding. Suspension of bioaerosols in the air can also be caused by activities such as sweeping floors and changing bed linens. There is evidence from a number of studies showing the spread of airborne microbes varies not only with location but also with seasonal environmental and meteorological parameters such as temperature, humidity, time of day, and wind speed. The quantities of airborne bacteria and fungi in a hospital ward were demonstrated by Augustwoska and Dutkiewicz (2006) to significantly vary both monthly and daily. The rate of airborne bacteria is significantly affected by hospital type, room type, and sampling time. It is widely believed that outdoor air is the single most important contributor to the microflora found within a building. Open windows and doors have been linked in numerous studies to an increase in microbial rates and a more uniform interior air environment.

There is a wide variety of bacteria in the air within hospitals. Multiple healthcare personnel have reported detecting airborne microorganisms in hospitals. *Micrococcus* spp., alpha-hemolytic Streptococci, Diphtheroid bacilli, Gram-negative bacilli, *Bacillus* spp., *Streptomyces* spp., and *Staphylococcus aureus* were identified from several hospital wards by Jaffal et al. (1997). The most common Gram-positive bacteria found in clinical settings were staphylococci and micrococci species (Gaur, 2013). They may be more resilient because of the prevalence of Gram-positive cocci in the air. Indoor air quality is also evaluated for the presence of Gram-negative bacteria, which can have harmful effects due to the endotoxins they produce. Potentially harmful Gram-negative bacteria were most commonly found to be *Enterobacter* spp. and *Pseudomonas* spp (Gaur, 2013). Moist surfaces are a common habitat for these microorganisms. The availability of a restroom likely accounts for their abundance in the area around the sample.

Normal operations can be jeopardized by the presence of bioaerosols in numerous settings, including hospitals. The risk of airborne infection is particularly high in confined spaces because infectious aerosols tend to be very tiny (5 μm) and can remain suspended and viable in the air stream for extended periods. A wide variety of viral and bacterial diseases, including measles and tuberculosis, can infect humans and spread from an infected person to a healthy person via the air. Bioaerosol particles typically range in size from 0.3 to 100 μm, with the respirable size fraction of 1 to 10 μm being the main focus of attention. Bio aerosols between 1.0 and 5.0 μm in size tend to float in the air for extended periods, while bigger particles tend to be deposited on surfaces. This means that places like hospitals and clinics, as well as confined areas like trains and airplanes, have an exceptionally high risk of airborne infection. An estimated 1 in 10 hospitalized patients will contract a nosocomial infection (i.e., an infection acquired in a hospital) during their stay (Kandle et al., 2015). Many of these illnesses are spread through direct human contact, but there is growing evidence that others can be spread through the air. It has been estimated that 10% to 20% of all endemic nosocomial illnesses are spread through the air. While the airborne route is known to play a significant role in the transmission of illnesses like tuberculosis (TB), measles, and pulmonary aspergillosis, its overall contribution to the burden of nosocomial infection remains unknown. Since the function of aerosolized microorganisms is still completely unknown, the significance of airborne microflora in hospitals has been generally overlooked for a considerable amount of time.

In hospitals, the majority of airborne germs are spread within by hospital personnel, patients, and visitors, rather than arriving from the outdoors.

2.8 BIOAEROSOL POLLUTION FROM WASTEWATER

It is well-known that bioaerosols, or aerosols harboring microbial pathogens, contribute to the spread of infection within buildings by being carried by air currents through a building's sewage and plumbing systems. A cross-transmission pathway is established when a flaw in the sanitary plumbing system compromises the system's integrity and allows bioaerosols to escape from the system and enter the building. Odors and microorganisms that are harmful to humans have been traced back to wastewater treatment plants (WWTPs). Numerous studies have shown that during the various phases of wastewater treatment, bioaerosols are produced that include significant numbers of microorganisms, even higher than those found in the wastewater or sludge stream. The four stages of wastewater treatment are as follows: (i) pretreatment, where coarse solids are removed from wastewater as it enters the treatment plants from the sewer systems; (ii) primary treatment, where suspended solids are removed using physical (sedimentation) and chemical methods (coagulation-flocculation); (iii) secondary treatment, where suspended and dissolved solids are removed using anaerobic and aerobic processes; and (iv) tertiary treatment, where dissolved particles are (Lou et al., 2021). Exposure to bioaerosols in wastewater treatment plants (WWTPs) creates occupational health concerns for workers and neighboring people since bioaerosols can include a wide variety of bacterial, viral, and fungal species. The air quality was found to be worse around the raw sewage inlet and the biological reactor (Banchón et al., 2021). Most analyses showed that the air quality in the wastewater treatment plant had a higher microbial load than the control location. According to the findings, the treatment facility suffers from air pollution caused by microorganisms, including pathogens. The produced bioaerosol can have detrimental effects on human health and the environment by contributing to the worsening of atmospheric air quality. Liquid media (such as a WWTP aeration basin, when a toilet is flushed, when a shower is taken, and any time polluted water is stirred by stirring, splashing, etc.) are another common source of bioaerosols.

Every year, almost 1.7 million immunocompromised people in the United States develop a healthcare-associated infection. These infections contribute to higher rates of illness, death, and healthcare expenditures. The production of bioaerosols during patient care is an understudied transmission route. Bioaerosols pose a health risk when they are inhaled or come into contact with the skin (Knowlton et al., 2018). Patients' feces being flushed down the toilet could produce bioaerosols. There has been no research done on the concentration of bioaerosols produced by flushing fecal wastes during patient care. Particle concentrations recorded before and after the flush were found to be significantly different (0.3-10 μm), as reported by Knowlton et al. (2018). Flushing fecal waste results in bioaerosol concentrations that are statistically substantially higher than background values (p-value = 0.005). It's possible that the aerosols produced stick around for longer than 30 minutes after a flush, as there was no significant difference in bioaerosol concentrations with time (p-value = 0.977) or distance (p-value = 0.911) from the toilet. When flushed, toilets release aerosol particles, the majority of which have a diameter of 0.3 μm or less. Past-use bacteria and feces are among the aerosolized particles. Toilet flushing may contribute to the spread of disease-causing germs since bioaerosol concentrations vary depending on environmental factors. Aerosolization in a sewage system occurs normally as a result of the irregular flow of wastewater into the system. The aerosolization process relies heavily on the surge waves produced by these discharges. The toilet is a major source of water pressure fluctuations in a plumbing system for human use.

It's been known for a long time that the aerosolization of biomatter induced by flushing toilets might spread infectious microorganisms. Depending on the toilet's design, water pressure, or flushing energy, it may release a significant number of microbe-containing aerosols when flushed.

The Legionella bacterium, which causes Legionnaire's disease, is just one of many waterborne and waste-borne diseases that can be widely spread through aerosolization (Gleason and Cohn, 2022).

A study by Gormley et al. (2021) confirms that aerosol particles, typically less than 5 μm in diameter, are generated (Figure 2.6) as a result of a toilet flush and characterizes the bioaerosol emissions from sanitary plumbing and sewerage systems into the interior of a building under defective conditions (when U-traps become empty). Aerosolized microorganisms were confirmed by slit-to-agar air sampling. Their research applies to the study of any microorganism, including viruses, because the processes they investigate are universal. It is vital to evaluate room ventilation at times of increased danger, as the number of particles released into a room (a bathroom, for example) by a single 6 L toilet flush can be equal to a person talking loudly in that room for nearly 6 and a half minutes. When investigating vertical transmission within a structure, as Gormley et al. (2021) did, the quantity of particles, their size, and the ease with which aerosolization occurs all imply that the sanitary plumbing system should always be considered as a potential transmission pathway for pathogens.

Over two million people in over 100 countries have contracted the Coronavirus COVID-19 virus recently (Mohammadi et al., 2020). Since there are many potential entry points for infectious illnesses, they have posed a serious danger to human health and the economy in recent decades. Airborne transmission (Usman et al., 2021) is a particularly problematic mode of spread since it is so widespread (Lai et al., 2018; Gormley et al., 2021). Transmission at a distance greater than 1 m is considered airborne, as is transmission through aerosols (particle diameter, 5 μm).

Because of its own stability, aerosols can both house and lengthen the retention period of airborne germs and viruses. The danger of serious infection and subsequent illnesses for humans exposed to air full of inhalable bioaerosols colonized by viruses and bacteria is high (Lou et al., 2021). It has been established that two locations—the restroom and the sewage treatment facility—are major sources of infectious bioaerosols that pose a threat to human health.

In the first study of bioaerosols, Jessen found that flushing the toilet releases germs that then colonies the area immediately surrounding the bowl (Johnson et al., 2013). Disease transmission

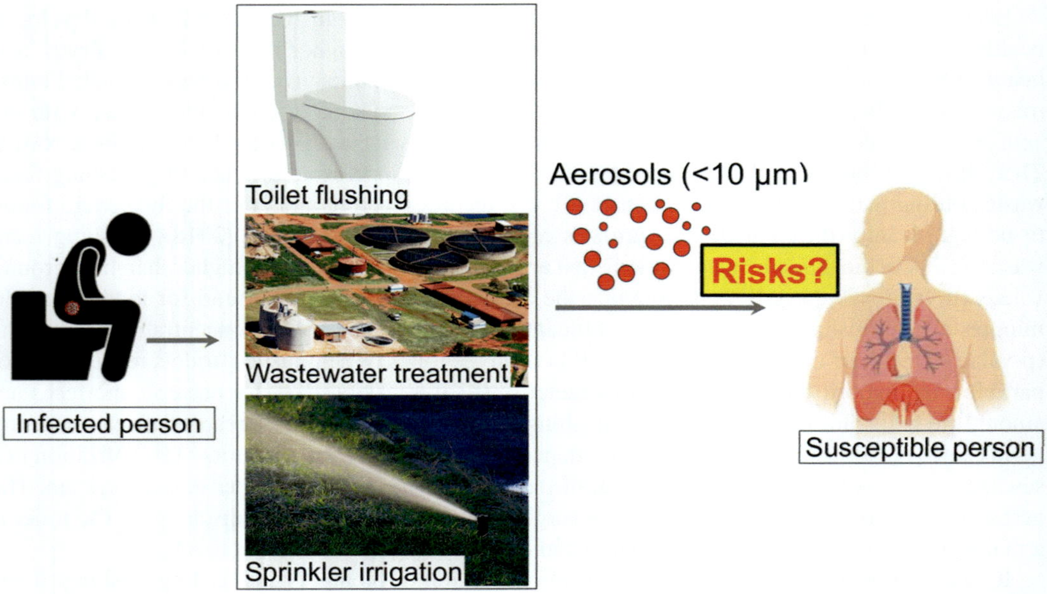

FIGURE 2.5 Risk of Airborne transmission from toilet plume, Bio aerosols (Usman et al. (2021), https://doi.org/10.3390/w13040436. Open access.)

pathways in the built environment and the mechanisms/processes that generate aerosols are depicted in Figure 2.5. As a result of this study, there is now more interest in toilet safety than ever before. Patients' feces are a big contributor to the microbial population in the bathroom. While flushing a toilet can reduce the concentration of bacteria, it also creates a large quantity of bioaerosols due to the turbulence and fluctuation of the water in the bowl. These bioaerosols may raise the risk of infection for those who use the restroom after someone has flushed. There are essentially three stages to the creation and spread of pathogenic bioaerosols through the air. For instance, *Clostridioides difficile* enters the toilet bowl through the emission of feces, which it has previously shed from the gastrointestinal system. Bioaerosols containing *Clostridium difficile* can be created by atomizing water droplets in a toilet's jet stream at high velocities. Depending on the air's upward velocity after being flushed, bioaerosols can be released from the bowl of the toilet and spread through the building via air currents. Some flight attendants on a journey from London to the Philippines became ill with norovirus after using the plane's lavatory (Li et al., 2020).

There is some evidence to suggest that flushing the toilet can disperse bioaerosols that linger in the bathroom after use (Verani et al., 2014). Multiple interactions between liquid and air, flushing techniques, and toilet design all have a role in the formation of bioaerosols during a toilet flush. To be more specific, the flushing process begins when the flushing valve is activated. At that point, some water enters the bowl of the toilet along the wall, while the rest splashes in with a great deal of velocity. In just a few seconds, the syphon effect takes place. Fine droplets or droplet films containing bacteria are formed as a result of the intense turbulence, and are then aerosolized when they burst at the air-liquid interface, becoming suspended in the air. The pressure and weight of a combined liquid likewise rise due to the syphon phenomenon. When liquid splashes on a bowl's side, vortices occur close to the surface (Lou et al., 2021). Inertia drives an upward airflow vortex within the toilet bowl, propelling the vortices upward in a continuous motion. Because of the upward circulation vortex, bioaerosols can be ejected from the bowl and into the air above the toilet, where they can spread more quickly (Li et al., 2020). It was first postulated by Johnson et al. (2013) that flushing the toilet could cause bioaerosols to be released into the air. Before flushing, researchers in that study poured a concentrated solution of *Serratia marcescens* into the toilet, and then they positioned sedimentation plates around the bowl at various distances to collect the resulting bioaerosols. Their findings revealed that bacterial colonies established themselves on the sedimentation plates that had been set down on the floor. This finding provided strong evidence that the little drops of liquid could be transporting germs and other pathogens to the toilet seat. There were also microorganisms found in the air, suggesting that aerosols may contribute to the microbes' prolonged retention time in the air.

Many wastewater treatment plants (WWTPs) have been built in China as a result of the country's growing urbanization in recent years. Wastewater and sludge include a wide variety of microorganisms, some of which are capable of being aerosolized into the air and producing bioaerosols when subjected to the various external forces (such as mechanical aeration, bubble aeration, and sludge dewatering) that are common in WWTPs. Bacteria, fungi, viruses, pollen, and other antigens, as well as the pieces and compounds created by microbes, are all examples of bioaerosols. Particles of this bioaerosol range in size from 1 nm to 102 nm and have widely varying concentrations. Bioaerosol concentrations in WWTPs are typically between 102 and 104 CFU/m^3. Aerator tanks and residual sludge storage yards in WWTPs are common locations for high bioaerosol emission (Szyak-Szydowski et al., 2016). In addition, bioaerosol particles can be carried great distances by wind, posing a risk to the health of not just those immediately in the vicinity but also those further off.

2.9 RESPIRATORY DISEASES

On the one hand, plants are a major contributor to atmospheric bioparticles by releasing pollen, endospores, leaf fragments, and other biological components into the air. However, animals that shed tiny hairs, feathers, or droplets of various body fluids can also be considered biogenic particle sources. When people sweat heavily, cough, or sneeze, they release bioaerosols into the air. Microorganisms of various kinds, including bacteria, fungi, protozoa, viruses, and algae, make up a third class of bioparticles that play a crucial role here (Seltzer, 1995). For decades, scientists have been trying to understand how ambient air pollution affects people's respiratory health. Particulate matter refers to both solid and liquid components of air pollution. Even though particulate matter (PM) might have components from human, geological, and biological origins (particles that originate from microbial, fungal, animal, or plant sources are distinguished from PM as bioaerosols). The lung is the biological system that is most directly exposed to the outside world (Herr et al., 2023). The lung's surface area is roughly 25 times bigger than the skin's; thus, it actually receives more direct exposure to the air's many gases, liquids, and particles than the skin does. Inhaling 12–14 times per minute at an average tidal volume of 500 mL/breath, the average adult takes in around 10,000 L of air every day. On average, a human breathes in 106 airborne germs per day. Pneumonia can be caused by some microbes (such as nontuberculous mycobacteria (NTM), *Legionella*, and *Mycoplasma* species). Nitrogen, oxygen, and hydrogen are the three gases that make up the air we breathe. However, with every inhale comes the introduction of chemical aerosols into the respiratory system. It is estimated that there are 2,000 to 3,000 particles (both aqueous and solid) per milliliter of air, even in comparatively clean conditions. Even though the lungs have several defense mechanisms, the number of particles that actually make it there might worsen or even trigger new respiratory diseases. Inhalable particles are those that are captured by the nose and throat and deposited there, whereas respirable particles are those that make it to the lungs (https://atm.amegroups.com/article/view/10635/html).

Dusty industries (such as coal and granite mining and founding) have significantly higher rates of respiratory diseases than the general population, leading many researchers to suspect that airborne particulate matter (PM) from sources such as household pollutants, wood smoke, bioaerosols, and diesel exhaust may be to blame (Mack et al., 2019). Non-workplace exposure to airborne particles and the development of respiratory and cardiovascular illness became clearer in the latter half of the 20th century.

Various occupational lung diseases can be brought on by exposure to bio dusts and bioaerosols in the workplace. Bioaerosols are a heterogeneous collection of microbes, plant and animal cells, dispersal units (spores and pollen), and allergenic proteins. Bio dusts include comparable biological components that, when released into the air, can be carried by the wind before slowly settling due to the force of gravity. Workers in the agricultural sector, waste management, woodworking, and the food processing industry are all potentially exposed to pathogens (https://dhs.wisconsin.gov/publications/p02188.pdf). Many different types of occupational lung diseases can be caused by exposure to bio dusts and bioaerosols. Both sensitization and exposure to bio dusts and bioaerosols are associated with increased sensitivity and inflammation in the airways. Vegetable dust, plant fibers from cotton, hemp, or flax processing, and wood dusts (especially western red cedar) are also potential culprits.

Hypersensitivity pneumonitis (HP) is an immunologically mediated inflammatory illness of the lung induced by a wide range of organic substances; it is also known as extrinsic allergic alveolitis. Farmers' lung, cheese workers' lung, and grain workers' lung are all caused by bacteria or fungi, while bird breeders' lung and animal handlers' lung are caused by proteins in animal urine and feces. Symptoms of HP include a persistent cough, a pounding headache, and difficulty breathing, and are often misdiagnosed as a viral infection or pneumonia. If left untreated, chronic symptoms might cause lung fibrosis, which is permanent. Organic toxic dust syndrome (OTDS) is a form of inhalation fever characterized by acute yet severe flu-like symptoms that often subside after a few days. Unlike HP, OTDS can appear in people who have never been exposed to the virus. Bioaerosols containing fungi, bacteria, or endotoxin are among the culprits. Bio dusts from cotton, grain, and wood can induce chronic inflammation of the bronchial tree, resulting in coughing and sputum production. Smoking may aggravate symptoms.

Allergies and inflammations are two of the most prominent ways in which bioaerosols contribute to disease development. Toxic responses and malignant alterations of certain target tissues may also be noticed, but rarely. Lung clearance refers to the collection of intrinsic defense mechanisms that the lung has developed to reduce the likelihood of insufficiencies caused by bioaerosols (Amarloei et al., 2020). The upper bronchial and extra-thoracic airways are often clear of accumulated particle aggregates within a few hours. In most cases, it takes many days to weeks for the respiratory tract to clear itself of particles that travel mostly through the lower bronchial and bronchiolar airways. Lastly, particles with a preference for alveolar deposition may be lodged there for a long time, months or even years, depending on their bio-solubility and the likelihood of their uptake by alveolar macrophages. Bioaerosol load from various sources (Qiu et al., 2023) and the factors that affect it are depicted in Figure 2.6.

Particles smaller than 10 μm can enter and settle in the pulmonary region, while particles between 10 and 100 μm in size are inspired but settle in the upper airways and tracheobronchial system. Particles of 20 μm were estimated to take 4 minutes to settle on the ground from a height of 3 m, while particles of 3 μm practically do not settle due to regular resuspension by air currents, leading some to believe that particles smaller than 20 μm are critical for aerosol transmission. Building-dwelling bacteria have an impact on human health from the moment of birth, via aerosol deposition, surface contact, and human-animal interactions. Since bioaerosols are so small, even a minute amount of them can enter the respiratory tract and increase the risk of contracting an infectious disease that is spread through direct contact with a biological agent (such as licking, touching, or biting), indirect contact (such as coughing or sneezing), vector-borne transmission, or airborne transmission. Source, aerosolization, and climatic conditions at the location can deteriorate bioaerosols of varying masses and structures (Jiayu et al., 2019). By binding to specific particles, they can change their aerodynamic and antigenic properties and so penetrate deeper into the lungs, where they can cause symptoms such as a runny nose, itchy eyes and throat, a heightened asthmatic response, and even weariness in those who are exposed. Exposure to aerodynamic diameters less than 2.5 μm over an extended period is associated with an increased risk of developing life-threatening chronic diseases like cancer.

There is a higher risk of respiratory infections among agricultural workers because of the large concentrations of airborne germs they are exposed to. Selecting and using the right respirators is essential for reducing exposures in agricultural settings. Due to the presence of several sources of biological material potentially linked to lung disorders, complex environments are employed to examine the causative impact of bioaerosol agents on respiratory health. Jiayu et al. (2019) and their research team found evidence that humans can be exposed to archaea through the airborne pathway, specifically through bioaerosols collected from swine barns. The immunogenic potential of these archaea and their constituent parts was found to be quite high. Molecular biology is the most effective method for assessing potential exposure to immunogenic archaea and other nonculturable microbes. Recent studies have shown that a sizable percentage of dairy barn employees develop humoral responses against *S. rectivirgula*, the bioaerosol agent responsible for farmers' lung, despite contemporary hay baling and management measures reducing exposure.

2.10 HEALTH EFFECTS OF BIOAEROSOLS

Composting organic waste has many advantages, but some people worry that the bioaerosols produced during the process could be harmful to their health. Bioaerosols can be ingested through hand-to-mouth contact or inhaled directly (Tang et al., 2018). Respirable bioaerosols have a diameter of less than 10 μm, which means they pass unimpeded through the hairs and specialized cells lining the nasal cavity and into the deepest parts of the lungs. This leads to contagious infections,

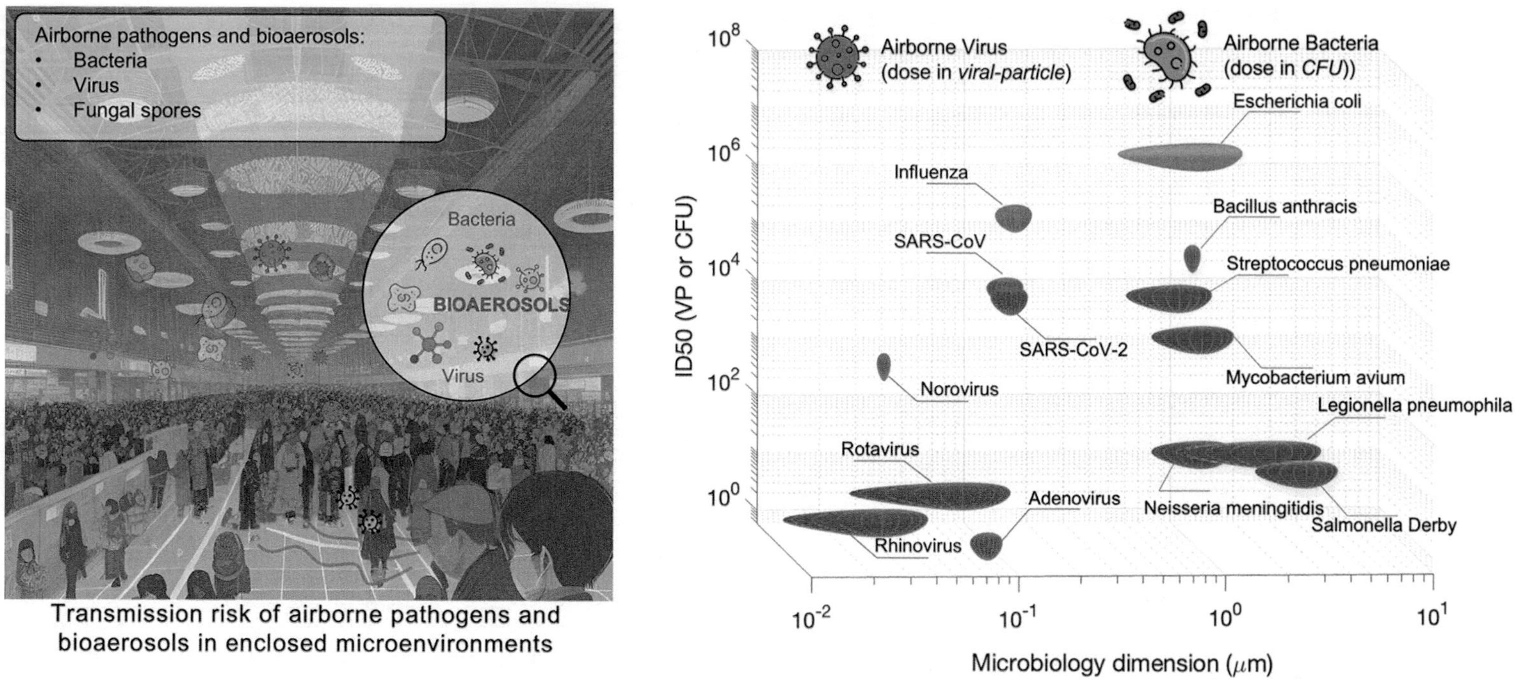

FIGURE 2.6 Airborne pathogens emission and transmission risk in enclosed indoor environments (Qiu et al. (2023), Creative Commons Attribution 3.0 Unported License.)

respiratory issues including asthma and coughs and bronchitis, digestive problems, eye irritation and dermatitis, gastrointestinal distress, acute toxic consequences, allergies, and even cancer. Airborne bacteria and fungi are abundant because they are ubiquitous in the environment. Although concentrations might vary widely, it's recommended that the ambient levels of gram-negative bacteria, total bacteria, and *Aspergillus fumigatus* not exceed 300, 1,000, and 500 CFU/m^3 respectively.

No detailed summary of the evidence regarding bioaerosols in dental and hospital settings is available. Few studies reported the microbial composition of dental clinics. Dental handpieces were found to generate aerosols in the dental settings. Another 30 sources of human activities, interventions, and daily cleaning performances in the hospital also generate aerosols. Fifty-five bacterial species, 45 fungal genera, and ten viruses were identified in a hospital setting, and 16 bacterial and 23 fungal species in the dental environment. None of the studies conducted in dental practice have used methods to identify the presence of viruses in the generated aerosols (Zemouri et al., 2017).

Composting relies on a species of bacteria known as thermophilic actinomycetes, which are spore-forming bacteria found in plant matter and essential for decomposing cellulose and lignin. To the point where thermophilic actinomycetes can thrive, a succession of microorganisms grows in actively composting material as the temperature rises. Therefore, you can expect to find more than 10 million spores per gram of it. Again, found naturally in vegetation and all over the environment, *Aspergillus fumigatus* is a fungus that thrives in compost piles, thanks to its ability to withstand higher temperatures.

Farmers' lung disease and mushroom workers' lung disease are two examples of occupational allergic lung disorders that have been linked to *Aspergillus fumigatus* and other thermophilic actinomycete species due to gross exposure, typically greater than tens of millions of spores per cubic meter of air. Chronic bronchitis, asthma, or alveolitis develops when the immune system is repeatedly triggered by such allergens, leading to the production of histamines, the narrowing of airways, and a decrease in lung capacity. Those who have had significant organ transplantation or who suffer from immune deficiency illnesses are at a higher risk than others because of their weakened immune systems.

2.11 THE RESPIRATORY INFECTIONS OF COVID-19

Diseases caused by the coronavirus family range from the common cold to severe conditions, including Middle East Respiratory Syndrome (MERS) and severe acute respiratory syndrome (SARS). Wuhan City, Hubei Province, China, had an increase in occurrences of respiratory illness, such as pneumonia, in December 2019, and the culprit was determined to be a newly described coronavirus called 2019-nCoV. The coronavirus subfamily Coronavirinae includes SARS-CoV-2. The family Coronavirinae consists of four different genera: C, C, C, and C. Of these, C and C are the most common mammalian hosts for the seven different coronavirus types that cause disease in humans. The "common cold" is caused by the a-coronaviruses 229E and NL63 and the b-coronaviruses OC43 and HKU1 between 10% and 30% of the time. Since SARS-appearance CoV-2 in 2002, another extremely virulent coronavirus, Middle Eastern Respiratory Virus (MERS-CoV), surfaced in 2012. The SARS-CoV virus likely originated in horseshoe bats and spread to humans. The agent that eventually adapted to efficient human-to-human transmission likely originated in an animal-to-human interspecies transmission around 2002–2003. In November 2002, the coronavirus that causes severe acute respiratory syndrome (SARS-CoV) was identified for the first time in humans.

Multiple SARS-CoV-2 superspreading occurrences and the recent COVID-19 pandemic point to the importance of aerosols in spreading the coronavirus. Droplets and aerosols containing the virus are expelled from the mouth and nose during normal breathing, talking, coughing, and sneezing. They are the primary vectors for the spread of respiratory illnesses like COVID-19. The fate of these droplets is determined by inertia, gravity, and evaporation, as described by Bourouiba (2020). The contamination of nearby surfaces is caused by drops that are too big to evaporate before they settle. Droplets below this size have a higher evaporation rate than settling rate, creating droplet nuclei that can float in the air for hours and be carried great distances. Flow-induced particle dispersion and sedimentation, as well as air-mucous contact, liquid sheet fragmentation, turbulent jets, droplet evaporation and deposition, and so on, all play a role in the transmission process. As a result, the physics of flows is pivotal in the spread of COVID-19. It is well-established that the fluid lining of the respiratory tract gives rise to respiratory droplets.

Numerous studies have sought to quantify the number density, velocity, and size distributions of droplets expelled by expiratory events because of the consequences they have for transmission (Asadi et al., 2019). Physicians use everything from hand washing and masks to fogging equipment, ventilation, and social distancing to combat the spread of disease (Mittal et al., 2020). Small droplets suspended in air currents are related to the airborne transmission channel. While the vapor-rich, buoyant, turbulent expiratory flow can delay evaporation, the majority of these droplets evaporate within a few seconds producing droplet nuclei (Bourouiba, 2020). Given that the SARS-CoV-2 virus has a half-life of around an hour in the human body, each droplet nucleus potentially contains numerous virions. The final step in airborne transmission is the host's inhalation of virus-infected particles or droplets, which deposits the virus in the host's respiratory mucosa. The mucus layer provides some protection against virus penetration and subsequent infection; hence, inhalation of infected droplets does not always result in infection. Mucus's physical characteristics are crucial to droplet production in the respiratory system. To reduce infection rates, it is possible to temporarily alter the physical features of the mucus lining through material delivery, to increase mucus stability.

Although 5 μm has traditionally been used to differentiate aerosols from droplets, 100 μm should be used instead because it represents the largest particle size that can remain suspended in still air for more than 5 s from a height of 1.5 m, typically reach a distance of 1 to 2 m from the emitter (Figure 2.7) (depending on the velocity of airflow carrying the aerosols), and can be inhaled (Wang et al., 2021; https://en.wikipedia.org/wiki/Airborne_transmission).

The Residence time of aerosols of varying size in still air can be estimated from Stokes' law for spherical particles (Figure 2.8) (Wang et al., 2021). Airborne transmission of respiratory viruses and their details are given by Wang et al. (2021).

Studies have demonstrated that viruses are abundant in small aerosols (< 5 μm), suggesting that they may be present in aerosols produced by an infected person. Temperature, relative humidity, UV light, airflow, and ventilation are only a few of the environmental elements that might influence the spread of virus-laden aerosols. Aerosols containing viruses can lodge in many locations in the respiratory system after being breathed. Although smaller particles can also be deposited in the upper airway, they are more likely to reach the alveolar portion of the lungs. There is strong and unmistakable evidence for airborne transmission, as shown by animal experiments and airflow simulations, the frequency of indoor superspreading events of SARS-CoV-2, , and the observed transmission of SARS-CoV-2 despite the use of masks and eye protection (Smets et al., 2016). However, droplet transmission is much more common when people are within 0.2 meters of each other, making fomite transmission of SARS-CoV-2 far less efficient. People with the virus can expel both aerosols and droplets during coughing and sneezing, but the latter quickly settle to the ground or surfaces, leaving behind a greater concentration of aerosols.

Hands are the most common vector for spreading infections from infected surfaces contaminated with respiratory droplets carrying viruses, making frequent hand washing with soap the best defense against this risk. One side of a soap molecule is hydrophilic and polar, attracting water, whereas the other is hydrophobic, attracting oils and lipids. Therefore, the lipid content of the substance clinging to the hand is emulsified into the bulk fluid during hand washing, and the material is subsequently washed away. It is, therefore, more crucial than ever to understand the physics behind face masks' efficacy as a protection against airborne viruses. Face masks offer "inward" protection by preventing

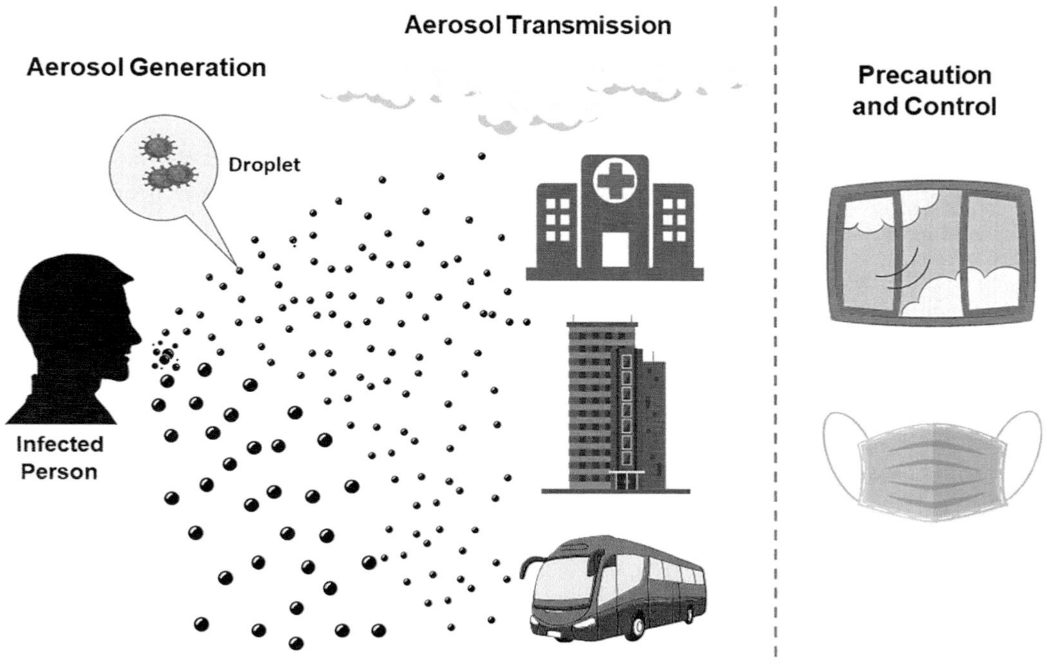

FIGURE 2.7 Infected people generate larger droplets and aerosols, which can infect over longer distances (https://en.wikipedia.org/wiki/Airborne_transmission.)

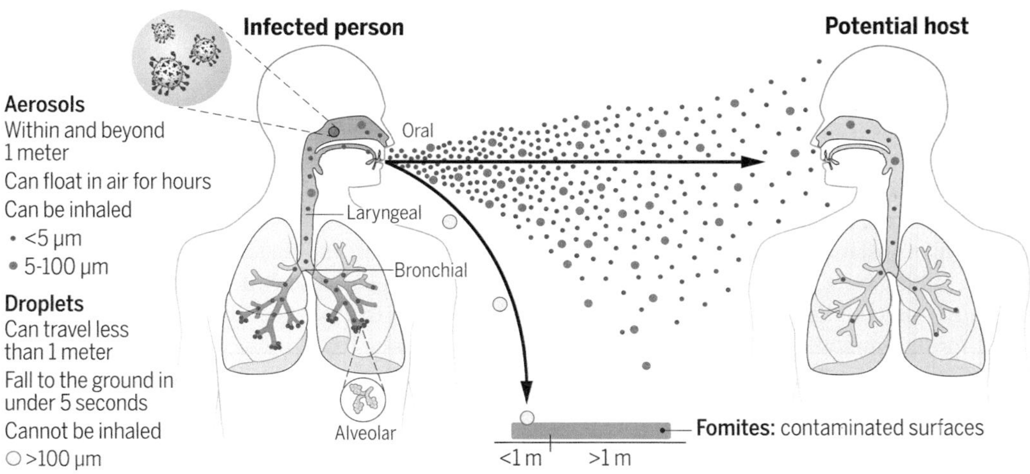

FIGURE 2.8 Airborne transmission of respiratory Viruses and Represents the Residence time of aerosols of varying size in still air can be estimated from Stokes' law for spherical particles (Wang et al. (2021). Creative Commons Attribution License 4.0 (CC BY).)

inhalation of aerosolized particles contaminated with viruses. Because of diffusion, inertial impaction, interception, and electrostatic attraction, the face-mask material effectively entraps droplets and particles. If using face masks becomes commonplace after the spread of COVID-19, it will be crucial that the filters be designed with this delicate balancing act in mind.

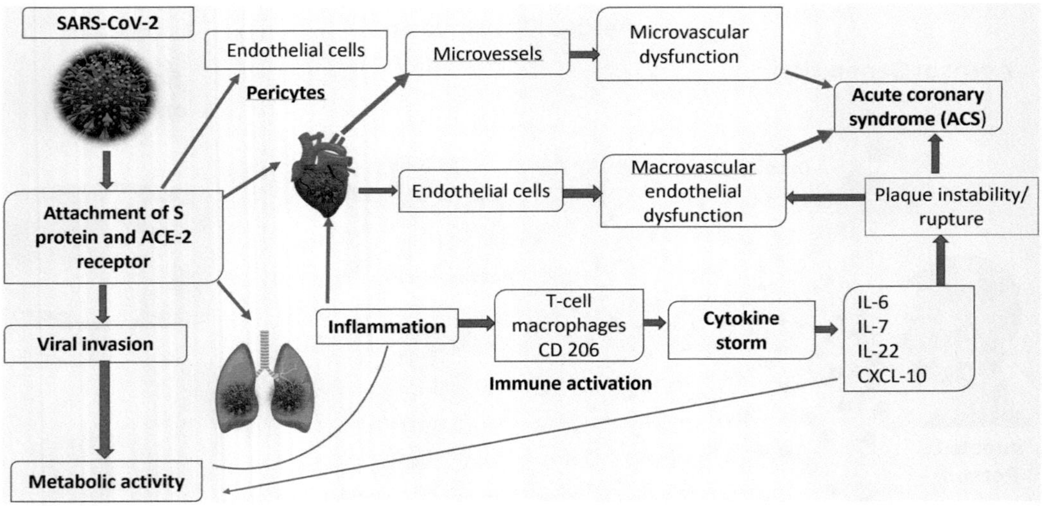

FIGURE 2.9 Effects of SARS-CoV-2 on the heart and lungs, which result in acute coronary syndrome (ACS) (Chowdhury et al. (2021). doi.org/10.22207/JPAM.15.2.14, https://creativecommons.org/licenses/by/4.0/.)

The COVID-19 pandemic has highlighted major knowledge gaps in areas such as the transmission channels of respiratory (https://utswmed.org/medblog/lifecycle-of-a-coronavirus/) infections and the techniques used to mitigate their spread (Figure 2.9; Chowdhury et al., 2021). Droplet generation in the lungs, two-phase flows during exhalation, the movement of evaporated droplets, and the aerodynamics of a face mask are only a few of the topics covered by Mittal et al. (2020) and Oliveira et al. (2021).

2.12 COVID-19'S IMPACT ON HEALTH

Nearly 4.9 million individuals had lost their lives to the pandemic by mid-October 2021, according to sources. The coronavirus causes an illness that most commonly affects the upper respiratory system, but can also manifest itself on the skin. The skin and mucous membranes are the first contact points for the coronavirus, making it particularly contagious. Numerous studies are looking into the skin manifestations of coronavirus, which might include rashes, blisters, irritation, dermatitis, and other conditions. Excessive use of soaps and disinfectants can also irritate the skin. The use of PPE might potentially aggravate existing skin and mucosal irritation. Mild respiratory infections are the most common outcome of exposure to any of the four endemic human CoVs (HCoV-NL63, HCoV-229E, HCoV-OC43, and HCoV-HKU1). Not only can severe pneumonia be caused by these four endemic HCoV, but also by the two epidemic CoV, severe acute respiratory syndrome (SARS-CoV) and Middle East respiratory syndrome (MERS-CoV) (www.mayoclinic.org/diseases-conditions/coronavirus/in-depth/coronavirus-long-term-effects/art-20490351).

Post-COVID-19 syndrome is characterized by the following most-reported symptoms: Fatigue, signs and symptoms that worsen with exertion, symptoms related to the lungs (the respiratory system) such as a high temperature, a cough, and/or shortness of breath. Some other symptoms could be: Cognitive or concentration issues, insomnia, vertigo on standing, numbness or tingling in the extremities, a loss of smell or taste, depression, or anxiety are all examples of neurological and mental health issues. Tension in the muscles or tendons, conditions, or symptoms related to the heart can manifest in a variety of ways. Gastrointestinal distress, such as nausea, bloating, and gas, circulatory problems, such as a clot in the deep veins of the legs that travels to the lungs and limits blood flow to the lungs, and other vascular problems (pulmonary embolism), rash, and menstrual cycle abnormalities are other symptoms.

Chronic health issues due to COVID-19 are as follows. Organ failure might be a factor. Organ damage affecting the heart, kidneys, skin, and brain has been reported in people with severe COVID-19 sickness (Fröhlich-Nowoisky et al., 2016). It's also possible to experience inflammation and immune system issues. The duration of these effects is unknown. The impacts may also contribute to the onset of preexisting illnesses like diabetes, heart disease, and neurological disorders. Neurological manifestations include brain fog, the inability to think clearly or focus on a task at hand, headache, sleep disorders, vertigo upon standing (lightheadedness), nervous anticipation, subtle shifts in flavor or aroma, sadness, or nervousness.

The SARS-CoV-2 virus can target many systems simultaneously, wreaking havoc on the lungs, heart, neurological system, kidneys, liver, and more. After receiving intensive care unit therapy, patients may experience mental health issues due to things like grief and loss, unresolved pain or weariness, or post-traumatic stress disorder (PTSD) (ICU). Shortness of breath, palpitations, and rapid heartbeat are all possible side effects of COVID-19, and one study found that 60% of people who recovered from the virus showed signs of ongoing heart inflammation (www.hopkinsmedicine.org/health/conditions-and-diseases/coronavirus/covid-long-haulers-long-term-effects-of-covid19). Even people who just experienced a moderate episode of COVID-19 and who were otherwise healthy showed signs of inflammation. More than 110 million cases of SARS-CoV-2 infection and over 2.1 million fatalities have been documented across the 38 OECD countries as of mid-October 2021.

As a result of COVID-19, there was a rise in diabetes. COVID-19 has a complicated role in the development of diabetes, especially type 2. Some people who recover from a severe case of COVID-19 appear to acquire symptoms of type 2 diabetes later in life, which is consistent with the fact that type 2 diabetes is a risk factor for severe cases of COVID-19. Whether or not children who have received COVID-19 are more prone than adults to develop long-term symptoms is unknown at this time. However, COVID-19 can have a lasting effect on children, manifesting itself in symptoms including exhaustion, headaches, school difficulties, emotional issues, shortness of breath, and more. Isolation, economic disadvantage, lack of access, and altered daily routines further complicate the complexities of long-term COVID-19 care, and the fact that millions more people have had COVID-19 than SARS or MERS only adds to the magnitude of the potential problem of lingering health problems in the context of the pandemic (Singh et al., 2021). More than 11 million people in India had contracted COVID-19 as of March 11, 2021, with an estimated 158,000 deaths as a result. People in rural areas and underserved urban populations had more trouble gaining access to healthcare, saw their diabetes or hypertension symptoms deteriorate, and lost large amounts of money and jobs as a result.

Some of the short-term and long-term lung difficulties brought on by the new coronavirus are explained by Johns Hopkins' Panagis Galiasatos (www.hopkinsmedicine.org/health/ conditions-and-diseases/coronavirus/what-coronavirus-does-to-the-lungs). To what extent does chronic obstructive pulmonary disease affect lung function? Pneumonia and the life-threatening acute respiratory distress syndrome (ARDS) are just two of the lung consequences that can be brought on by COVID-19. Another potential complication of COVID-19 is sepsis, which can permanently damage the lungs and other organs. More severe cases of bronchitis and other diseases of the airways may be caused by newer coronavirus strains, which may necessitate hospitalization in some. In severe cases of COVID-19, a significant pro-inflammatory condition can lead to various important diseases, problems, and syndromes, as Galiatsatos explained, based on what we've learned about SARS-CoV-2 and the subsequent COVID-19.

Pneumonia due to COVID-19: Pneumonia causes trouble breathing because fluid and inflammation build up in the lungs. When people have trouble breathing, they may need to be hospitalized for treatment with oxygen or a ventilator. COVID-19 pneumonia typically spreads to both lungs.

Fluid builds up in the air sacs of the lungs, reducing their capacity to take in oxygen and leading to symptoms like shortness of breath and coughing. While most patients who develop pneumonia make a full recovery, the pneumonia caused by COVID-19 can be quite severe. Damage to the lungs can linger long after the sickness has cleared up, making it difficult to breathe for a while.

Bronchitis due to CMV: According to Galiatsatos et al. (2005), the problem with COVID-19-related bronchitis is that too much sputum is being produced in the airways, which causes coughing and chest congestion. Sputum also causes airway constriction, which makes it harder to breathe. He says, "Patients may develop a cough that remains with them for months after the initial infection" due to bronchitis. A person's quality of life may be diminished due to their chronic cough and chest congestion.

Shock Due to Acute Respiratory Failure (ARDS): In advanced cases of COVID-19 pneumonia, fluid might seep from the lungs' tiny blood vessels and fill more of the air sacs. Acute respiratory distress syndrome (ARDS), a kind of lung failure, can develop from chronic shortness of breath. Patients with ARDS often need ventilator support since they are unable to breathe on their own. Deadly ARDS can happen anywhere, at home or in the hospital. Those who recover from ARDS and COVID-19 may be left with permanent lung scarring.

Sepsis is a potential consequence for patients with a severe case of COVID-19. When an infection enters the bloodstream and travels through the body, it is said to have entered a state known as sepsis. According to Galiatsatos, "the lungs, the heart, and the other body systems all function together like instruments in an orchestra." The breakdown of interorgan communication is a hallmark of sepsis. The lungs and the heart, among other organs, can shut down one-by-one. Even if a person survives sepsis, their lungs and other organs may be permanently damaged. Galiatsatos said (Galiatsatos et al., 2005), the immune system is actively attempting to combat the invader when someone has COVID-19. Because of this, a superinfection involving both COVID-19 and another pathogen (bacteria or virus) is possible. An additional infection may cause more lung damage. Galiatsatos added that it is remarkable that around one in four individuals with severe COVID-19 also have a superinfection, which means that these patients would take longer to recover.

After a severe case of COVID-19, Galiatsatos said (Galiatsatos et al., 2005), a patient's lungs can heal, but it won't happen overnight. Lung injury recovery is a lengthy process. First, there's the lung damage, and then there's the scarring. The infected person's lungs will eventually recover, but it may take anywhere from three months to a year or more (www.hopkinsmedicine.org/ health/conditions-and-diseases/coronavirus/what-coronavirus-does-to-the-lungs) for lung function to return to pre-COVID-19 levels. Symptoms may arise simply from the process of the lungs recovering. It's like getting a cast off after being immobilized for a few months because your leg broke. After having a leg bone set, no one would anticipate being able to start running again immediately. Patients will feel pain as their legs mend and muscular tissue is rebuilt. Yes, this is what our lungs have to endure. Both doctors and patients, Galiatsatos emphasizes, need to be ready for long-term care. Repeated administrations of the COVID-19 vaccine, including booster shots, are essential not only for protection against the virus itself but also, maybe, for preventing irreversible lung damage. COVID-19 survivors should keep in touch with their doctors so they may get the most out of the time they have while their lungs are mending. When the pandemic is gone, there will be a new population of people who require medical attention. Medical professionals, including doctors and respiratory therapists, must assist these patients in regaining as much lung function as possible.

Consequences of COVIDs through Time, in Relation to Other Respiratory Viruses: According to Baskett et al. (2023), COVID-19 at the time of the index encounter revealed to be a significant predictor of diagnoses related to hair loss, palpitations, chest discomfort, dyspnea, obesity, and joint pain between around 30 and 365 days after the index encounter, in comparison to the generic VRIs. The only symptoms that were substantially higher in the VRI group compared to the COVID-19 group or the control group compared to the COVID-19 group, were dyspnea and chest discomfort.

Patients infected with COVID-19 were significantly more likely to experience pulmonary embolism, various respiratory failures, hypoxemia, oxygen dependency, and pneumonia than patients infected with VRI or SARS-CoV-2.

Long-term neurological effects were not shown to be strongly linked to COVID-19 in this study. Although there was no statistically significant difference between the categories, patients with COVID-19 and generic VRI appeared to have much greater rates of anosmia. Notably, the researchers discovered no strong link between COVID-19 and emotional distress. When compared to a control group, those with SARS-CoV-2 infection had significantly higher rates of cognitive impairment compared to those with VRI. Heart palpitations were found to be specifically linked to COVID-19. Compared to the overall population of VRI patients, COVID-19 patients had a higher prevalence of tachycardia, heart failure, unexplained anemia, hyperlipidemia, and hypertension; however, this difference was not statistically significant. Particularly associated with the first COVID-19 were fatigue and joint aches. Type 1 diabetes was diagnosed in a higher proportion of COVID-19 patients than VRI patients, although there was no statistically significant association between COVID-19 patients and controls.

2.13 CONCLUSION

Human diseases, including infectious/respiratory symptoms and cancer, have been linked to the presence of bioaerosols. Due to their pervasiveness in nature, bioaerosols can be discovered almost anywhere. Because of a variety of internal sources typically connected with human activities, indoor bioaerosol concentrations may be even higher than outdoor values. As a result, improving indoor air quality can be greatly aided by increasing the pace of mechanical or natural ventilation and engaging in routine cleaning and repair. Microbial aerosol inhalation can cause respiratory illnesses, allergic reactions, infections, and inflammation, among other harmful health effects. Infectious bacteria, fungi, and viruses can settle in the respiratory tract host cells after inhalation. Some spread and infect other tissues, including the gastrointestinal tract. A gathering of biological material in the air is called a bioaerosol. Bacterial cells and cell fragments, fungal spores and hyphae, viruses, and byproducts of microbial metabolism can all be found in bioaerosols. Health problems like infection, allergic reactions, inflammation, and respiratory diseases can be brought on by bioaerosols. High-temperature techniques are useful for eliminating bioaerosols, but they need higher temperatures, use more energy, and are ineffective in reducing the emissions of allergens like spores.

REFERENCES

Alsved, M., A. Matamis, R. Bohlin, M. Richter, P.E. Bengtsson, C. J. Fraenkel, and P. J. Medstrande. 2020. Exhaled respiratory particles during singing and talking. *Aero. Sci. Technol.* 54(11): 1245–1248. https://doi.org/10.1080/02786826.2020.1812502.

Amarloei, A., M. Fazlzadeh, A.J. Jafari, A. Zarei, and S. Mazloomi. 2020. Particulate matters and bio aerosols during middle east dust storms events in Ilam, Iran. *Microchem. J.* 152: 104280. https://doi.org/10.1016/j.microc.2019.104280

Ariya P.A., J. Sun, N.A. Eltouny, E.D. Hudson, C.T. Hayes, and G. Kos. 2009. Physical and chemical characterization of bio aerosols – Implications for nucleation processes. *Intern. Rev. Physical Chem.* 28 (1): 1–32. https://doi.org/10.1080/01442350802597438

Asadi, S., A.S. Wexler, C.D. Cappa, S. Barreda, N.M. Bouvier and W.D. Ristenpart. 2019. Aerosol emission and super emission during human speech increase with voice loudness. *Sci. Rep.* 9 (1): 2348.

Augustowska, M., and J. Dutkiewicz. 2006. Variability of airborne microflora in a hospital ward with in a period of one year. *Ann. Agric. Environ. Med.* 13: 99–106. www.airqualitydirect.com/bio-aerosols.htm

Banchón, C., T. Vivas, A. Aveiga, and L. Diaz. 2021. Airborne bacteria from wastewater treatment and their antibiotic resistance: A meta-analysis. *J. Ecol. Eng.* 22 (10): 205–214. https://doi.org/10.12911/22998993/142207

Baskett, W.I., A.I. Qureshi, D. Shyu, J.M. Armer, and C.-R. Shyu. 2023. COVID-specific long-term sequelae in comparison to common viral respiratory infections: An Analysis of 17 487 infected adult patients. *Open Forum Infect. Dis.* 10 (1): ofac683. https://doi.org/10.1093/ofid/ofac683

Bourouiba, L. 2020. Turbulent gas clouds and respiratory pathogen emissions: Potential implications for reducing transmission of COVID-19. *J. Amer. Med. Assoc.* 323 (18): 1837–1838. doi:10.1001/jama.2020.4756

Bonifait, L., R. Charlebois, A. Vimont, N. Turgeon, M. Veilette, Y. Longtin, J. Jean, and C. Duchaine. 2015. Detection and quantification of airborne Norovirus during outbreaks in healthcare facilities. *Clin. Infect. Dis.* 61 (3): 299–304. https://doi.org/10.1093/cid/civ321

Brown, G.D., D.W. Denning, N.A.R. Gow., S.M. Levitz, M.G. Letia, M.G., and T.C. White. 2012. Hidden killers: Human fungal infections. *Sci. Transl. Med.* 4 (165): 165rv13. https://doi.org/10.1126/scitranslmed.3004404

Cai, J., Q. Ren, F. Chen, L. Chen, J. Wu, Z. Wu, L. Chen, R. Liu, and G. Zhgang. 2019. Microbiology community structure in bio aerosols and the respiratory diseases. *J. Environ. Sci. Public Health* 3 (3): 347–357. https://doi.org/10.26502/jesph.96120068

Choi, H., N. Schmidbauer, and C.G. Bornehag. 2017. Volatile organic compounds of possible microbial origin and their risks on childhood asthma and allergies within damp homes. *Environ. Int.* 98: 143–151. https://doi.org/10.1016/j.envint.2016.10.028

Chowdhury, M.R., A.M. Mas-ud, M.R. Ali, M. Fatamatuzzohora, A. Sultana Shimu, M. Anamul Haq, M. Ashikul Islam, M. Firose Hossain, M. Hosenuzzaman, M. Mominul Islam, M. Faruk Hasan, and M.N. Matin. 2021. Harmful effects of COVID-19 on major human body organs: A review. *J. Pure Appl. Microbiol.* 15 (2): 500–511. https://doi.org/10.22207/JPAM.15.2.14

Chua, M.H., W. Cheng, S.S. Goh, J. Kong, B. Li, J.Y.C. Lim, L. Lu Mao, S. Wang, K. Sue, L. Yang, E. Ye, K. Zhang, W.C.D. Cheong, B.H. Tan, Z. Li, H.T. Ban, B.H., and X.J., Xu. 2020. Face masks in the new COVID-19 Normal: Materials, testing, and perspectives. *Research* ID 7286735: 40. https://doi.org/10.34133/2020/7286735

Després, V.R., J.A. Huffman., S.M. Burrows, C. Hoose, A.S. Safatov, G. Buryak, J. Fröhlich-Nowoisky, W. Elbert, M.O. Andreae, U. Pöschal, and R. Jaenicke. 2012. Primary biological aerosol particles in the atmosphere: A review. *Tellus* B64: 1. https://doi.org/10.3402/tellusb.v64i0.15598

Douwes, J., W. Eduard, and P.S. Thorne. 2008. Bioaerosols. *International Encyclopedia of Public Health*, 287–297. https://doi.org/10.1016/b978-012373960-5.00281-1

Eduard, W., D. Heederik, C. Duchaine and B.J. Green. 2012. Bio aerosol exposure assessment in the workplace: The past, present and recent advances. *J. Environ. Monit.* 14 (2): 334–339. https://doi.org/10.1039/c2em10717a

Feazel, L.M., L.K. Baumgartner, K.L. Peterson, D.N. Frank, J.K. Harris, and N.R. Pace. 2009. Opportunistic pathogens enriched in showerhead biofilms. *Proc. Natl. Acad. Sci.* USA, 106: 16393–16399.

Fröhlich-Nowoisky, J., C.J. Kampf, B. Weber, J.A. Huffman, C. Pöhlker, M.O. Andreae, N. LangYona, S.M. Burrows, S.S. Gunthe, W. Elbert, H. Su, P. Hoor, E. Thines, T. Hoffmann, V.R. Després, and U. Pöschl. 2016. Bioaerosols in the Earth system: Climate, health, and ecosystem interactions. *Atmospheric Res.* 182: 346–376.

Galiatsatos, P., I. Shirer, E. Lamoureux and A. Szilagyi A. 2005. Meta-analysis of outcome of cytomegalovirus colitis in immunocompetent hosts. *Dig. Dis. Sci.* 50: 609–616. https://doi.org/10.1007/s10620-005-2544-6

Gaur, J. 2013. *Study of microbial air quality in public places of Jodhpur, India.* PhD Thesis, J.N. Vyas University, Jodhpur, Rajasthan, India.

Ghimire, P.S., L. Tripathee, and S. Kang. 2022. Modification and coupled use of technologies are an essential envisioned need for bio aerosol study – An emerging public health concern. *Fundamental Res.* 2 (2): 218–221. https://doi.org/10.1016/j.fmre.2021.10.012

Ghosh, B., H. Lal, and A. Srivastava. 2015. Review of bio aerosols in indoor environment with special reference to sampling, analysis and control mechanisms. *Environ. Int.* 85: 254–272. https://doi.org/10.1016/j.envint.2015.09.018

Gleason, J.A. and P.D. Cohn. 2022. A review of legionnaires' disease and public water systems – Scientific considerations, uncertainties and recommendations. *Int J Hyg Environ Health.* 240: 113906. https://doi.org/10.1016/j.ijheh.2021.113906

Gormley, M., T.J. Aspray, and D.A. Kelly. 2021. Aerosol and bioaerosol particle size and dynamics from defective sanitary plumbing systems. *Indoor Air* 31: 1427–1440.

Han, Y., T. Yang, Xu. Yan, L. Li, and X. Liu. 2020. Effect of aeration mode on aerosol characteristics from the same wastewater treatment plant. *Water Res* 170: 11532.

Herr, C.E., A. Zur Nieden, A. Jankofsky, M., N.I. Stalinskis, R.H. Boedeker, and T.F. Eikmann. 2023. Effects of bioaerosol polluted outdoor air on airways of residents: A cross-sectional study. *Occup. Environ. Med.* 60: 336–342. doi: 10.1136/oem.60.5.336

Hospodsky, D., J. Qian, W.W. Nazaroff, N. Yamamoto, K. Bibby, H. Rismani-Yazdi, and J. Peccia. 2012. Human occupancy as a source of indoor airborne bacteria. *PLoS ONE* 7 (4): e34867. https://doi.org/10.1371/journal.pone.0034867

Jaffal, A.A., H. Nsanze, A. Bener, A.S. Ameen, I.M. Banat, and A.A. El Mogheth. 1997. Hospital airborne microbial pollution in a desert country. *Environ. Intern.* 23: 167–172.

Jiayu, C., R. Qiaoqiao, C. Feilong, L. Chen, W. Jiguo, W. Zhendong, C. Lingyun, R. Liu, and Z. Guoxia. 2019. Microbiology community structure in bioaerosols and the respiratory diseases. *J. Environ. Sci. Public Health* 3: 347–357.

Johnson, D., R. Lynch, C. Marshall, K. Mead and D. Hirst. 2013. Aerosol generation by modern flush toilets. *Aerosol. Sci. Technol.* 47: 1047–1057. https://doi.org/10.1080/02786826.2013.814911

Joung, Y., Z. Ge, and C. Buie. 2017. Bioaerosol generation by raindrops on soil. *Nat. Commun.* 8: 14668. https://doi.org/10.1038/ncomms14668

Kallawicha, K., and H.J. Chao. 2022. Bioaerosols: An unavoidable indoor air pollutant that deteriorates indoor air quality. In: Saini, J., M. Dutta, G. Marques, and M.N. Halgamuge (eds.) *Integrating IoT and AI for Indoor Air Quality Assessment.* Internet of Things. Springer. https://doi.org/10.1007/978-3-030-96486-3_3.

Kandle, K., S. Kandle, R. Kandle, and V. Jahagirdar. 2015. Bioaerosols: Role in healthcare institutions. *The Indian Practitioner.* 68 (10).

Kaur, N., and C. Hans. 2007. Air bacterial isolations from operation theaters in a tertiary care hospital in India. *J. Clin. Diagn. Res.* 2: 87–89.

Kavita, N., G. Jyoti, and C. Ritika. 2014. Bioaerosols in healthcare settings – A brief review, Int. *J. Geol. Earth Environ. Sci.* 4 (3): 59–64.

Kim, Ki-H., E. Kabir, and S. A. Jahan. 2018. Airborne bioaerosols and their impact on human health, *J. Environ. Sci.* 67, 23–35.

Krishnamurthy, S., M. Akila, and P. Hema. 2020. On distinguishing the natural and human-induced sources of airborne pathogenic viable bio aerosols: Characteristic assessment using advanced molecular analysis. SN *Appl. Sci.* 2: 1162. https://doi.org/10.1007/s42452-020-2965-z

Knowlton, S.D., C.L. Boles, and E.N. Perencevich. 2018. CDC Epicenters Program. Bioaerosol concentrations generated from toilet flushing in a hospital-based patient care setting. *Antimicrob. Resist. Infect. Control.* 26 (7): 16. https://doi.org/10.1186/s13756-018-0301-9

Lai, A.C.K., T.-F. Tan, W.-S. Li, and D.K.M. Ip. 2018. Emission strength of airborne pathogens during toilet flushing. *Indoor Air* 28: 73–79. https://doi.org/10.1111/ina.12406

Li, Y-Y., J.-X. Wang, and X. Chen. 2020. Can a toilet promote virus transmission? From a fluid dynamics perspective. *Phys. Fluids*. 32: 065107. https://doi.org/10.1063/5.0013318

Lou, M., S. Liu, C. Gu, H. Hu, and Z. Tang. 2021. The bio aerosols emitted from toilet and wastewater treatment plant: A literature review. *Environ. Sci. Pollut. Res. Int.* 28 (3): 2509–2521. https://doi.org/10.1007/s11356-020-11297-8

Mack, S.M., A.K. Madl, and K.E. Pinkerton. 2019. Respiratory health effects of exposure to ambient particulate matter and bioaerosols. *Compr Physiol.* 10 (1): 1–20. https://doi.org/10.1002/cphy.c180040

Mbareche, H., L. Morawska, and C. Duchaine. 2019. On the interpretation of bioaerosol exposure measurements and impacts on health. *J. Air Waste Manage. Assoc.* 69 (7): 789–804. https://doi.org/10.1080/10962247.2019.1587552

Mittal, R., R. Ni and J.-H. Seo. 2020. The flow physics of COVID-19. *J. Fluid Mech.* 893: F2.

Mohammadi, M., M. Meskini, and A.L. do Nascimento Pinto. 2020. Novel coronavirus (COVID-19) overview. *J. Public Health*. https://doi.org/10.1007/s10389-020-01258-3

Morawska, L. 2006. Droplet fate in indoor environments, or can we prevent the spread of infection? *Indoor Air* 16 (5): 335–347. https://doi.org/10.1111/j.1600-0668.2006.00432. x

National Academies of Sciences, Engineering, and Medicine. 2005. *Sensor Systems for Biological Agent Attacks: Protecting Buildings and Military Bases.* The National Academies Press. https://doi.org/10.17226/11207

Nazaroff, W.W. 2016, Indoor bioaerosol dynamics. *Indoor Air.* 26 (1): 61–78. https://doi.org/10.1111/ina.12174

Oliveira M.T.N. de, F.S.I. Figueiredo, da Silva, L.C.M., K.S. da Silva, L.R. Bueno, B.B. da Luz, C.R. Corso, M.F. de Paula Werner, E.S. Fernandes and D. Maria-Ferreira. 2020. Tissue proteases and immune responses: Influencing factors of COVID-19 severity and mortality. *Pathogens* 9: 817. https://doi.org/10.3390/pathogens9100817

Qiu, G., X.C. Zhang, A.J. DeMello, M. Yao, J. Cao, and J. Wang. 2023. On-site airborne pathogen detection for infection risk mitigation. *Chem. Soc. Rev.* 52: 8531–8579. https://doi.org/10.1039/D3CS00417A

Sadigh, A., E. Fataei, M. Arzanloo, and A.A. Imani. 2021, Bacteria bioaerosol in the indoor air of educational microenvironments: Measuring exposures and assessing health effects. *J. Environ. Health Sci. Eng.* 19 (2): 1635–1642. https://doi.org/10.1007/s40201-021-00719-5

Schenkel, D., M.C. Lemfack, P. Birgit, and R. Splivallo. 2015. A meta-analysis approach for assessing the diversity and specificity of belowground root and microbial volatiles, *Front. Plant Sci.* 6: 707. https://doi.org/10.3389/fpls.2015.00707

Seltzer, J.M. 1995. Biologic contaminants. *Occup. Med.* 10: 1–25.

Shammi, M., M.M. Rahman, and S.M. Tareq. 2021. Distribution of bio aerosols in association with particulate matter: A review on emerging public health threat in Asian megacities. *Front. Environ. Sci.* 9: 698215. https://doi.org/10.3389/fenvs.2021.698215

Singh, K., D. Kondal, S. Mohan, S., S. Jagannathan, M. Deepa, N.S. Venkatesh Murthy, P. Jarhyan, R.M. Anjana, K.M.V. Narayan, V. Mohan, N. Tandon, M.K. Ali, D. Prabhakaran and K. Eggleston. 2021. Health, psychosocial, and economic impacts of the COVID-19 pandemic on people with chronic conditions in India: A mixed methods study. *BMC Public Health* 21: 685. https://doi.org/10.1186/s12889-021-10708-w

Smets, W., S. Moretti, S. Denys, and S. Lebeer. 2016. Airborne bacteria in the atmosphere: Presence, purpose, and potential. *Atmospher. Environ.* 139: 214–221.

Stetzenbach, L.D. 2007. Introduction to aerobiology. In: C.J. Hurst, R.L. Crawford, J.L. Garland, D.A. Lipson, A.L. Mills, and L.D. Stetzenbach (eds.) *Manual of Environmental Microbiology.* Wiley. https://doi.org/10.1128/9781555815882.ch73

Szyłak-Szydłowski, M., A. Kulig and E. Miaśkiewicz-Pęska. 2016. Seasonal changes in the concentrations of airborne bacteria emitted from a large wastewater treatment plant *Intern. Bi deteriorat. Biodegrad.* 115: 11–16.

Tang, K., Z. Huang, and J. Huang. 2018. Characterization of atmospheric bio aerosols along the transport pathway of Asian dust during the Dust-Bioaerosol 2016 Campaign. *Atmosphere. Chem. Phys.* 18 (10): 7131–7148.

Tjalvin, G., Ø. Svanes, R.J. Bertelsen, B.E. Hollund, T.B. Aasen, C. Svanes, and J. Kirkeleit. 2018. Hypersensitivity pneumonitis in fish processing workers diagnosed by inhalation challenge. *ERJ Open Res.* 4 (4): 00071–2018. https://doi.org/10.1183/23120541.00071-2018

Verani, M., R. Bigazzi, and A. Carducci. 2014, Viral contamination of aerosol and surfaces through toilet use in health care and other settings. *Am. J. Infect. Control.* 42: 758–762. https://doi.org/10.1016/j.ajic.2014.03.026

Usman, M., M. Farooq, M. Farooq, and I. Anastopoulos. 2021. Exposure to SARS-CoV-2 in Aerosolized wastewater: Toilet flushing, wastewater treatment, and sprinkler irrigation. *Water* 13: 436. https://doi.org/10.3390/w13040436

Wang, C.C., K.A. Prather, J. Sznitman, J. Jimenez, S.S. Lakdawala, Z. Tufekci, and L.C. Marr. 2021. Airborne transmission of respiratory viruses. *Science* 373: eabd9149. https://doi.org/10.1126/science.abd9149

Wei, M., C. Xu, J. Chen, C. Zhu, J. Li, and G. Lv. 2017. Characteristics of bacterial community in cloud water at Mt Tai: Similarity and disparity under polluted and non-polluted cloud episodes. *Atmos. Chem. Phys.* 17: 5253–5270. https://doi.org/10.5194/acp-17-5253-2017

Xie, W., Y. Li, W. Bai, J. Hou, T. Ma, X. Zeng, L. Zhang, and T. An. 2021. The source and transport of bio aerosols in the air: A review. *Front. Environ. Sci. Eng.* 15 (3): 44. https://doi.org/10.1007/s11783-020-1336-8

Zemouri, C., H. de Soet, W. Crielaard, and A. Laheij. 2017. A scoping review on bio-aerosols in healthcare and the dental environment. *PLoS ONE* 12 (5): e0178007. https://doi.org/10.1371/journal.pone.0178007

3 Impacts of Black Carbon on the Environment and Health

3.1 INTRODUCTION

A major constituent of soot, black carbon is the most solar energy-absorbing component of particulate matter and can absorb one million times more energy than CO. — Columbia Climate School.

Black carbon, also known as soot, is the second-leading cause of global warming after carbon dioxide (CO_2). Black carbon may account for as much as half of Arctic warming, and on top of that, it makes people sick; it is a leading cause of respiratory illness and premature death. A by-product of inefficient combustion, black carbon is emitted from diesel engines, forest fires, and residential cooking and heating (Shrestha et al., 2010). In many places around the world, black carbon from inefficient indoor cookstoves causes millions of premature deaths every year. Through the distribution and use of existing technology, black carbon emissions from cookstoves and other sources can be significantly reduced, having a positive impact on public health and the climate. Black carbon is emitted from gas and diesel engines, coal-fired power plants, and other sources that burn fossil fuel (https://bezerocarbon.com/insights/the-black-carbon-black-box-introduction-to-black-carbon-and-short-lived-climate-pollutants). It comprises a significant portion of particulate matter or PM, which is an air pollutant. Black carbon absorbs solar energy, warming the atmosphere. When it falls to earth with precipitation, it darkens the surface of snow and ice, reducing their albedo (the reflecting power of a surface), warming the snow, and hastening melting (de la Sota et al., 2019). Black carbon is released as particulate matter, which has a diameter of less than 2.5 micrometers, also known as PM2.5. It is one of several short-lived climate pollutants (SLCPs) emitted from low-efficiency cooking technologies. Black carbon persists from a few days to a couple of weeks (consequently termed "short-lived"). In contrast, GHGs can remain in the atmosphere for up to hundreds of years. Black carbon emissions (Seinfeld, 2008) can therefore have more immediate effects and, as such, are reported to contribute to up to 35% of global climate forcing, whereby elements on the earth's surface or in the atmosphere change the global energy flux. As the atmospheric carbon dioxide (CO_2) levels continue to increase from pre-industrial concentrations of 280 ppm (parts per million volume) to an average value of 385 ppm (Shrestha et al., 2010), BC can influence the environment at local, regional, and global scales in different ways. Maximizing BC formation (as charcoal and char) during combustion processes can decrease CO_2 emission on one hand, whereas on the other hand, it can increase atmospheric pollution with excess soot BC yield. The global dimming effect is another consequence of aerosol BC emissions. The emission and transport of BC in the environment are also

DOI: 10.1201/9781003618164-3

associated with the fate of other organic substances, particularly polycyclic aromatic hydrocarbons (PAHs), whose isomers can be utilized to trace the sources of BC in soils and sediments.

Black carbon emissions have become a focus of attention due to their effects on the near-term warming of the atmosphere and on human health. Reducing black carbon emissions is of particular interest in polar regions, such as the Arctic, which are especially sensitive to the effects of black carbon. When suspended in the air, black carbon turns solar radiation into heat, consequently contributing to air warming, regional cloud formation, and precipitation patterns. When black carbon particles settle on snow and ice, they darken the surface, reducing their albedo and enhancing absorption of solar radiation, thus indirectly increasing the rate of melting (www.canada.ca/en/environment-climate-change/services/air-pollution/publications/black-carbon-inventory-emissions-2024/chapter-1.html).

Black carbon is released as a particle and, therefore, falls in the category of particulate matter, not greenhouse gas. All particle emissions from a combustion source are broadly referred to as particulate matter (PM) and usually delineated by sizes less than 10 micrometers (PM10) or less than 2.5 micrometers (PM2.5). Black carbon is an ultrafine particle or PM0.1, and poses a strong health risk. They also strongly absorb light and convert that energy to heat. Soot is composed of various pollutants, including chemicals, acids, metals, and soils, suspended in the air after emission. Soot can come in solid, liquid, or gaseous aerosol states capable of penetrating your lungs when inhaled and, subsequently, your bloodstream. "Soot" has a more technical term known as "black carbon." Combustion-generated aerosols, however, also have a light-absorbing black carbon (BC) component.

3.2 BLACK CARBON AS A CONTRIBUTOR AND ITS ROLE

Black Carbon (BC) has recently emerged as a major contributor to global climate change, possibly second only to CO_2 as the main driver of change. BC particles strongly absorb sunlight, giving soot its black color. BC is produced both naturally and by human activities as a result of the incomplete combustion of fossil fuels, biofuels, and biomass. Primary sources include emissions from diesel engines, cook stoves, wood burning, and forest fires. Reducing CO_2 emissions is essential to avert the worst impacts of future climate change, but CO_2 has such a long atmospheric lifetime that it will take several decades for CO_2 concentrations to begin to stabilize after emissions reductions begin. In contrast, BC remains in the atmosphere for only a few weeks, so cutting its emissions would immediately reduce the rate of warming, particularly in the rapidly changing Arctic. Moreover, reduced exposure to BC provides public health co-benefits, especially in developing countries. Technologies that can reduce global BC emissions are available today. When you hear the word "soot," an image probably immediately comes to mind of an early 20th-century factory exhausting a dark substance into the air or coal miners caked in thick black dust. "Soot" is a term for a type of particle pollution called PM2.5, attributed to fine particles that are smaller than dust and mold particles. Black carbon is created from factories, cooking pots, and ships that burn heavy fuel oil (Figure 3.1).

Jurányi et al. (2023) pointed out that a factor of four higher black carbon mass concentration (21.6 ng m^{-3} average, 14.3 ng m^{-3} median) was found in spring, compared to summer (4.7 ng m^{-3} average, 3.9 ng m^{-3} median). In spring, much higher inter-annual and geographic variability prevailed compared to the stable situation in summer. Fortunately, black carbon is a short-lived climate pollutant with a lifetime of only days to weeks after release in the atmosphere. However, do not let its short duration make it appear any less dangerous than CO_2. Black carbon can significantly affect the climate, agriculture, and human health (Niranjana and Thakur, 2017). To put this in perspective,

FIGURE 3.1 Black carbon emitters. ((a). Black carbon at a site in Indonesia, https://en.wikipedia.org/wiki/Black_carbon; (b). Biomass Burning: Li et al (2021); (c). Ships that burn heavy fuel oil emit dirty exhaust containing black carbon. Image: Cyprien Hauser. https://dialogue.earth/en/climate/black-carbon-the-low-hanging-fruit-for-cleaner-shipping/. https://creativecommons.org/licenses/by-nd/2.0/.)

black carbon has a warming impact on the climate 460–1,500 times stronger than CO_2. Black carbon can negatively affect the health of ecosystems by depositing on plants, thus increasing their temperature, dimming the sunlight that reaches forests, and causing disruption by modifying rain patterns. Any change in rain patterns can have devastating and far-reaching impacts on ecosystems across the globe (Grant-Jacob and Mills, 2022). Black carbon and its co-pollutants are elements of fine particulate matter known as PM2.5, the leading environmental cause of chronic lung diseases such as bronchitis, aggravated asthma, acute respiratory infections (such as pneumonia), and premature death (National Academies of Sciences, Engineering, and Medicine, 2016). Black carbon is a component of fine particulate matter (PM2.5) air pollution, the leading environmental cause of poor health and premature deaths, as shown in Figure 3.2.

3.2.1 Properties of Black Carbon

* Black carbon warms the Earth by absorbing heat in the atmosphere and by reducing albedo (the ability to reflect sunlight) when deposited on snow and ice.
* BC is the strongest absorber of sunlight and heats the air directly.
* It emits infrared radiation that increases the temperature.
* In addition, it darkens snowpack and glaciers through deposition and leads to the melting of ice and snow (black carbon is contributing to the melting of the Himalayan Glaciers).
* Regionally, BC disrupts cloudiness and monsoon rainfall.
* Black carbon stays in the atmosphere for only several days to weeks.

Thus, the effects of BC on the atmospheric warming and glacier retreat disappear within months of reducing emissions.

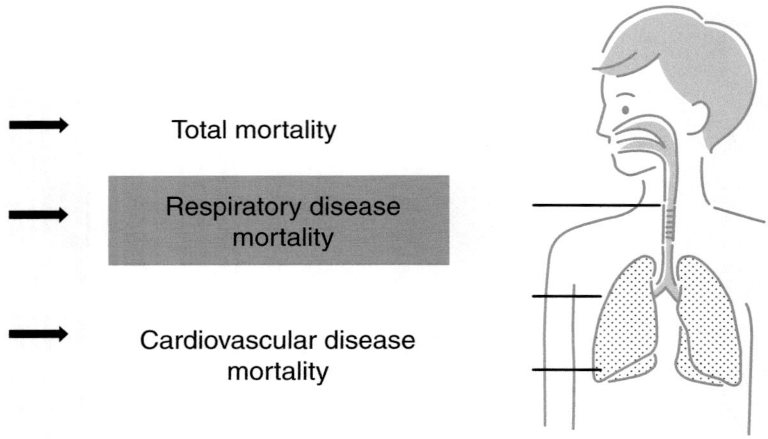

FIGURE 3.2 The major health problems due to soot and carbon black (CB) (Niranjana and Thakur,2017. Front. Immunol. 8:763. doi: 10.3389/fimmu.2017.00763.)

3.2.2 Adverse Effects of Black Carbon

The impact of black carbon is as much a local problem as a global one. The effects depend on the source and location of emissions, with biomass sources of black carbon having localized effects on human health, while fossil fuel sources can contribute to more global problems, such as increasing the risk of natural disasters and global warming.

Black carbon is a short-lived climate pollutant with a lifetime of only days to weeks after release into the atmosphere.

It warms the earth 460–1,500 times more powerfully than CO_2.

During this short period, black carbon can have significant direct and indirect impacts on the climate, glacial regions, agriculture, and human health, landing on plant leaves and increasing their temperature, altering rainfall patterns, and dimming sunlight reaching the Earth

Several studies have demonstrated that measures to prevent black carbon emissions can reduce near-term warming of the climate and prevent premature deaths.

Black carbon absorbs solar energy, warming the atmosphere.

When it falls to earth with precipitation, it darkens the surface of snow and ice, reducing their albedo (the reflecting power of a surface), warming the snow, and hastening melting.

Scientists use color to classify carbon at different points in the carbon cycle based on carbon function, characteristics, and location. This creates a more descriptive framework than traditional "organic" and "inorganic" labels (www.usgs.gov/media/images/carbon-rainbow).

Purple – Carbon captured through the air or industrial emissions
Blue – Carbon stored in ocean plants and sediments
Teal – Carbon stored in freshwater and wetland environments
Green – Carbon stored in terrestrial plants
Black – Carbon released through the burning of fossil fuels
Gray – Carbon released through industrial emissions
Brown – Carbon released by the incomplete combustion of organic matter
Red – Carbon released through biological particles on snow and ice that reduces albedo

3.2.3 Source Apportionment of Black Carbon Aerosols

In general, black carbon (BC) is an aerosol product of incomplete combustion of fossil fuels, biofuels, or biomass, generally originating from anthropogenic activities such as emissions from vehicles, industries, and burning of solid waste, among others (Bond et al., 2013). It is mostly used as an indicator of combustion sources since its physical properties and airborne concentration vary depending on the type of fuel used, combustion characteristics, and meteorology. Light-absorbing carbonaceous (LAC) aerosols are ubiquitous in the atmosphere. They are emitted and formed from, e.g., the combustion processes of fossil fuels and biomass, biogenic sources, and via heterogeneous/multiphase reactions (Andreae and Gelencsér, 2006). "Black carbon" (BC) belongs to this group of LAC compounds, and it is considered to be the most important light-absorbing aerosol component in the atmosphere due to the positive radiative forcing it imposes on the climate [Intergovernmental Panel on Climate Change (IPCC). 2007, 2014]. In addition to various climate impacts, BC deposited on snow may lead to a reduction of snow albedo, and BC has adverse effects on public health. BC sources in urban areas are characteristically dominated by combustion processes, typically from anthropogenic sources, such as transportation, industry, and residential combustion (Bond et al., 2013). It has been estimated that, globally, 24% and 60% of anthropogenic BC emissions are from transport and residential combustion sources, respectively.

Black carbon has been linked to increased frequency of droughts in northern China and floods in southern China, as well as to increased intensity of tropical cyclones originating in the Arabian Sea. Wildfires inject large amounts of black carbon (BC) particles into the atmosphere, which can reach the lowermost stratosphere (LMS) and cause strong radiative forcing. Figure 3.3 highlights the black carbon emissions from different Sources.

From October through December, a thick and deadly haze hangs over Delhi, a metro area of roughly 30 million residents. The smoke is a product of coal-burning power plants and tens of thousands of field fires (www.nature.org/en-us/magazine/magazine-articles/india-agriculture/). Black carbon (BC) emissions from open biomass burning (BB) are known to have a considerable impact on the radiative budget of the atmosphere at both global and regional scales. Crop residue burning is a leading contributor to air pollution and ill health in India. Despite current bans to curtail agricultural fires, burning persists because of a lack of alternatives that are both effective and politically viable (Sarkar et al., 2018; Konovalov et al., 2018).

Crop residue burning (CRB) is a recurring problem, during October–November, in the northwestern regions (Punjab, Haryana, and western Uttar Pradesh) of India. The emissions from the CRB source regions spread in all directions through long-range transport mechanisms, depending upon the meteorological conditions. Although the intentional use of fire to transform land has decreased globally, regulatory strategies have been less effective in southern and eastern Asia. This is particularly true in northwestern India, where rice residue burning makes a substantial contribution to air pollution and short-lived climate pollutants. Pullabhotla and Souza (2022)'s and Singh et al. (2021, 2022)'s results show that a standard deviation increase of exposure to upwind fires increases the incidence of hypertension by 1.8% – relative to the average hypertension incidence among individuals not exposed to upwind fires. We estimate that the exposure to fires contributes to nearly 14% of total mortality due to hypertension in India annually.

A spatial and monthly inventory has been constructed for carbonaceous particles emitted by boreal and temperate wildfires in forests, shrublands, and grasslands, with burned area data statistics, fuel load maps, fire characteristics, and particle emission factors. The period considered is 1960-1997, and an important year-to-year variability was observed. On average, boreal and temperate vegetation fires represent 4% of global biomass burning, but during extreme years, their contribution may reach 12%, producing 9% and 20% of black carbon (BC) and particulate organic matter (POM), respectively, emitted by worldwide fires (Lavoué et al., 2000).

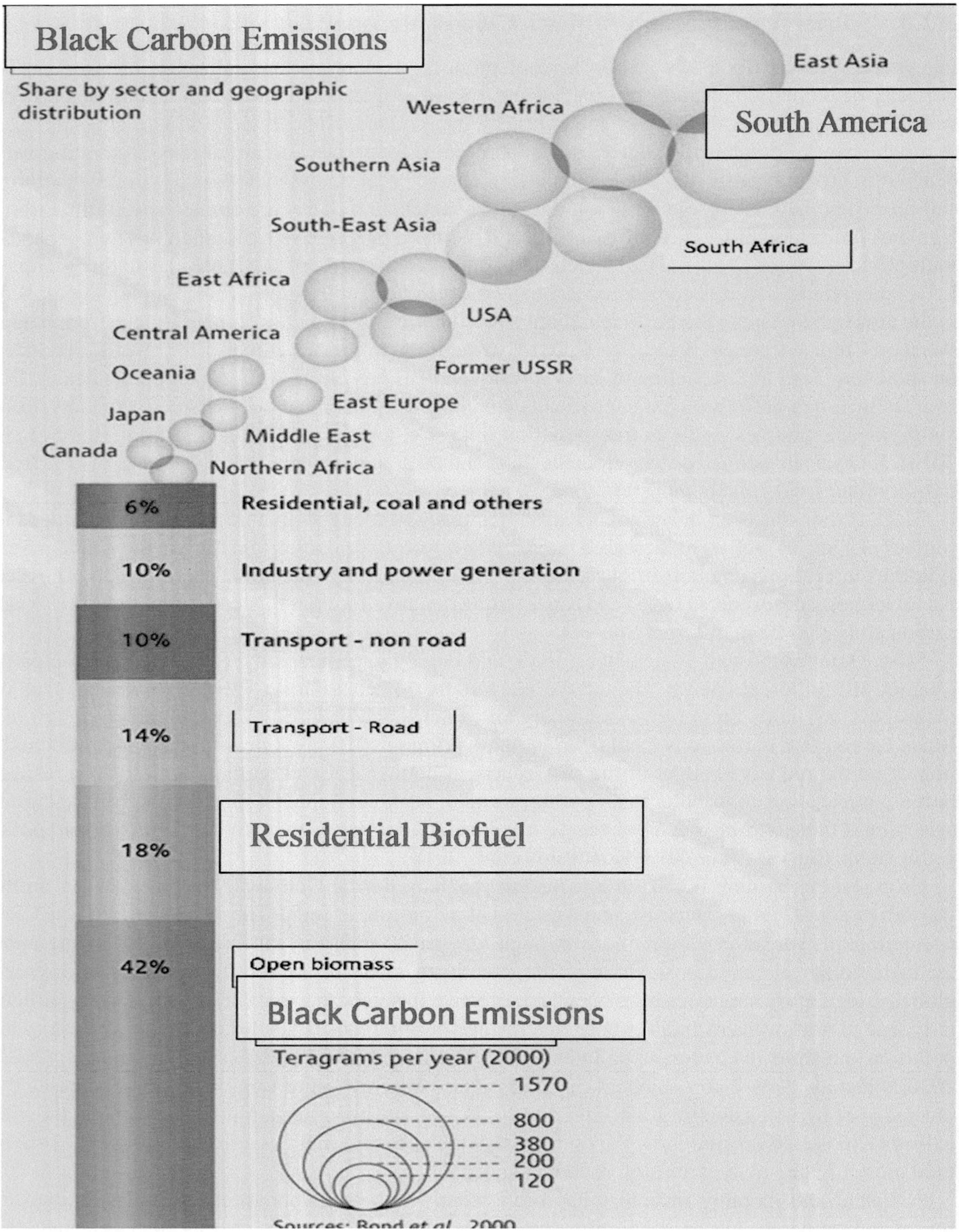

FIGURE 3.3 The Black carbon emissions (Black carbon emissions: www.grida.no/resources/7340. Creator: Riccardo Pravettoni. For any form of publication, please include the link to this page.)

Black carbon has a warming impact on the climate 460-1,500 times stronger than CO_2 per unit of mass. The average atmospheric lifetime of black carbon particles is 4-12 days. The black carbon processes in the climate system are given in Figure 3.4 (https://commons.wikimedia.org/wiki/File:Carbon_cycle.jpg).

Black carbon (BC) aerosol is a vital air pollutant throughout the surface earth system and has attracted great concern regarding its multiple impacts on human health, climate change, and atmospheric visibility (Chen et al., 2020a,b). As the most important light-absorbing component of PM2.5 (particulate matter with an aerodynamic diameter less than 2.5 μm), BC exerts a key and unique role in the climate system by absorbing solar radiation, affecting chemical/physical properties of cloud, and influencing snow and ice cover. BC is even found to be the second most important climate-warming agent after carbon dioxide, with a positive climate forcing of 1.1 W m^{-2}, greater than that of methane. BC also has impacts on urban weather conditions and may play a key role in extreme weather. In polluted environments, BC has a significant influence on pollution development. In addition, BC leads to visibility impairment because of its strong absorption of visible light (Watson, 2002) and has adverse impacts on human health due to its adsorption capacity (Colicino et al., 2017). Nevertheless, due to a lack of observational constraints and uncertainties in emission inventories, large uncertainties still exist in BC emissions, including absolute fluxes and relative

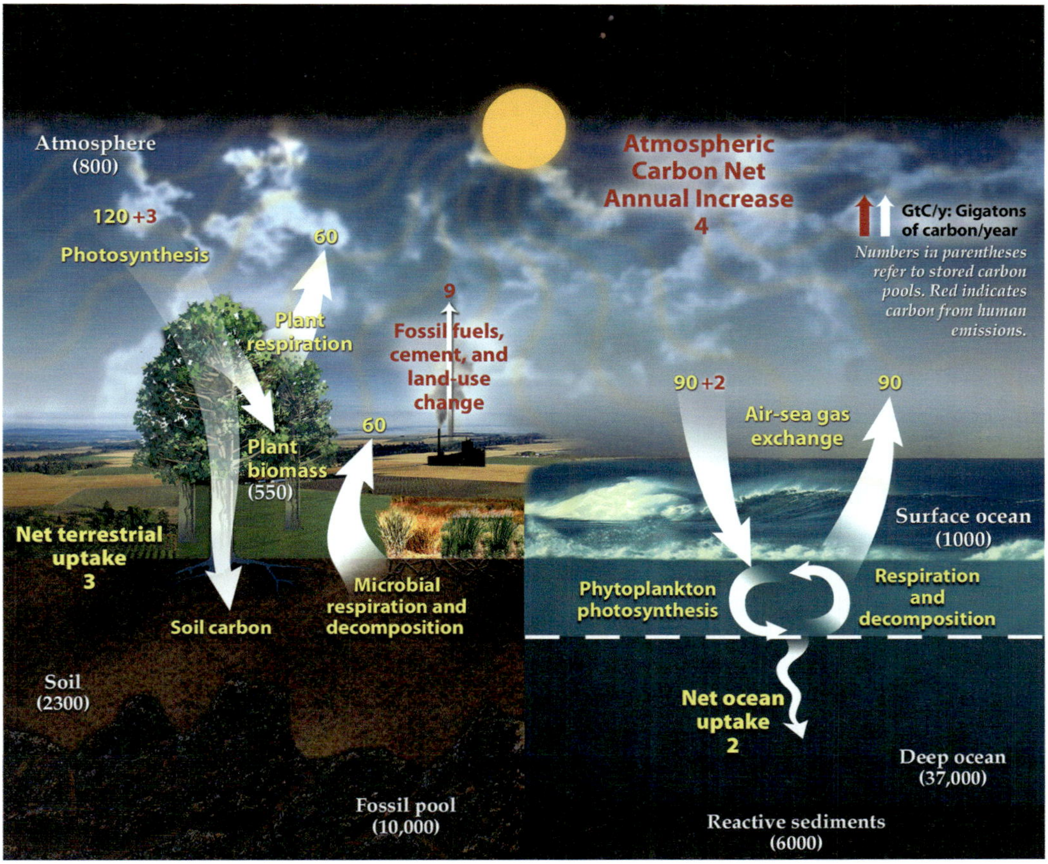

FIGURE 3.4 Black carbon processes in the climate system. (https://commons.wikimedia.org/wiki/File:Carbon_cycle.jpg.)

source contributions of fossil versus biomass combustion, which will complicate our knowledge of the multiple BC effects. It is also necessary to clarify the contributions of different sources to BC to determine efficient emission mitigation strategies. At a global scale, BC emission sources can be attributed to fossil fuels (~ 40 %), open biomass burning (~ 40 %), and biofuels (~ 20 %) (Ramanathan and Carmichael, 2008). However, these fractions vary significantly because of the substantial spatial and temporal variations in BC emissions.

Even though the chemical constituent of PM from biomass burning is characterized by extensive spatiotemporal heterogeneity, black carbon (BC) is one of the major components and is primarily a pollutant emitted from the incomplete combustion process (Bond et al., 2013). In addition to the open burning of biomass, BC could also be produced from diesel engines, industrial coal combustion, and residential solid fuels (wood, crop residue, and coal). In general, BC has attracted increasing scientific interest in recent decades because it negatively impacts human health, climate change, and the environment. The study of BC concentration and its control depends on the evidence of the negative impact that BC has on people's health in terms of the alterations it causes in respiratory and cardiovascular functions. BC aerosol particles are present mostly in the fine mode (PM2.5) and can be inhaled and deposited in the lungs quite efficiently. BC is considered to be more harmful to people than other particle components, such as PM10 and PM2.5 particulate materials. BC also has an important impact on the climate as it is responsible for a positive radiative forcing (warming) of 0.6 W/m^2 (Blanco-Donado et al., 2022). This value is about one-quarter of the CO_2 radiative forcing, a very significant figure. BC also has an impact on the increase in snow melting when it gets deposited over the snow. Adopting these measures would have major positive co-benefits for public health, especially in the developing world (www.ccacoalition.org/en/slcps/black-carbon).

3.3 BLACK CARBON VERSUS ELEMENTAL CARBON

The term "black carbon" is often used interchangeably with the term "elemental carbon" in the atmospheric sciences. Both are light-absorbing carbon compounds, but whereas BC is formed from incomplete combustion in reduced or anoxic environments, the term "elemental carbon" is used for carbon fractions measured after oxidative combustion in the presence of oxygen above a certain temperature threshold (https://escholarship.org/uc/item/1br0q643). Another light-absorbing carbon is "brown carbon", which is a non-soot organic carbon aerosol originating from humic-like substances (HULIS), bioaerosols, and tar.

3.4 BC IN THE ATMOSPHERE

Black carbon (BC), which can linger in the environment for years to millennia, is created when biomass and fossil fuels are not completely burned. Under anthropogenic warming, BC production is predicted to rise in tandem with rising fire activity and may serve as a counterbalance to climate change. BC is frequently separated into dissolved and particulate forms, each of which may have distinct residence durations, environmental transport modes, and outcomes (Coppola et al., 2022).

According to BC aerosol emission inventories, burning biomass releases 2–11 Tg BC annually, while burning fossil fuels releases 4.5–12.6 Tg BC annually. The PM2.5 size-class BC aerosol is first released as soot, but as it matures, photodegrades, and coagulates with hydrophilic aerosol species, it becomes soluble in raindrops. BC aerosols can be deposited far from the fire source since submicrometric-sized BC can stay in the atmosphere for up to 14 days. Dry deposition, which involves dropping from suspension in the atmosphere, or wet deposition, which involves dropping raindrops, are the two ways that BC exits the sky and lands on land or the ocean's surface.

3.5 CLIMATE IMPACTS OF BLACK CARBON EMISSIONS – EFFECTS ON GLACIERS

In the atmosphere, black carbon affects the Earth's temperature by absorbing solar energy and releasing it as heat. Warm air flows from lower latitudes to the Arctic. When it falls out of the atmosphere onto ice or snow, black carbon warms the surface and dramatically increases the rate of melting. Because snow and ice reflect solar energy back out to space, reductions in their extent lead to more warming. This feedback loop is one reason why the Arctic is warming more than two times faster than the rest of the world. The snow and ice reflect solar energy back out to space (https://eua-bca.amap.no/background).

Snow-covered regions are the most vulnerable to the warming effects of black carbon, and any particles reaching them are of concern if they are darker than snow, because they can reduce reflectivity and speed melting. Glaciers are melting in the Andes, the Rocky Mountains, the Canadian Rockies, the Alps, the Himalayas, and around the world (https://news.climate.columbia.edu/2016/03/22/the-damaging-effects-of-black-carbon/). As glaciers melt and retreat, the ice melt, which during the dry season feeds the rivers that supply irrigation systems, will dwindle. The glaciers of the Himalayas and the Tibetan Plateau sustain the rivers of China and India and are critical to the fresh water and food supplies of these two countries.

Black carbon, like all particles in the atmosphere, also affects the reflectivity, stability, and duration of clouds and alters precipitation. Depending on how much soot is in the air and where black carbon sits in the atmosphere, it has different effects. If it absorbs heat at the level where clouds are forming, they will evaporate. When it lies above lower stratocumulus clouds that block the sun, it stabilizes them and thus has a cooling effect. Because black carbon interacts with other components of particulate matter, such as sulfates and nitrates that reflect sunlight and cool the atmosphere, scientists do not know exactly how much black carbon itself directly contributes to global warming.

Black carbon, which often causes glaciers to recede, is emitted in the atmosphere due to the incomplete combustion of fossil fuels in motor vehicles and aircraft exhausts, biofuels, and biomass. Both India and China are the largest emitters of black carbon in the world, with emissions of up to 25-35 percent (www.downtoearth.org.in/blog/climate-change/black-carbon-is-a-threat-to-himalayan-glaciers-74542).

The Indo-Gangetic plains will become the largest contributor of black carbon, with about 20 percent from bio-fuels, 40 percent from fossil fuels, and about 40 percent from biomass burning. Black materials absorb light and emit infrared radiation that increases the temperature. So, when there is an increase in black carbon in the high Himalayas, it will contribute to faster melting of Himalayan glaciers.

In the long run, changes in the atmospheric composition of the high Himalayas will affect rain and snow precipitation patterns (Amjad et al., 2022). Accordingly, natural resources and socio-economic activities of Himalayan communities will also be affected. The GB Pant Institute of Himalayan Environment and Development found that black carbon concentration over the Parbati Glacier in Himachal Pradesh ranged between 0.34 micrograms per cubic meter ($\mu g / m^3$) and 0.56 $\mu g / m^3$. In the foothills of the Parbati, Hamta, and Beas Kund glaciers, the black carbon concentration was 796 $\mu g / m^3$, 416 $\mu g / m^3$, and 432 $\mu g / m^3$. Meanwhile, the Wadia Institute of Himalayan Geology found that black carbon concentration on the Gangotri glacier ranged from 0.01-4.62 $\mu g / m^3$. Thus, one can conclude that high-altitude glaciers of Uttarakhand are more prone to further climate change disasters as compared to others (www.downtoearth.org.in/blog/climate-change/black-carbon-is-a-threat-to-himalayan-glaciers-74542). In areas with year-round ice, such as glaciers, the presence of black carbon increases flood risk. Glacial melt from the Himalayas increases the flood risk for 78 million people living in the basins of the Ganges and the Brahmaputra rivers. Black carbon deposits due to human activities are increasing glaciers and snow melt in the Himalayan ranges and are changing temperatures and precipitation patterns, the World Bank has said in a study (www.hindustantimes.com/india-news/heres-how-black-carbon-deposits-causing-glaciers-snow-melt-in-himalayan-ranges-101622789118845.html).

Himalayan glaciers provide water to hundreds of millions of people, but they're melting faster because of climate change and a threatening level of pollution, which is likely to get only worse if actions are not taken in time (Mani, 2021). Melting glaciers and the loss of seasonal snow pose significant risks to the stability of water resources in South Asia. The 55,000 glaciers in the Himalaya, Karakoram, and Hindu Kush (HKHK) mountain ranges store more freshwater than any region outside of the North and South Poles (Mani, 2021). Their ice reserves feed into three major river basins in South Asia—the Indus, Ganges, and Brahmaputra—that are home to 750 million people. One major regional driver of the accelerating glacier melt is climate change, which is altering the patterns of temperature and precipitation. A second driver may be deposits of anthropogenic black carbon (BC), which increase the glaciers' absorption of solar radiation and raise air temperatures. BC is generated by human activity both inside and outside of South Asia, and it may be meaningfully reduced by policy actions taken by the South Asian countries themselves. Mani et al. (2021) investigated the extent to which the BC reduction policies of South Asian countries may affect glacier formation and melt within the context of a changing global climate. It assesses the relative impact of each source of black carbon on snow and glacier dynamics. The study Glaciers of the Himalayas provides new evidence on the extent to which black carbon reduction policies by South Asian countries have an impact on glacier formation and melt in the Himalaya, Karakoram, and Hindu Kush Mountain ranges within the context of the changing global climate. Recent evidence suggests that in addition to changing temperatures and precipitation patterns, anthropogenic black carbon (BC) deposits — soot — are further accelerating glacier and snow melt in these mountain ranges. It is generated by human activity both inside and outside South Asia. It is part of a larger basket of aerosols that impact climate change directly and indirectly, according to Hartwig Schafer, vice president, South Asia Region, the World Bank.

Black soot deposited on Tibetan glaciers has contributed significantly to the retreat of the world's largest non-polar ice masses, according to new research by scientists from NASA and the Chinese Academy of Sciences (www.nasa.gov/topics/earth/features/carbon-pole.html). Soot absorbs incoming solar radiation and can speed glacial melting when deposited on snow in sufficient quantities. Temperatures on the Tibetan Plateau – sometimes called Earth's "third pole" – have warmed by 0.3°C (0.5°F) per decade over the past 30 years, about twice the rate of observed global temperature increases. New field research and ongoing quantitative modeling suggest that soot's warming influence on Tibetan glaciers could rival that of greenhouse gases. "Tibet's glaciers are retreating at an alarming rate," said James Hansen, coauthor of the study and director of NASA's Goddard Institute for Space Studies (GISS) in New York City. "Black soot is probably responsible for as much as half of the glacial melt, and greenhouse gases are responsible for the rest." Reduced black soot emissions, in addition to reduced greenhouse gases, may be required to avoid the demise of Himalayan glaciers and retain the benefits of glaciers for seasonal fresh-water supplies.

The Gangotri glacier holds a place of great significance for Indians. Located in the Uttarkashi District of Uttarakhand, this glacier is not only among the largest in the Himalayas, but also one of the primary sources of water for the river Ganga. Moreover, it is a pilgrimage site for devout Hindus, while also being popular among the trekking circles. The glacier has faced a constant threat of excessive melting, as global warming continues unabated. Now, the Ministry of Science and Technology has said that factors like agricultural burning and forest fires could trigger large-scale glacial melting in Gangotri come summer (https://weather.com/en-IN/india/science/news/2020-03-06-black-carbon-melting-gangotri-glacier-water-security).

Scientists from the Wadia Institute of Himalayan Geology (WIHG) conducted a study at Chirbasa station near Gangotri Glacier. The study found that the concentration of black carbon in the region increased by 400 times during the summer of 2016. Black carbon—a component of fine particulate matter formed due to the incomplete combustion of fossil fuels, biofuels, and biomass—is capable of triggering glacial melting due to its light-absorbing nature. The presence of black carbon aerosols in the eco-sensitive zone, such as the Himalayan glacier valleys, is a matter of serious concern and needs to be meticulously monitored, says the ministry in a statement. According to the study, the seasonal increase in black carbon concentration in the region is dueto agricultural burning in western India and forest fires along the Himalayan slopes. Both these activities undergo a spike during summer—the former due to farmers' practice of burning crop residue to prepare their field for seeding the next season, and the latter because forests get littered with dry components that ignite by the slightest of sparks. Further, the long-range transportation of pollutants in winter also amplifies the black carbon content around the glacier, thereby posing a direct threat to its existence.

The melting of Gangotri could lead to enormous short- as well as long-term consequences for the entire country. The immediate effect of the melting is likely to be in the form of floods, as glacial melt will lead to the accumulation of excess water in glacial lakes. Flooding of such lakes, along with rivers and other reservoirs, could wreak havoc in multiple north Indian states. Such intense inundation will not only damage the environment but also force the regional residents to migrate southward. Once the flooding phases out, droughts will set in due to the disappearance of the water source, which could further pose severe challenges to water and food security in the region. Ganga's stream would possibly narrow down, and the river may very well become a seasonal one. Its basins could also undergo desertification due to the lack of water year-round, causing great harm to flora, fauna, and human settlements that dwell around the river. Multiple societal and cultural changes across India could follow.

Dust is not to be confused with black carbon, which is generated by human activities. Black carbon is also present within High Mountain Asia, emitted by burning materials such as crop stubble and worn tires on the agricultural Indus and Gangetic plains. However, black carbon is carried within the surface layers of the atmosphere and is therefore deposited at lower elevations, between 0 and 3,000m altitude. The main finding of the study was that dust-induced snow darkening had a greater effect above 4,000m altitude than that of black carbon, which has long been seen as the dominant threat. As temperatures rise, snow at lower elevations disappears, but dust's role remains with high elevation snow. With global warming, the snow line rises and the snowpack in low elevations disappears.

Both black carbon and natural dust affect albedo—the reflectance of solar radiation from a surface. Snow, being shiny and white, has a high albedo. It reflects a large proportion of radiation from its surface and thereby keeps cool. But surfaces covered in black carbon and dust, being dark-colored, absorb more solar radiation, warming the surface and causing melting (https://news.climate.columbia.edu/2020/12/03/dust-speeds-snowmelt-himalayas/). The research signifies the role of long-distance transported dust in snow melting in the Himalayas.

A Parliamentary Standing Committee on Water Resources has recommended a detailed study on the impact of atmospheric pollution, including black carbon, on the Himalayan glacier system (www.deccanherald.com/national/house-panel-suggests-study-on-impact-of-black-carbon-on-himalayan-glacier-1206459.html). Black carbon absorbs light and heat, which causes an increase in temperature, causing glaciers to melt faster. To assess the extent and scope of the adverse impact of atmospheric pollution, including black carbon, on the fragile and sensitive Himalayan glacier system, the Department of Water Resources in consultation with other ministries concerned should commission a research project for a comprehensive examination of this aspect and submit the research findings within a time-bound period, the committee said in its report.

A study conducted by the National Centre for Polar and Ocean Research and Space Physics Laboratory at their Himansh station in the Lahaul-Spiti region has revealed that the total Suspended Particulate Matter (SPM) showed significant variations with the dominance of mineral dust components therein (~67%). It has also been stated that black carbon reportedly contributes (~4%) to near-surface composite aerosol mass concentrations and also absorbs light and emits infrared radiation that increases the temperature. Hence, an increase in black carbon in the high Himalayas is stated to contribute to the faster melting of glaciers, said the report (www.deccanherald.com/national/house-panel-suggests-study-on-impact-of-black-carbon-on-himalayan-glacier-1206459.html). Black carbon particles consist of nearly pure elemental carbon with some oxygen and hydrogen bound into a layered, hexagonal structure, which corresponds to a somewhat disordered graphitic crystal structure. Black carbon absorbs light and heat, causing an increase in temperature, causing glaciers to melt faster. It is formed by the incomplete combustion of fossil fuels, biofuels, and biomass. This has worsened the air quality since black carbon is emitted directly into the atmosphere in the form of fine particles and is a light-absorbing component of particulate matter.

The reports published (www.newindianexpress.com/nation/2020/feb/03/high-levels-of-black-carbon-in-glaciers-warn-scientists-2098214.html) reveal that if wildfires continue for a longer period, it can affect the glaciers too, warned scientists. The report mentions that black carbon was found up to 4.62 cubic micrograms in over 900 glaciers during the wildfire season in Uttarakhand, especially from April to June in 2020. Imagine wearing a black shirt or a white one during the summer; the dark cloth absorbs the sun's energy, making the wearer warmer, while the white reflects. Similarly, ice and snow reflect the sun's heat back to the atmosphere while darker surfaces absorb the sun's rays. This is known as the albedo effect. Black carbon particles do just that; they absorb the sun's radiation and heat up what is around them, melting ice and exposing darker seawater or land underneath. This can lead to a vicious cycle of warming and melting that gets worse over time (https://clearseas.org/en/blog/black-carbon-in-the-arctic-what-you-need-to-know/).

3.5.1 Glacier's Role in Managing Climate Change

Glaciers begin to form when snow remains in the same place year-round, where enough accumulation takes place to transform into ice. As more snow gets added in each cycle, previous layers get compressed to recrystallize and harden. An increase in the size of snow crystals reduces air pockets, resulting in a dense mass of ice. A glacier is a large accumulation of ice and snow that slowly moves over land. At lower temperatures, snow accumulates due to low melting. The surplus of accumulated snow begins to move in the form of glaciers. Glaciers play a strategic role as a vital national resource in ensuring water security. Hence, there is a need to carry out new strategies for combating the challenges posed by climate change and global warming in glacier management. Various parts of glaciers move at different speeds. Most glaciers move very slowly, but a few glaciers can move 50 meters per day. Further, glaciers are of two types. One is an alpine glacier, and the other is an ice sheet glacier. Alpine glaciers form near mountainsides and move through valleys.

Parliament's Standing Committee on Water Resources has recommended setting up an apex body for glacier management in India (Read more at: www.studyiq.com/articles/glacier-management/). In India, the Department of Science & Technology will be responsible for studying, monitoring, and providing early warning responses to reduce the potential glacier-related risks. It will coordinate the activities of all the departments/agencies involved in monitoring and research work of Himalayan glaciers (www.studyiq.com/articles/glacier-management/)

Global warming: Global warming is increasing the global temperatures, thereby reducing the ability of snow to accumulate and form ice. Ocean warming is also melting the polar glaciers.

Carbon Pollution: Pollutants such as black carbon can accumulate over snow cover and increase the temperature of the snow, causing it to melt.

Climate change: Climate change has affected the precipitation cycle across the globe. In the absence of precipitation in the form of snow, glaciers tend to reduce in size (www.studyiq.com/articles/glacier-management/).

Forest fires and crop burning, in addition to pollution due to emissions from other sources like coal-burning power stations, transportation, and brick kilns, are causing high soot depositions in the Himalayas. Black carbon, a component in soot that absorbs light more strongly than any other substance in the atmosphere, is now settling over Himalayan glaciers and disturbing the principle on which these glaciers have survived over centuries, namely that light-colored surfaces reflect sunlight and darker colored ones absorb it. Recent research has shown once again that the many fires in the lower Himalayas and the foothills are forcing the glaciers in the higher reaches to melt much faster (https://indiaclimatedialogue.net/2020/03/17/soot-on-himalayan-glaciers-accelerates-melting/).

When black carbon particles settle over glaciers, they darken the surface and enable higher absorption of light and heat, eventually causing the snow or ice to melt. In addition to such direct impacts, black carbon particles that remain suspended in the atmosphere absorb sunlight and infrared radiation and warm the air, which also contributes to glacial melt, among other impacts of warming, like altered monsoon patterns.

3.5.2 Significant Effects of Temperature Rise on Glaciers

Due to the rise in the air temperature, ice shelf collapse has increased in the last few decades. Around 87% of glaciers in the Antarctic Peninsula are shrinking. According to recent research, twelve glaciers around Antarctica showed near-frontal surface lowering since the 1960s (www.earthreminder.com/impact-of-climate-change-on-glaciers/). Due to the thinning of glaciers, the acceleration of glaciers has increased. Because thin glaciers float faster, we can conclude that rapid warming has resulted in glaciological changes. According to a report of a study by NASA and the European Space Agency, from 1992 to 2018, about 3.8 trillion tons of ice have melted from Greenland's ice sheets. If this continues, about 3 to 5-inch sea level rise will cause natural disasters like floods until 2100. Glacier melting has impacted society directly by affecting landscapes. This is a sign to us to improve the climate. But if we ignore the warning, then we have to face the consequences. As the temperature is increasing, there is no other way to reduce this rapid melting rate. The only way is to control the temperature and take strict action against this.

Due to the rapid melting of glaciers, the melted water covers land , and the sea level is rising. A rise in sea level increases soil erosion in coastal areas. It raises storms and floods as warming air and sea temperature start more frequent and exceptional seaside storms like hurricanes and typhoons. The melting rate is rapid; if this rapid melting rate continues, then many lands will be covered by water in the future. If all the ice in Greenland melts one day, then the ocean level will rise by 20 feet globally. This will disrupt the typical pattern of ocean circulation, and we face more storms. This will directly affect the aquatic animals, and the businesses related to the sea will face a significant loss. Then, the economy is also going to be involved.

Originally, scientists began studying glaciers only for the clues they offered about Earth's climate during past ice ages and the role they played in shaping the landscape. Today, they are also trying to understand how quickly human-caused climate change will cause them to disappear altogether.

The repercussions of melting glaciers are multifaceted. For instance, the drinking water supplies of the millions of people who depend on meltwater rivers would be impacted by the melting of the Himalayan glaciers. Additionally, homes at sea level will flood for those who live there as sea levels rise. The hundreds of millions of people who live in Bangladesh on low-lying ground would be significantly impacted by this. Since Asia is the most populous of the four regions—Africa, the Arctic, the Amazon, and Asia—the effects of climate change and glacier melting are probably going to be felt there the most. In certain areas, glaciers supply both people and wildlife with water that is essential to life. One significant river in Central Asia, the Amu Darya, for example, gets some of its water from glacier melt. Although that melt only contributes 8% of the river's yearly flow on average, it is responsible for 27% of the flow in late summer, when snowmelt supplies comparatively little water.

3.6 IMPACT OF BLACK CARBON IN THE ARCTIC

The effect of black carbon is complex, as black carbon particles emitted from outside the Arctic and transported into the region via weather systems may interact with cloud formation, initially having a cooling impact. Then, black carbon absorbs solar energy and warms the atmosphere while floating in the air. However, some airborne black carbon particles will eventually land on ice or snow. Conditions in the Canadian Arctic, with vast tundra, snow-covered landscapes, and expansive sea ice, are particularly susceptible to the effects of black carbon. When black carbon particles settle on ice or snow, they change the surface's color and its ability to reflect light (albedo). This reduced albedo is one reason why the Arctic is warming more than two times faster than the rest of the globe.

Now covered in a layer of black carbon, the ice or snow surface absorbs more of the sun's radiant energy, accelerating melt. This direct warming impact adds to the atmospheric warming created by floating black carbon particles. As the warming ice melts, it exposes dark water that absorbs more of the sun's heat energy than ice or snow, further accelerating the warming process. In the Arctic, the warming effects of black carbon outweigh the cooling effects from upper atmosphere particles, as the majority of black carbon particles are emitted from sea-level sources within the region, such as ships that are operating close to glaciers, snow-covered coasts, or sea ice edges (https://clearseas.org/en/blog/black-carbon-in-the-arctic-what-you-need-to-know/).

Black carbon comes from natural and human sources. In the Arctic, some black carbon is produced naturally through wildfires. Arctic wildfires come in many forms, but the most problematic are "zombie fires." Zombie fires can thaw permafrost while smoldering under snow or ice and emit CO_2 and black carbon particles once the snow and ice melt and the fire erupts on the surface (https://clearseas.org/en/blog/black-carbon-in-the-arctic-what-you-need-to-know/). Various human activities produce black carbon, from household cooking fires to commercial fuel combustion and resource extraction processes such as flaring waste gas. Commercial ship exhaust accounts for a growing share of black carbon particles in the Arctic.

3.7 BLACK CARBON AND CLOUD FORMATION

Black carbon causes warming of the atmosphere by a number of different processes. These particles absorb visible light due to their dark color. This absorption leads to a disturbance of the planetary radiation balance and eventually to warming. Black carbon aerosols have a large impact on regional circulation and rainfall patterns as they cause significant asymmetry in heating patterns over a region. While not fully quantifiable, the impact of black carbon on regional weather patterns and regional warming is more certain than its impact on global warming (Zhou et al., 2020). It was mentioned that the earlier start time of heavy rainfall related to absorbing aerosols (BC) is due to the radiative heating of absorbing aerosols, while the longer rainfall duration is probably caused by both the cloud effect of sulfate-like aerosols and the increased moisture supply (Figure 3.5). This is because, at the global scale, organic carbon may offset warming due to black carbon. Black carbon is a warming agent, and organic carbon is a cooling agent. They are always co-emitted, but in different proportions, depending on the source. At the regional scale, changes are more closely related to atmospheric heating, which is dominated by black carbon, and co-emitted species have a lesser impact. Black carbon particles also influence cloud formation, cloud lifetime, rainfall, and weather patterns (William et al., 2002). The formation of cloud droplets typically cannot occur without aerosols. Clouds strongly affect the Earth's energy balance by trapping heat near the surface and reflecting solar radiation back to space.

The soot particles aged by sulfuric acid, on the other hand, cause more ice crystals to form and make cirrus clouds optically thicker, i.e., they are less permeable to radiation. They extend as far as the tropopause, which is located at an altitude of 10 to 18 kilometers, and also linger longer in higher regions of the atmosphere (www.nature.com/articles/s41561-020-0631-0). In aerosol hot spots around the globe, solar radiation is dimmed down on its way to the Earth's surface. The

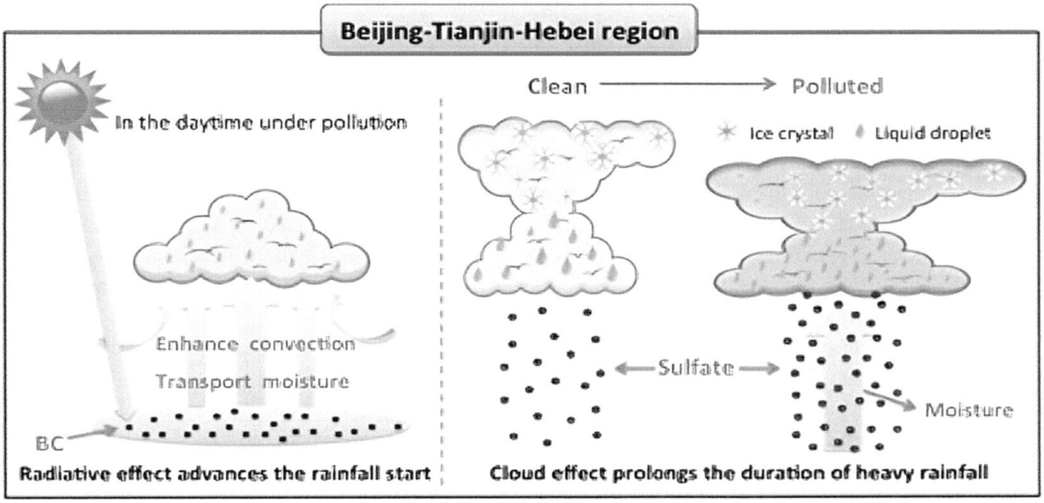

FIGURE 3.5 A schematic diagram for black carbon and other aerosol impacts on clouds (Zhou et al., 2020. Under the Creative Commons Attribution 4.0 License.)

resulting surface cooling turns out to be almost in balance with the heating of the atmosphere due to black carbon.

They can also change wind patterns by affecting the regional temperature contrasts that drive the winds, influencing where rain and snow fall. These effects can be both local and distant. Another regional impact of short-lived climate forcers is their effect on accelerating the melting of ice and snow. Firstly, ozone and black carbon warm the atmosphere and therefore promote melting of ice and snow in the Arctic and heavily glaciated parts of elevated regions, such as the Himalayas. A UNEP/WMO assessment concluded that by 2040, atmospheric warming could be reduced by about 0.7°C in the Arctic by implementing black carbon and methane measures, two-thirds of the warming projected for that region in that year. It also concluded that this benefit was larger in the Arctic compared to the average global climate benefit (https://svs.gsfc.nasa.gov/10387).

Aerosol type plays an important role in determining how aerosols affect clouds. Whereas reflective aerosols tend to brighten clouds and make them last longer, the black carbon from soot can have the opposite effect. Studies of pollution over the Indian Ocean and biomass burning smoke in the Amazon have shown that the black carbon warms the surrounding atmosphere and can cause cloud droplets to evaporate. This process, called the "semi-direct effect," turns clouds into a smoky haze that suppresses precipitation. Aerosols in brown clouds are made up primarily of black carbon and organic carbon. These aerosols, especially the black carbon component, absorb solar radiation, and this absorption results in enhanced solar heating of the atmosphere. Other aerosols, such as sulfates and nitrates, scatter solar radiation back to space.

In addition, pollution from atmospheric brown clouds is a threat to human health. Particulate matter, such as soot and dust, has been linked in epidemiological studies to cardiovascular problems, chronic respiratory problems, and mortality. Brown clouds also contain ozone and other hazardous pollutants. Ozone can irritate lung tissue, exacerbate asthma, and reduce lung function. Ozone has also been linked to reduced crop yields.

Soot particles change precipitation by absorbing solar radiation and heating the surrounding air. The atmosphere rapidly adjusts to this added warming by changing relative humidity, clouds, and precipitation. Soot particles at higher altitudes stabilize the atmosphere and increase cloud cover located below. Given all the processes that soot particles influence in the atmosphere, the similarities in underlying processes between the two climate models are striking. The resulting change in precipitation and temperature differs quite substantially (https://agupubs.onlinelibrary.wiley.com/doi/full/10.1029/2019JD032239).

Both the magnitude and sign of changes in surface temperatures depend on the vertical distribution of BC. BC at low altitudes can warm the surface (and the surrounding air), while BC at high altitudes has a weak positive or negative surface temperature change, by absorbing and reducing downwelling solar radiation. Depending on the altitude, BC can either reduce or increase the cloud cover. For instance, by warming the surrounding air, BC can evaporate nearby clouds or inhibit cloud formation (Ackerman et al., 2000). BC located above clouds, on the other hand, may increase the static stability in the atmosphere and increase stratocumulus clouds below. BC located below clouds can promote convection and thereby increase cloud cover.

Atmospheric black carbon is suspended particulate matter formed by the incomplete burning of fossil fuels, biofuels, and biomass. Black carbon is short-lived; it remains suspended in the atmosphere for only one to four weeks, thus its effects on the climate are largely regional. It affects the climate in several ways:

- When present in the atmosphere, black carbon absorbs sunlight and generates heat; it warms the air above and shadows the valleys and lowland areas below.
- During winter, when the atmospheric concentration of black carbon is greatest, there is increased fog, and temperatures in the river valleys of the northern Ganges Plains are notably lower.

During winter in the USA, black carbon significantly impacts fog formation by acting as a nucleus for water vapor condensation, leading to denser and more persistent fog, especially in urban areas with high emissions; this is because black carbon readily absorbs sunlight, warming the surrounding air and increasing humidity, creating conditions favorable for fog development.

In Europe during winter, black carbon significantly impacts fog by acting as a key component in fog formation, enhancing its density and persistence due to its ability to absorb sunlight, leading to warmer air near the ground, which traps moisture and creates thicker fog, particularly in areas with high pollution levels from combustion sources like traffic and heating systems.

- Atmospheric black carbon can also affect regional cloud formation and precipitation patterns by altering how much sunlight is absorbed.
- When deposited on snow and ice at higher elevations, black carbon absorbs heat, leading to accelerated melting of the snow and ice.

3.8 BLACK CARBON EMISSIONS AND THEIR IMPACT ON THE MONSOON RAINFALL

Strong radiative qualities allow black carbon, a major short-lived climatic pollutant, to absorb solar energy and cause localized warming. Therefore, the presence of BC in the atmosphere has the potential to alter cloud formation, temperature patterns, and atmospheric circulation, which in turn could affect the region's monsoon rainfall distribution and intensity. Unlike carbon dioxide, short-lived climate pollutants (SLCPs) have comparatively short air lives, spanning from hours to months, making them powerful climate forcers. They consist of gases like methane, ozone, and halogenated chemicals, as well as aerosols like sulphate and nitrate. SLCPs affect the climate within the first 20 years of emission and can be either direct (like methane and HFCs) or indirect (like NO_x and CO) (Bond et al., 2013). They have a significant impact on climate change and air pollution, and their regional effects affect temperature, the hydrological cycle, and weather patterns (Bond et al., 2013; Ramanathan et al., 2001). Addressing SLCP emissions presents a timely chance to fight global warming because of their brief lifespans.

Because of their significant contribution to particulate matter (PM) in the context of air quality and their potent radiative forcing characteristics, carbonaceous aerosols—including black carbon (BC) and organic carbon (OC)—have attracted a lot of interest in both scientific and policy circles. Assessing the climate effects of carbonaceous aerosols requires an understanding of how their many components partition, mix, coat, and age. One of the most important SLCPs among them is black carbon, which has significant regional effects and a short atmospheric lifetime of a few days to weeks. After carbon dioxide (CO_2) and methane, black carbon emissions are the third-largest cause of the current climate warming (Ramanathan and Carmichael, 2008).

As soot particles develop in the atmosphere, they pick up a layer of chemically translucent or transparent materials, which makes them bigger and increases the possibility that they will absorb sunlight. As a result, the air is heated more by these "aged" coated soot particles than by newly released, uncoated particles. However, there is significant temporal and regional fluctuation in the distribution of black carbon concentrations, which results in uneven global effects. Additionally, cloud formation is influenced by black carbon, which affects precipitation and regional weather patterns (Andreae and Gelencsér, 2006). When it is deposited over snow and ice, the surface's albedo decreases, making it less able to reflect sunlight and warming the surface. This changes weather patterns and ecological cycles by intensifying heat in areas with high concentrations of black carbon. Consequently, black carbon has important direct and indirect effects on agriculture, human health, snow and ice dynamics, and climate. The simulated BC concentrations are significantly impacted by the 99% drop in BC emissions, not only over India but also over large areas of south and eastern China.

It is interesting to note that burning biomass from agricultural waste typically results in higher BC concentrations in China. As a result, the simulated BC concentration in this region has been significantly impacted by the decrease in BC emissions, resulting in a drop in values. Through indirect and semi-direct effects on aerosols, the variations in the simulated BC concentration produce a complicated feedback process that affects cloud microphysical characteristics the increased activation of cloud condensation nuclei (CCNs) into cloud droplets, which is made possible by the atmosphere's higher moisture content. Simulated rainfall patterns over the Indian subcontinent have changed significantly as a result of the reduced BC scenario. We see a large improvement in simulated rainfall throughout peninsular India, especially in the eastern coastal regions and the northern portion of the peninsula.

3.9 EFFECT OF BLACK CARBON ON CROP/VEGETATION

Black carbon (BC) is most often released through the incomplete combustion of fossil fuels, the burning of biomass, or as diesel exhaust. It consists of highly condensed carbonaceous by-products (e.g., soot) and residues (e.g., charcoal). As the main constituent of soot, it is BC that is almost exclusively responsible for light absorption by particles in the atmosphere, thus changing the radiation balance of the Earth and contributing to global warming. The report from IPCC (2007) gave a consensus estimate of the warming influence of BC as 0.34 ± 0.25 W m^{-2}, but Ramanathan and Carmichael (2008) used data from satellites, aircraft, and surface instruments to calculate that the warming effect of BC in the atmosphere is three to four times greater than this.

There is a broad consensus that human activities are changing the global climate through anthropogenic emissions of CO_2. This realization has highlighted the importance of conserving and enhancing existing carbon reservoirs. While terrestrial forests have received most of the attention, it is being increasingly acknowledged that blue carbon ecosystems, such as seagrass, mangrove, and salt marsh, may also play a significant role. Climate change is caused by long-lived greenhouse gases, black carbon, and tropospheric ozone, which are short-lived climate pollutants that exist locally in the atmosphere for weeks to months. Black carbon, mostly caused by rural cookstoves, affects the plants by reducing the amount of sunlight they receive. 51% of black carbon is emitted from household energy. It alters the regional rainfall patterns and accelerates the melting of glaciers; it also reduces sunlight, affecting plant health and productivity.

The direct impacts of BC on radiation and crop growth are straightforward: BC is an absorbing aerosol that reduces both direct and diffuse light available to plants, and—all else equal—should therefore lower yields. However, this effect is difficult to isolate because BC is usually committed or mixed in the atmosphere with other scattering aerosols to create compound particles of varying radiative properties. Anthropogenic climate change is a result of both global emissions of long-lived greenhouse gases (LLGHGs) and other short-lived climate pollutants (SLCPs). Two potent SLCPs, tropospheric ozone and black carbon, have direct effects on crop yields beyond their indirect effects through climate; emissions of black carbon and ozone precursors have risen dramatically in India over the past three decades. Jennifer and Ramanathan (2014) presented their results of the combined effects of climate change and the direct effects of SLCPs on wheat and rice yields in India from 1980 to 2010. Their statistical model suggests that, averaged over India, yields in 2010 were up to 36% lower for wheat than they otherwise would have been, absent climate and pollutant emissions trends, with some densely populated states experiencing 50% relative yield losses. SLCPs have indirect effects on agricultural productivity through their impacts on temperature (all) and precipitation (BC). However, BC and ozone are of particular interest for agriculture because they also have direct impacts on crop growth. BC aerosols alter the quantity and nature of the solar radiation reaching the surface.

Yield increases from the reduction of air pollution could help offset anticipated future expected yield losses resulting from temperature and precipitation changes. In the short term, this is an appealing option because SLCP mitigation will produce immediate results that can help counter the impacts of climate change and sea level rise already "locked in" from historical LLGHG and SLCP emissions. In the long term, although farmers may select/breed more pollution-resilient cultivars or alter management practices to help minimize such losses, air pollution mitigation—particularly of ozone precursors—will become an ever-more important food security measure.

3.10 BLACK CARBON IS SUPERCHARGING EL NIÑO

Black carbon is a dangerous, sooty material produced by the incomplete burning of biomass and fossil fuels—it poisons the air we breathe and accelerates the melting of glaciers and polar ice caps. It was reported that with the Arctic Sea ice cover shrinking rapidly, polar bears are forced onto land for longer periods, risking health degradation and starvation. Ice sheets melting and a threat to polar bears (Figure 3.6). As sea ice melts, polar bears are moving onto land for at least part of the year. Polar bears rely heavily on sea ice for many activities such as travelling, resting, mating, and, of course, hunting. Their diet consists mainly of seals, as they require large amounts of fat to survive the cold climate. The Arctic is heating up twice as fast as anywhere else on the planet, shrinking sea ice cover by 14% per decade.

The polar ice caps are melting (Ifrah Amjad et al., 2022) due to climate change caused by global warming. According to NASA, Arctic ice is melting at a rate of 13% per decade, and the oldest and thickest Arctic ice has declined by 90% in the last 30 years. Rapid melting of ice makes diversity more vulnerable to extinction (Jenssen et al., 2015), as their habitat is under threat and loss of ecological niche.

Black carbon is the second-most-significant pollutant driving climate change after carbon dioxide. But as new research suggests, black carbon emissions in the northern hemisphere might also be increasing the frequency of extreme episodes of one of the world's most important climate phenomena: the El Niño-Southern Oscillation. Yang suggests that reducing black carbon emissions could help mitigate extreme El Niño events (https://hakaimagazine.com/news/black-carbon-is-supercharging-el-nino/).

FIGURE 3.6 Persistent organic pollutants' (POPs) (high levels of these harmful compounds) effect on polar bears and possibly enhance the effects of climate warming in polar bears (Jenssen et al., 2015, https://doi.org/10.3389/fevo.2015.00016. https://creativecommons.org/licenses/by/4.0/.)

3.11 THE BLACK CARBON CYCLE AND ITS ROLE IN THE EARTH SYSTEM

Black carbon (BC) is produced by the incomplete combustion of biomass by wildfires and the burning of fossil fuels. BC is environmentally persistent over centuries to millennia, sequestering carbon in marine and terrestrial environments. Tiny air pollution particles, commonly called soot, but also known as black carbon, are in the air and on the move throughout our planet. Black carbon enters the air when fossil fuels and biofuels, such as coal, wood, and diesel, are burned. Since black carbon readily absorbs heat from sunlight, the particles can affect Earth's climate, especially on a regional scale. Though the global distribution of soot remains difficult to measure, NASA researchers use satellite data and computer models to better understand how these short-lived particles influence Earth's climate, cryosphere, and clouds. However, its production, storage, and dynamics, and therefore its role in the broader carbon cycling during global change, are poorly understood (https://ui.adsabs.harvard.edu/abs/2022NRvEE...3.516C/abstract). Wildfires are the main source of BC, producing 128 ± 84 teragrams per year. Negative climate-BC feedbacks could arise as wildfire increases with anthropogenic warming, producing more BC, which in turn will sequester carbon, but the magnitude of these effects is unknown. Most BC is stored in terrestrial systems, with some transported to the ocean via rivers and the atmosphere. However, the oceanic BC budget is not balanced, with known BC removal fluxes exceeding BC inputs. Coppola et al. (2022) demonstrated these observed inconsistencies using a simple ocean box model, which highlights key areas for future research. Black carbon alone is estimated to have a 20-year Global Warming Potential (GWP) of 4,470, and a 100-year GWP of 1,055–2,240. Fossil fuel soot, as a result of mixing with cooling aerosols and particulate matter, has a lower 20-year GWP of 2,530 and a 100-year GWP of 840–1,280.

In understanding the global carbon cycle, "black carbon"—decay-resistant carbon molecules altered by exposure to fire or combustion—has long been presumed to originate on land and work its way to the ocean via rivers and streams. Black carbon aerosol alters atmospheric circulation patterns due to strong absorption of solar radiation. Adjustments in circulation are important in driving regional changes in precipitation, convection, and cloud. Circulation adjustments have little impact on global-mean precipitation and effective radiative forcing (https://agupubs.onlinelibrary.wiley.com/doi/abs/10.1029/2019JD030568). There's a stubborn, heat-absorbing particle that floats along in Earth's atmosphere: It initially doesn't like water, it absorbs light, and it takes its time moving on. Black carbon in the atmosphere tends to linger until it finally absorbs enough water to fall from the sky. In the meantime, black carbon absorbs the Sun's energy and heats up the surrounding air, creating a radiative effect. Fresh, young black carbon tends to be resistant to water. Over time, the particles age and become more hygroscopic, or able to absorb water from the air. But when does black carbon start absorbing water, act as cloud nuclei, and remove itself from the atmosphere? Researchers previously investigated the hygroscopic conditions of black carbon in the lab, with limited conditions on chemical sources and water vapor conditions. In all of these studies, the cloud nucleation values of black carbon were indirect measurements (https://eos.org/research-spotlights/how-long-do-black-carbon-particles-linger-in-the-atmosphere). Hu et al. (2021) noted that their work can help improve estimates of the longevity of suspended black carbon particles in the atmosphere and, therefore, the radiative impacts those particles can have.

3.12 CONCLUSION

It is clear that deep and immediate carbon dioxide reductions are required to protect the long-term climate. Besides, our understanding of the impact of aerosols on climate is not as good as our understanding of the impact of greenhouse gases, such as CO_2, on the atmosphere. This is because the radiative properties of aerosols vary widely and also have large temporal and spatial variations. In addition, there is a complex interaction between natural aerosols, such as dust and salt, and anthropogenic aerosols such as black carbon, organic carbon, and sulphate. Filters are available to trap 90 percent of the black carbon emitted from diesel engines. Eliminating unnecessary idling of diesel engines also reduces the black carbon emissions. Any mechanism targeted toward reducing soot BC should aim at all combustion activities within humankind's control and at large geographic scales, since its climate and pollution effects are far-reaching.

Because black carbon is very short-lived in the atmosphere, its negative effects can be quickly eliminated if emissions are reduced. Government policies to reduce black carbon emissions could yield results very quickly. Open fires and inefficient cookstoves are major polluters; they account for over 20% of global black carbon emissions. A localized solution for reducing black carbon emissions is the use of improved cookstoves that enhance the efficiency of biofuels, reduce total emissions, and have health benefits.

Improved brick manufacturing methods, less open burning in agriculture, better fuel and vehicle standards, cleaner household energy, easier access to financing and technology, and a global push for clean air are all contributing to the decline in black carbon emissions, which is predicted to continue as efficiency gains and the electrification of automobiles and other vehicles continue. Co-emitted chemicals can occasionally obscure the effects of black carbon. The kind and number of co-pollutants vary depending on the source, and some have a cooling influence on the environment. The most promising sources for mitigation and short-term climate and health benefits are those that emit a high ratio of warming to cooling pollutants. Similar to how asphalt surfaces produce heat islands in cities, black carbon, when hung in the atmosphere, contributes to warming by transforming incoming solar energy into heat. In addition, black carbon affects local weather and precipitation patterns and cloud formation. The primary environmental cause of ill health and early mortality is fine particulate matter (PM2.5) air pollution, of which black carbon is a component. These particles can enter the lungs deeply, are several times smaller than a grain of table salt, and help carry harmful substances into the circulation.

Long-term exposure to PM2.5 air pollution is linked to over 4 million deaths annually. Ecosystem health is impacted by black carbon in a number of ways. When the fine particles land on plant leaves, they raise the temperature of the plant's surface, reduce the amount of sunlight that reaches the ground, and disrupt the natural flow of rainwater. The effects of shifting rainfall patterns on ecosystems and human livelihoods can be profound. The Climate and Clean Air Coalition, or CCAC, is in Favor of control measures that might cut global black carbon emissions by up to 80% if they are implemented globally by 2030. Net cost savings could be used to accomplish a number of these emission reductions. Public health would greatly benefit from implementing these policies, particularly in poorer nations. Large volumes of black carbon are produced by open burning in agriculture, although no-burn waste management methods are easily substituted. The fossil fuel industry can readily and affordably reduce its black carbon emissions. Although the manufacture of bricks is a significant source of black carbon emissions, kiln design modifications can significantly lower these emissions.

REFERENCES

Ackerman, A.S., O.B. Toon, D.E. Stevens, A.J. Heymsfield, V. Ramanathan, and E.J. Welton. 2000. Reduction of tropical cloudiness by soot. *Science*, 288(5468): 1042–1047. https://doi.org/10.1126/science.288.5468.1042

Amjad, I., R. Noor, A. Khalid, and S. Riaz. 2022. Polar ice caps melting–A hallmark to vanishing ice along with a global climate change- and addressing solutions to it. *J. Environ. Sci. Public Health*, 6: 135–144.

Andreae, M.O., and A. Gelencsér. 2006. Black carbon or brown carbon? The nature of light-absorbing carbonaceous aerosols. *Atmospheric Chem. Phys.*, 6: 3131–3148.

Blanco-Donado, E., P. Ismael, L. Schneider, P. Artaxo, J. Lozano-Osorio, L. Portz, and M.L.S. Oliveira. 2022. Source identification and global implications of black carbon. *Geoscience Front.*, 13(1): 101149. https://doi.org/10.1016/j.gsf.2021.101149

Bond, T.C., J. Doherty, W. Fahey, M. Forster, T. Berntsen, J. DeAngelo, G. Flanner, S. Ghan, B. Kärcher, S. Koch, Y. Kinne, K. Kondo, C. Quinn, G. Sarofim, M.G. Schulz, H. Venkataraman, S. Zhang, N. Zhang, S.K. Bellouin, P.K. Guttikunda, M. Hopke, Z. Jacobson, J.W. Kaiser, Z. Klimont, U. Lohmann, J.P. Schwarz, D. Shindell, T. Storelvmo, S.G. Warren and C.S. Zender. 2013. Bounding the role of black carbon in the climate system: A scientific assessment. *J. Geophys. Res. Atmos.*, 118: 5380–5552.

Chen, L., F. Zhang, P. Yan, X. Wang, L. Sun, Y. Li, X. Zhang, Y. Sun, and Z. Li. 2020a. The large proportion of black carbon (BC)-containing aerosols in the urban atmosphere, *Environmental Pollution*, 263(Part B): 114507. https://doi.org/10.1016/j.envpol.2020.114507

Chen, H., B. Zhuang, J. Liu, S. Li, T. Wang, X. Xie, M. Xie, M. Li and M. Zhao. 2020b. Regional climate responses in East Asia to the black carbon aerosol direct effects from India and China in summer, *J. Climate*, 33: 9783–9800. https://doi.org/10.1175/JCLI-D-19-0706.1

Colicino, E., A. Wilson, and M.C. Frisardi. 2017. Telomere length, long-term black carbon exposure, and cognitive function in a cohort of older men: The VA normative aging study, *Environ. Health Persp.*, 125: 76–81. https://doi.org/10.1289/EHP241

Coppola, A.I., S. Wagner, S.T. Lennartz, M.S. Nicholas, D. Ward, T. Dittmar, C. Santín and M.W. Jones. 2022. The black carbon cycle and its role in the Earth system. *Nat. Rev. Earth Environ.*, 3: 516–532. https://doi.org/10.1038/s43017-022-00316-6

de la Sota, C., M. Viana, M. Kane, I. Youm, O. Masara, and J. Lumbreras. 2019. Quantification of carbonaceous aerosol emissions from cook stoves in Senegal. *Aerosol and Air Quality Res.* 19: 80–91. https://doi.org/10.4209/aaqr.2017.11.0540

Grant-Jacob, J.A and B. Mills, 2022. Deep learning in airborne particulate matter sensing: A review. *J. Phys. Commun.* 6: 122001. https://doi.org/10.1088/2399-6528/aca45e

Hu, D., D. Liu, S. Kong, D. Zhai, Y. Wu, S. Li, S. Ding, S. Zheng, Y. Cheng, K. Hu, Z. Deng, Y. Wu, P. Tian, Q. Liu, M. Huang and D. Ding. 2021. Direct quantification of droplet activation of ambient black carbon under water supersaturation. *JGR Atmospheres*, 126(13): e2021JD034649.

IPCC, Intergovernmental Panel on Climate Change. 2007. Climate change 2007: The scientific basis. In: Solomon, S. et al. (eds.) *Contribution of Working Group I to the Fourth Assessment Report of the Intergovernmental Panel on Climate Change*. New York: Cambridge University Press.

IPCC, R.K. Pachauri, and L.A. Meyer (eds.) 2014. Climate change: Synthesis report. *Contribution of Working Groups I, II and III to the Fifth Assessment Report of the Intergovernmental Panel on Climate Change.*

Jennifer, B., and V. Ramanathan, 2014. Recent climate and air pollution impacts on Indian agriculture. *Proc. Natl. Acad. Sci.*, 111(46): 16319–16324.

Jenssen, B.M., G.D. Villanger, K.M. Gabrielsen, J. Bytingsvik, T. Bechshoft, C. Sonne, and R. Dietz. 2015. Anthropogenic flank attack on polar bears: Interacting consequences of climate warming and pollutant exposure, *Front. Ecol. Evol., Sec. Interdisciplinary Climate Studies*, 3. https://doi.org/10.3389/fevo.2015.00016.

Jurányi, Z., M. Zanatta, M.T. Lund, B.H. Samset, R.B. Skeie, S. Sharma, M. Windisch and A. Herber. 2023. Atmospheric concentrations of black carbon are substantially higher in spring than summer in the Arctic. *Commun. Earth Environ.*, 4: 91. https://doi.org/10.1038/s43247-023-00749-x.

Konovalov, I.B., D.A. Lvova, M. Beekmann, H. Jethva, E.F. Mikhailov, J-D. Paris, B.D. Belan, V.S. Kozlov, P. Ciais, and M.O. Andreae. 2018. Estimation of black carbon emissions from Siberian fires using satellite observations of absorption and extinction optical depths, *Atmos. Chem. Phys.*, 18: 14889–14924. https://doi.org/10.5194/acp-18-14889-2018

Lavoué, D., C .Liousse, H. Cachier, B .Stocks and J. Goldammer. 2000. Modeling of carbonaceous particles emitted by boreal and temperate wildfires at northern latitudes. *JGR Atmospheres*, 105 (D22): 26871–26890. https://doi.org/10.1029/2000JD900180ff

Li, W., P. Ge., M. Chen., J. Tang., M. Cao., Y. Cui., K. Hu., and D. Nie. 2021. Tracers from biomass burning emissions and identification of biomass burning. *Atmosphere*, 12: 1401. https://doi.org/10.3390/atmos12111401.

Mani, M.S. 2021. *Glaciers of the Himalayas: Climate Change, Black Carbon, and Regional Resilience*. World Bank. https://openknowledge.worldbank.org/entities/publication/8b7f8f18-f83a-5d2f-bbf5-0bd2fd805027

National Academies of Sciences, Engineering, and Medicine. 2016. Climate and health challenges posed by Black carbon. *Proceedings of a Workshop–in Brief.* The National Academies Press. https://doi.org/10.17226/23648.

Niranjana. R and A.K. Thakur, (2017) The toxicological mechanisms of environmental soot (black carbon) and carbon black: Focus on oxidative stress and inflammatory pathways. *Front. Immunol.* 8: 763. https://doi.org/10.3389/fimmu.2017.00763

Pullabhotla, H.K., and M. Souza. 2022. Air pollution from agricultural fires increases hypertension risk. *J. Environ. Econ. Manag.* 115: 102723 https://doi.org/10.1016/j.jeem.2.102723

Ramanathan, V., and G. Carmichael. 2008. Global and regional climate changes due to black carbon. *Nat. Geoscience*, 1: 221–222. www.sciencedaily.com/releases/2008/03/080323210225.htm

Ramanathan, V., P.J. Crutzen, J.T. Kiehl, and D. Rosenfeld. 2001. Aerosols, climate, and the hydrological cycle. *Science*, 294: 2119–2125.

Sarkar, S., R.P. Singh, and A. Chauhan .2018. Crop residue burning in northern India: Increasing threat to Greater India. *JGR Atmospheres*, 123: 6920–6934. https://doi.org/10.1029/2018JD028428

Seinfeld, J. 2008. Black carbon and brown clouds. *Nature Geosci.* 1: 15–16. https://doi.org/10.1038/ngeo.2007.62

Shrestha, G., S.J. Traina and C.W. Swanston, 2010. Black carbon's properties and ole in the environment: A comprehensive review. *Sustainability*, 2: 294–320. https://doi.org/10.3390/su2010294

Singh, P., A. Roy, D. Bhasin, M. Kapoor, S. Ravi, and S. Dey. 2021. Crop fires and cardiovascular health – A study from North India. *SSM – Population Health*, 14: 100757.

Singh, D., S.K. Dhiman, V. Kumar, R.Babu, K. Shree, A. Priyadarshani, A. Singh, L. Shakya, A. Nautiyal and S. Saluja, 2022. Crop residue burning and its relationship between health, agriculture value addition, and regional finance. *Atmosphere*, 13: 1405. https://doi.org/10.3390/atmos13091405.

Watson, J.G. 2002. Visibility: Science and regulation. *J. Air Waste Manag. Assoc.*, 52(6): 628–713. https://doi.org/10.1080/10473289.2002.10470813

Williams, E., D. Rosenfeld, N. Madden, J. Gerlach, N. Gears, L. Atkinson, N. Dunnemann, G. Frostrom, M. Antonio, B. Biazon, R. Camargo, H. Franca, A. Gomes, M. Lima, R. Machado, S. Manhaes, L. Nachtigall, H. Piva, W. Quintiliano, L. Machado, P. Artaxo, G. Roberts, N. Renno, R. Blakeslee, J. Bailey, D. Boccippio, A. Betts, D. Wolff, B. Roy, J. Halverson, T. Rickenbach, J. Fuentes, and E. Avelino. 2002. Contrasting convective regimes over the Amazon: Implications for cloud electrification. *J. Geophys. Res.* 107: NO. D20. 8082, https://doi.org/10.1029/2001JD000380

Zhou, S., Yang, J., Wang, W.-C., Zhao, C., Gong, D., and Shi, P. 2020. An observational study of the effects of aerosols on diurnal variation of heavy rainfall and associated clouds over Beijing–Tianjin–Hebei. *Atmos. Chem. Phys.*, 20: 5211–5229. https://doi.org/10.5194/acp-20-5211-2020.

4 Perspectives on Surface Ozone Pollution on Crop Yields and Environment

4.1 HISTORY OF OZONE DEPLETION: A TEAM EFFORT BY SCIENTISTS

People have associated ozone with thunderstorms ever since the beginning of time. It was initially identified as a distinct chemical molecule by Christian Friedrich Schönbein, a chemistry professor at the University of Basel beginning in 1828. Using the Greek word for "to smell," ozein, Christian Friedrich Schönbein coined the term "ozone" in the 1830s to describe the gas. In 1913, French physicists Henri Buisson and Charles Fabry discovered the ozone layer in the upper atmosphere. Using the Fabry-Pérot interferometer, Fabry and Buisson determined the effect of Doppler on light in a controlled laboratory setting. In the 1913 experiment, the spectrum of the missing radiation was matched to a single molecule: ozone. The ozone layer, which protects Earth from damaging UV radiation, has been found by scientists to lie between 12 and 18 miles above the surface. The ozone layer protects us from dangerous levels of ultraviolet (UV) radiation that can cause skin cancer, cataracts, and other eye problems if allowed to reach the Earth's surface.

Guy P. Brasseur (2020) enthrallingly detailed the most current ozone research breakthrough. This interesting story will explain why ozone is so important in the modern world. Scientists have been covering up the fact that depletion of the ozone layer in the atmosphere is caused in large part by human activities. The survival of the ozone layer is crucial to the efficacy of any census. Spreading urbanization and growing manufacturing and vehicle industries will produce events characterized by high concentrations of ozone pollution, which has been linked to adverse health consequences and increased mortality.

After its discovery in 1839 (Schönbein, 1840), ozone was studied extensively by scientists for several decades. For a while, its chemical makeup remained a mystery. Soret (1865) in Geneva first noticed that ozone molecules contain three oxygen atoms. Houzeau (1858) was the first to show that atmospheric ozone exists as a chemically stable molecule. As could be expected, this has piqued the interest of the weather community. Several spectrometers were put to use by Dobson in the 1920s. Ozone is created when oxygen molecules are broken apart by sunlight in the stratosphere, as proposed by Chapman's hypothesis (1930). Paul Crutzen (1970) demonstrated that minute amounts of nitrogen oxides in the stratosphere can hasten the destruction of the ozone layer. No one had ever suspected before that nitrogen oxide existed in the stratosphere. Nitric oxide (NO) is completely destroyed by sun UV light at altitudes greater than 60 kilometers, according to scientific studies. Due to its inability to pass into the stratosphere, its impact on the ozone layer is minimal below 50 km in altitude (Brasseur and Nicolet, 1973). Airborne oxygen, according to Nicolet's (1971) theory, generates N_2O by interacting with a molecule generated by soil bacteria. According to a new model, nitrogen oxides are produced in the mesosphere and could be transported to lower altitudes in the polar regions throughout the winter (Brasseur, 1984).

DOI: 10.1201/9781003618164-4

In the spring, when the Sun returns to high latitudes and restarts photochemistry, this mechanism is responsible for a significant proportion of stratospheric ozone depletion. In 1974, when it was discovered by Molina and Rowland that chlorine in the stratosphere was destroying ozone, a new "ozone war" broke out (Molina and Rowland, 1974). Studies by Solomon et al. (1986) and Solomon (1999) show that the chlorine action on particles in the polar stratospheric cloud layer is the primary driver of ozone depletion. Essentially, this discovery means the "battle for the ozone layer" has been won. Extremely high quantities of ozone were measured in the troposphere during summer smog episodes in Los Angeles, as reported by Haagen-Smit (1952), causing damage to plant leaves.

Scientists were more worried about the potential loss of stratospheric ozone; therefore, they didn't pay enough attention to Paul Crutzen's work, even though he had built a model of tropospheric ozone as early as 1972. Photo-oxidation of methane results in the creation of ozone, as established by Chameides and Walker (1973), which is a major step forward. Several studies have shown that the OH radical is critically important to the atmosphere. It has been argued that basic research efforts and the quest for socially relevant answers to build a sustainable society may coexist (Brasseur, 2020).

For their contributions to atmospheric chemistry, especially in regards to ozone formation and degradation, Paul Crutzen, Sherwood Rowland, and Mario Molina were awarded the Nobel Prize in Chemistry in 1995 (NobelPrize.org). Important research released in 1974 established the connection between spray bottles carrying chlorofluorocarbons (CFCs) and ozone depletion. The discovery made scientists Paul Crutzen, Mario Molina, and F. Sherwood Rowland household names. Crutzen discovered that nitrogen pollution might deplete ozone in the atmosphere in 1970, which is crucial for protecting life from UV rays. The ozone layer's past could be the key to the future. According to Molina and Rowland, "human usage of CFCs was to continue at an unchanging rate," which scientists predicted would result in "several percent loss of the ozone layer" and the "hole" forming in some decades (1985). It's common knowledge that every year in the spring, as the weather warms up in Antarctica and chemical reactions between CFCs and ozone become more active, this phenomenon manifests itself. As the Antarctic winter progresses, the hole seals up as a result of the low temperatures. The Antarctic ozone hole is just one among several. In 2003 and 2005, a small hole was discovered over Tibet. Significant enough thinning to be classified as a hole has been confirmed by scientists above the Arctic. Scientists say the ozone hole is getting smaller: Scientists estimate that the chasm would have been 40% larger by 2013 had the Montreal Protocol never been implemented. Instead, by 2050, the hole should have totally closed. Johnson and other scientists see hope in the international effort to fix the ozone hole, despite the sometimes-bleak mood in the scientific community. Johnson claims that the ozone hole is closing and that this trend will continue.

4.2 PROMOTING OZONE AWARENESS AMONG THE GENERAL PUBLIC

Ozone is a gas that occurs naturally in the atmosphere. Ozone molecules, which are represented chemically as O_3, are composed of three oxygen atoms each. There are two main ozone pockets in the atmosphere. The troposphere, which extends from the ground up to a height of about 16 kilometers (10 miles), contains just around 10 percent of the atmosphere's total ozone. Roughly 90 percent of the planet's remaining ozone layer is found in the stratosphere, an atmospheric layer that extends from the tropopause to a height of about 50 kilometers (31 miles). Dobson units (DU), generally abbreviated as "DU," are frequently used to report total ozone column levels. Typical levels fall between 200 and 500 DU, with an average abundance of around 300 DU (https://ozone.unep.org/20-questions-and-answers). To achieve a total ozone of 300 DU, only a layer of pure ozone gas 3 millimeters (0.12 inches) thick at the Earth's surface is required. Astonishingly, a layer of pure ozone barely 3 millimeters thick protects life on Earth's surface from the sun's harmful UV radiation. Ninety percent of the Earth's ozone layer is located in the stratosphere, the layer of the atmosphere that extends from about 10 to 50 kilometers in altitude. The ozone layer protects us from the sun's harmful ultraviolet (UV) rays by acting as a screen. The ozone layer in Earth's stratosphere blocks harmful ultraviolet (UV) radiation from the sun. The ozone layer is considered "good" ozone because it protects against UV-B radiation, which can cause skin cancer and other skin problems (https://biologyreader.com/ozone-formation.html). Too much ultraviolet radiation B (UV-B) at the Earth's surface can be harmful to plant and animal life. Ozone, which is located 15 kilometers above the Earth's surface, acts as a shield against the sun's harmful ultraviolet rays.

Getting enough of this essential mineral is critical for maintaining good health, especially considering that vitamin D shortage is another potential health catastrophe (www.suzannegazdamd.com/blog/vitamin-d-deficiency-is-it) caused by internal factors or food intake, respectively. Most people can get enough vitamin D by spending time in the sun (UVB 290-315 nm). The provitamin 7-dehydrocholesterol in the human body is converted into vitamin D when exposed to UV radiation (280-305 nm). Because vitamin D insufficiency is so common (Morales-González et al., 2019), it is important to supplement the diet. The skin's ability to synthesize vitamin D in response to UV light is commonly regarded to play a crucial role in ensuring adequate vitamin D levels in the blood. Vitamin D production in human skin begins only at a certain amount of incoming UV radiation. Vitamin D intake from food and sun both contribute to overall vitamin D levels in the body. When determining vitamin D levels in a population, it is important to account for lifestyle, cultural, and behavioral factors. High amounts of ambient ozone may limit vitamin D production, leading to an increase in histologic osteocalcin and, eventually, bone fractures, especially among the elderly. When exposed to the sun's ultraviolet (UV) rays for an extended period, most people experience a painful sunburn. Cataracts have been linked to prolonged exposure to high levels of ultraviolet radiation. In the United States, cataracts account for about half of all cases of blindness (Bikle, 2014). Two of the health benefits of sun exposure include the skin's ability to produce vitamin D and the immune system's modulation by UV light. Psoriasis patients have benefited from the latter, and those with other systemic autoimmune diseases like MS should see similar results. The harmful effects of sun exposure can be mitigated by taking precautions such as seeking shade, using sunscreen liberally, and donning UV-blocking sunglasses and clothes (https://pubs.rsc.org/en/content/articlelanding/2019/pp/c8pp90060d).

4.3 OZONE LAYER EVOLUTION

For almost two billion years, blue-green algae have been harnessing the sun's energy to recombine H_2O and CO_2 molecules into organic compounds and atomic oxygen, making them one of the earliest aquatic plants. Photosynthesis is the process through which plants convert the energy from sunlight into chemical energy. A rise in atmospheric oxygen caused by photosynthesis likely contributed to the demise of the earliest anaerobic species (Baldwin and Dameris, 2007). The ratio of oxygen to carbon dioxide went up as oxygen levels rose. Some molecules of atmospheric oxygen (O_2) were broken down into their component atoms of oxygen (O) after being exposed to the Sun's ultraviolet (UV) radiation. When these elements combine with oxygen molecules, they produce ozone (O_3), which is a potent UV absorber (www.albany.edu/faculty/rgk/atm101/ozone.shtml). The ozone layer protects Earth from the sun's radiation. An estimated 600 million years ago, Earth's ozone layer was high enough to protect its people from ultraviolet (UV) light with wavelengths between 200 and 300 nanometers (nm). The percentage of oxygen in the air was just around 10% of what it is now. Before that point, all life forms existed in the water. Land-dwelling organisms benefited from ozone since it allowed them to evolve and thrive.

4.4 HOW IS STRATOSPHERIC OZONE PRODUCED IN THE ATMOSPHERE?

Chapman, a British scientist, proposed a method for producing ozone in the stratosphere (Chapman, 1930). The basic steps of the Chapman mechanism are as follows: 1) Above 30 kilometers in altitude, reactive oxygen species are produced when molecular oxygen is photo-decomposed by sunlight (wavelength 242 nm).After being struck by a photon, an oxygen molecule breaks down into its constituent reactive atoms hv. These atoms form ozone (O_3) when they react with an oxygen molecule. The ozone layer is created when UV rays strike oxygen molecules, converting them into ozone (O_3) and atomic oxygen (O). This method is known as the Chapman cycle. During its participation in the Chapman cycle, oxygen is photolyzed by UV light, creating oxygen radicals.

Ozone is naturally created and destroyed by the following set of reactions:

$$O_2 + UV(<242nm) \rightarrow O + O \ (1)$$

$$O + O_2 + M \rightarrow O_3 + M \ (2)$$

$$O_3 + UV/visible \rightarrow O + O_2 \ (3)$$

$$O_3 + O \rightarrow O_2 + O_2 \ (4)$$

$$O_2 + UV(<242nm) \rightarrow O + O$$

In 1930, Sydney Chapman explained the occurrence of ozone in the stratosphere. For millions of years, the Chapman cycle has been steadily producing and depleting ozone. This causes substantial variations in ozone layer thickness throughout geologic timescales. In addition, photosynthesis provides a steady supply of new O_2 to the atmosphere, increasing the prospect that the ozone layer will regenerate on its own.

4.5 OZONE DEPLETION AND THE OZONE HOLE FORMATION

In the mid-1980s, a "hole" was discovered in the ozone layer above Antarctica, identifying the region of the Earth's atmosphere where ozone depletion was greatest. Ozone depletion is mostly attributable to the production and distribution of synthetic halogen chemicals. These chemicals are known as ozone-depleting substances (ODSs) (https://helpsavenature.com/ozone-layer-depletion-effects-causes). Every day, individuals all around the world use ODSs in millions of products.

The most widely used ODSs, chlorofluorocarbons (CFCs), were formerly widespread in appliances like refrigerators, air conditioners, aerosol cans, and asthma inhalers. In addition to CFCs and HCFCs, chemicals like halon and methyl bromide are also harmful to the ozone layer. Ozone is naturally produced in the stratosphere by a two-step reactive process. In the first step, solar ultraviolet radiation (sunlight) breaks apart an oxygen molecule to form two separate oxygen atoms. In the second step, each atom then undergoes a binding collision with another oxygen molecule to form an ozone molecule.

When a CFC molecule absorbs UV light in the stratosphere, the chlorine atoms in it are freed. UV radiation frees a chlorine (Cl) atom from a CFC molecule; this atom reacts with ozone and produces an oxygen gas molecule and chlorine monoxide (ClO) (https://ecampusontario.pressbooks.pub/envirobiology/chapter/10-2-ozone-depletion/). ClO reacts with another oxygen atom, freeing up the Cl atom, which then proceeds to destroy another ozone molecule. The Destruction Process of the Ozone (O_3) layer that occurs on the Earth (Figure 4.1) is explained. This Figure clearly explains the reaction (https://commons.wikimedia.org/wiki/Category:Ozone#/media/File:Chlorine catalysis.GIF). Ozone layer depletion is the thinning of the ozone layer in the upper atmosphere. This happens when chlorine and bromine atoms in the atmosphere destroy ozone molecules.

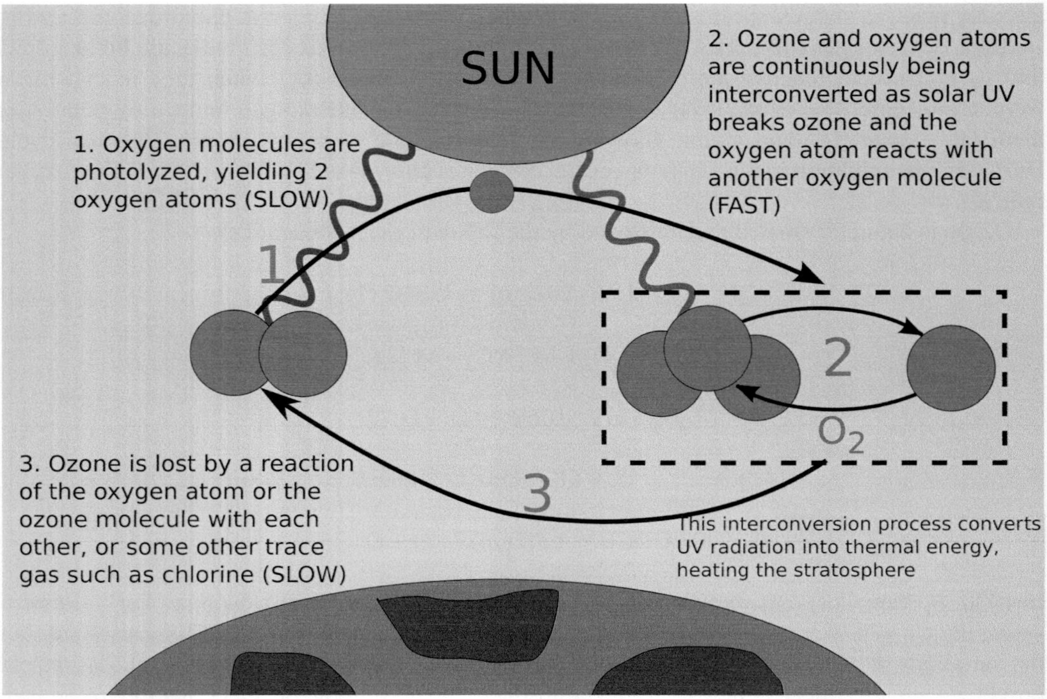

FIGURE 4.1 The destruction of O_3 on the Earth (Environmental Biology Copyright © 2017 by Matthew R. Fisher is licensed under a Creative Commons Attribution 4.0 International License, except where otherwise noted.)

A single chlorine atom can kill thousands of ozone molecules through interaction with just one. Ozone is depleted at a faster rate than it can be replenished due to chlorine and bromine interactions (www.whatarethe7continents.com/the-ozone-layer-what-how-formed-important/). Thanks to the Montreal Protocol, the use of chemicals that deplete the ozone layer has been prohibited. One worldwide model estimates that two million annual occurrences of skin cancer may be prevented by 2030 if the Montreal Protocol were fully implemented. One of the sustainable development objectives adopted in 2015 by all UN Member States was to enhance the health and happiness of all people. The Montreal Protocol has contributed significantly to accomplishing this goal. Plants can't protect themselves from UV-B rays because they need sunlight for photosynthesis. They have evolved defenses to avert or repair damage, such as pigments that absorb UV light. Unchecked ozone depletion (https://cimss.ssec.wisc.edu/wxwise/ozone/OZONE2.html) would have dramatically skewed this equilibrium in a negative way.

Increased UV radiation can have detrimental impacts on aquatic food chains, specifically on crustaceans and fish eggs. UV-B radiation interferes with many different types of biological processes in marine, freshwater, and riparian environments. Therefore, the world's food supply would have been at jeopardy if ozone depletion had been allowed to continue uncontrolled, threatening fisheries and other aquatic resources. If levels of ozone in the stratosphere were to be reduced by 10%, plant yields may fall by as much as 6%. Without the Montreal Protocol, ozone depletion would have impeded efforts to end world hunger as a sustainable development goal. These effects on ecosystems have the potential to alter the exchange of carbon dioxide between the atmosphere and the biosphere in response to global increases in UV-B. An increase in UV light also speeds up the decay of dead leaves and other organic matter. Overall, an increase in UV-B would reduce ecosystems' ability to sequester carbon dioxide, particularly carbon dioxide produced by humans. UV-B light is destructive to both organic and synthetic materials. Some solar panel materials, in addition to wood, plastic, and rubber, are vulnerable.

4.6 OZONE-DEPLETING COMPOUNDS

Chemicals that are harmful to the ozone layer have been used to clean many of the parts of our electronic devices and home appliances. Many commonplace items, such as car dashboards, building insulation, water boilers, and even shoe soles, are manufactured using chlorofluorocarbons (CFCs) and hydrofluorocarbons (HCFCs). Offices, data centers, military outposts, aircraft, and ships were just some of the many commercial and military structures that used Halon for fire protection. Many of the fruits and vegetables we ate had been fumigated with methyl bromide. As evidence accumulated showing that CFCs were degrading the ozone layer, and awareness of the different repercussions of uncontrolled depletion grew, scientists and policymakers urged countries to regulate their use of CFCs. Refrigeration and air conditioning made extensive use of chlorofluorocarbons (CFCs). CFCs were required to be phased out entirely per the Montreal Protocol. CFCs were gradually replaced by HCFCs and then HFCs. To protect the ozone layer, CFCs have been completely phased out in favor of hydrofluorocarbons (HFCs) and other alternatives. The rapid and complete elimination of CFCs in aerosols was one of the greatest achievements of the Montreal Protocol. Asthma and COPD inhalers containing CFCs were more challenging to replace. There are now effective, cost-effective, and CFC-free substitutes for all inhalation therapies on the market (https://ozone.unep.org/ozone-and-you). Once widely employed, the ODS methyl bromide has since been banned from use in agriculture and food storage due to its effectiveness against pests and diseases. As a result of the Montreal Protocol, more humane approaches to eradicating pests and diseases have developed. The successful restoration of the ozone layer not only sets an important precedent for the future but also shows how to address other worldwide environmental challenges.

4.7 OZONE AT THE SURFACE

Only a tiny fraction of Earth's total ozone is found in the troposphere, yet this small amount is considered "bad" due to its interaction with other molecules. It is well-known that exposure to tropospheric ozone is harmful to agricultural productivity, forest health, and human health. The photochemical smog that plagues many cities is caused in large part by atmospheric ozone. Breathing in ground-level ozone can cause a variety of unpleasant health effects, including chest pain, coughing, throat irritation, and congestion. Preexisting illnesses, including bronchitis, emphysema, and asthma, can be made worse by ozone's link to reduced lung function and inflammation of the lung lining.

4.7.1 The Production of Ozone in the Troposphere

Ground-level ozone is a very reactive secondary contaminant in the atmosphere. This type of pollution is the result of chemical interactions between major pollutants like hydrocarbons, nitrogen oxides, and sunshine. Photochemical smog is a major cause of respiratory irritation, and ozone is a major component of this pollution. Ground-level ozone is responsible for the unpleasant odor of photochemical smog and also poses a health danger, especially to the lungs of young children. Ground-level ozone is created when certain man-made and/or naturally occurring substances are exposed to sunlight and through a series of chemical reactions (https://csl.noaa.gov/assessments/ozone/2010/twentyquestions/Q2.pdf).

The lifetime of tropospheric ozone is only a few hours to a few weeks. It is created when sunlight reacts with hydrocarbons (including methane) and nitrogen oxides, but it is not directly caused by human activity. Efforts have been undertaken to reduce air pollution from human activities, including farming and the distribution of fossil fuels, which is a major contribution to the production of tropospheric ozone. Ozone is created when several different air contaminants react in the troposphere. These contaminants have several origins, some being natural and others being anthropogenic. Precursor pollutants result from human activity and include hydrocarbons and nitrogen oxides. The most common source of these chemicals is automobile exhaust, but they can also come from other places. Smog is full of ozone from the troposphere, and breathing it in can worsen asthma, bring on bronchitis, and even permanently damage lung tissue. Exposure to dangerous quantities of ozone in the environment is responsible for the deaths of approximately one million people annually. Everyone is at risk from ozone's negative effects, but infants, the elderly, and people with preexisting lung or heart conditions are most vulnerable.

Ozone in the troposphere, where crops are grown, is always present at a base concentration due to natural sources of the precursors. Complex photochemical interactions between oxides of nitrogen (NOx), carbon monoxide, and non-methane volatile organic compounds (NMVOCs) from human-caused sources create extra-tropospheric ozone. The result is an increase in hunger. Conditions for ozone formation are different in the troposphere and the stratosphere of Earth's atmosphere. When exposed to heat or sunlight, oxygen molecules undergo chemical changes or break down into individual atoms; these atoms can then combine with neighboring oxygen to produce stratospheric ozone. When VOCs, NO_X, and other pollutants are heated or exposed to sunlight, ozone is produced in the troposphere through a chemical process. Ozone is created when gaseous pollutants and natural gases in the troposphere react chemically, leading to the formation of a layer of ozone (Crutzen, 1970). Exposure to the trace amounts of ozone that exist naturally in the atmosphere has no health risks to humans. The soil and certain plants are responsible for producing this ozone. Because it contains three oxygen atoms, the chemical symbol for ozone is the number 3. Therefore, ozone formation is possible whenever air is present (Figure 4.2).

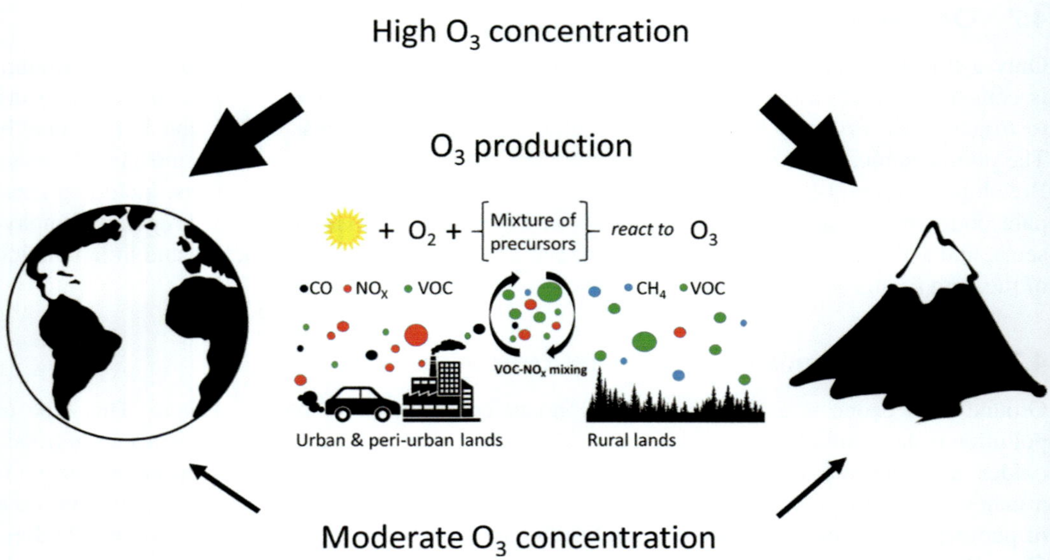

FIGURE 4.2 Scheme of tropospheric ozone (O_3) formation (Juráň et al., 2021., https://doi.org/10.3390/atmos12010082. Open access.)

Ozone is formed in the troposphere when pollutants like nitrogen oxides (NO_X) and volatile organic compounds (VOCs) undergo chemical reactions when exposed to heat or sunshine. Natural gases and gaseous pollutants such as hydrocarbons, nitrogen oxide gases, carbon monoxide, and hydroxyl radicals all contribute to the chemistry in the troposphere that leads to the formation of ozone. VOCs are released into the atmosphere through industrial processes, the application of paints and solvents, and the combustion of fuels like gasoline. Some of the most abundant natural sources of volatile organic molecules are trees and crops.

Plants absorb ozone through their leaves' numerous, microscopic stomatal pores, which normally open throughout the day to allow photosynthesis and transpiration. Stomata allow ozone to enter the plant; hence, the wider the stomata, the more ozone the plant is exposed to. Reactive oxygen species can kill plant cells and disrupt vital activities like photosynthesis once they have broken through a plant's cell walls and membranes. When ozone levels in the atmosphere rise above the "detoxifying threshold," the resulting harm to crops includes lower yields, lower quality products, and increased vulnerability to environmental pressures like drought.

Because of its ability to scatter the sun's ultraviolet rays when it is high in the stratosphere, ozone is good for life on Earth. For instance, a recent spike in skin cancer cases in Australia has been related to a hole in the stratospheric ozone layer. However, when ozone is present at ground level (in the troposphere), it can enter the lungs and induce a variety of unpleasant side effects. Remembering that ozone is "great up high but terrible nearby" should help.

Ozone molecules are absorbed into the body when you take a deep breath. Ozone triggers chemical changes in the airway lining, resulting in the separation and subsequent reassembly of tissues as it travels via the mouth and nose and into the lungs. Recombination compromises the lining's natural structure, making them less effective at keeping out bacteria and other pathogens. In response to this change, mucus production increases and the airways narrow, reducing breathing capacity. Besides shortness of breath, wheezing, and asthma, other clinical signs include lung infection and an increased risk of heart attack.

4.7.2 Risk from Ground-Level Ozone

The air we breathe contains ozone, and people with previous lung diseases may be more vulnerable to its effects. The effects of ozone on respiratory illnesses may lead to an increase in the need for medical care, including medication use, doctor and emergency department visits, and hospitalizations. There is evidence that ozone exposure contributes to the development of a number of lung ailments (www.trusens.com/guides/ultimate-guide-to-air-pollutants/types-of-ozone-and-their-effects/). Because their lungs are still developing, children and teenagers are at a higher risk of developing asthma. Not only do they spend more time outside, but they are also more susceptible to the weather. Ozone's effects are exacerbated in the elderly due to previous lung illness. People of all ages who spend time outside are more sensitive to ozone. It has already been proven that outdoor ozone levels are far higher than indoor ones. Exercising has been linked to deeper, faster breathing, which in turn increases ozone intake. Those who labor or exercise vigorously in the open air are included here. People whose diets are low in important nutrients are more vulnerable to the effects of ozone exposure. People who don't get enough vitamin C and E are more vulnerable to the effects of ozone exposure.

4.8 DIFFERENTIATE BETWEEN OZONE IN THE STRATOSPHERE AND OZONE IN THE TROPOSPHERE

The process begins in the high atmosphere when ultraviolet (UV) light from the sun breaks apart an oxygen molecule (O_2). Chemical interactions in the troposphere produce ozone, and they involve both naturally occurring gases and pollutants. Toxic ozone can be found in the troposphere, or lower atmosphere. Even at very low doses [in the ppb (parts per billion) range], air and water cleansing can begin. Since it forms when large pollutant emissions, such as NOx, break down. However, the protective ozone layer in the stratosphere helps to mitigate the effects of UV radiation from the sun. It occurs naturally but may be killed with a few simple steps and common household products like freezers and spray paint. One way to tell the difference between stratospheric and tropospheric ozone concentrations is to look at the percentage of ozone present in each layer. The stratosphere is the region of the atmosphere that exists between 10 and 50 kilometers above the surface of the Earth. Ozone reaches its highest density 20 kilometers above the Earth's surface. The concentration of ozone in the stratosphere is very high compared to that in the troposphere, which is very low. Ozone gas is created in the stratosphere when molecules of oxygen combine with atoms of oxygen.

4.9 PLANT SENSITIVITY TO OZONE

The concentration of ozone (O_3) in the atmosphere has been rising continuously since the beginning of the industrial period due to its role as a greenhouse gas and secondary air pollutant. Similarly, to how breathing in O_3 molecules can have harmful effects on humans and animals, the same is true for plants when O_3 molecules enter their leaves and induce phytotoxic damage. The decrease in crop output and quality is a major obstacle to the world's ability to feed its people (Jimenez-Montenegro et al., 2021).

As an oxidant, ozone in the air is harmful to living things and inanimate objects alike. Ozone deactivates Rubisco, an enzyme essential for photosynthesis. When the photosynthetic ability of plants was examined in vivo and in vitro under conditions of increased ozone exposure, a drop in Rubisco concentration was detected, but not in activation state. When a plant's stomata are open and exchanging gas, it can detect O_3. When ozone enters a plant, it can cause damage to the photosynthetic system and increase oxidative stress. When this occurs, the leaves' ability to do photosynthesis decreases, and they turn yellow. Its role in accelerating leaf senescence is also well established. Ozone has been shown to hasten senescence and, at high quantities, to trigger programmed cell death by interacting with DNA in the aqueous apoplast. In vitro and in vivo experiments have shown that C3 crops degrade Rubisco protein and reduce Rubisco function in response to O_3 stress. Antioxidant chemical levels increased in tropical maize in response to oxygen radical stress. This included phenolic compounds, flavonoids, and anthocyanin pigments. Since Rubisco is assumed to be localized in the bundle sheath cells of C4 plants, it is protected from O_3-induced ROS.

Research on Rubisco dates back to 1947, according to a study by Portis and Parry (2007). The electron transport rate was not connected with either the total amount of Rubisco or the activity of Rubisco activase (Rca), but it was correlated with the activation state of Rubisco (Perdomo et al., 2017). It was shown that lower levels of Rubisco activation at HT result from biological restrictions on the net CO_2 assimilation rate (AN). Based on these results, Rubisco should be a primary focus in efforts to improve the photosynthetic efficiency of C3 (wheat, rice) and C4 (maize) cereal crops in more varied and warmer locations. Reduced ATP availability decreases the CO_2 fixation rate since it depends on the regeneration of ribulose-1,5-bisphosphate (RuBP). Many different researchers have looked at how drought stress affects Rubisco activity, and the results have shown either a drastic reduction in activity, no change in activity, or a mild suppression.

Low chloroplast CO_2 has been linked to the deactivation of Rubisco in several Mediterranean-endemic taxa, as demonstrated by Perdoomo et al. (2017). To predict and mitigate the effects of expected climate change on agricultural and natural ecosystems, understanding how Rubisco activity is controlled under heat stress is vital. Water stress had no noticeable effect, but ozone reduced Rubisco activity by 30%. It was shown that the cumulative effects of the stressors on Rubisco activity were on par with those of ozone.

Under high-temperature and high-light circumstances, however, oxygen displays a considerable affinity for the photosynthetic enzyme Rubisco. To rephrase, oxygen hinders the efficiency with which C3 plants convert water and light into sugars (a process known as photorespiration). Materials for both students and educators can be integrated at https://serc.carleton.edu/integrate/teaching-materials-food-supply-student-materials/1167, where Edwards and Walker (1983) discussed how some plants have evolved C4 photosynthesis to thrive in high-temperature, high-light conditions where soil moisture is often scarce. To maximize photosynthesis and water use efficiency, C4 plants have evolved a specialized leaf anatomy and biochemistry that binds carbon dioxide upon entry, forming a 4-carbon molecule that transports and concentrates CO_2 in specific cells near the Rubisco enzyme. Therefore, C4 plants are typically more productive than C3 plants under optimal conditions of high light and heat. Corn, sorghum, sugarcane, millet, and switchgrass are all examples of C4 plants.

Rubisco is the most abundant and critically important protein on Earth. Photosynthesis, the process through which plants increase their growth and yield, requires this enzyme as its fuel. The plant needs sugar for development and metabolism, and Rubisco does this by converting carbon dioxide (CO_2) into sugar. Although Rubisco is essential for plant growth, its poor activity can compromise photosynthesis. This is because the protons raise the concentration of CO_2 within the Rubisco compartments (https://foundationfar.org/news/ripe-project-rubisco-production-can-enhance-co2-acquisition/), which speeds up the Rubisco reaction. To convert bicarbonate into carbon dioxide, protons are used. Photosynthetic species that use bicarbonate as a source of carbon dioxide (CO_2) have much to teach us about how to make our agricultural plants hardier, and aquatic environments rely significantly on this molecule.

According to NASA climate scientist and crop modeler Jonas Jägermeyr, "even under hopeful climate change scenarios, world agriculture is facing a new climate reality." The Rubisco was tweaked. Photosynthesis, the process by which plants produce their own food from carbon dioxide (CO_2), light (photons), and water, requires all three. It is possible to increase crop yields while using the same amount of land if plant photosynthesis can be enhanced. Modern plants' Rubisco enzymes are quite inefficient at drawing CO_2 from the air. The Rubisco enzymes in today's plants have evolved to significantly lower CO_2 levels. Reproducing the ancient form of the enzyme, which evolved at far greater CO_2 concentrations, may theoretically speed up photosynthesis. There is scientific consensus that we can improve crop yields by adapting Rubisco to today's growing conditions (www.freethink.com/science/rubisco).

At the turn of the 21st century, annual increases in background O_3 concentrations in the Northern Hemisphere ranged from 0.52 percent to 2 percent (Vingarzan,2004). Mean O_3 concentrations are expected to rise by 20–25 percent by 2050 and by 40–60 percent by 2100, respectively. This has been proven in several contexts (Wagg et al., 2012). The increasing industrialization of the twentieth century is to blame for this. Emissions of O_3 precursors like NOx or VOCs have increased, particularly in the Northern Hemisphere, due to increased industrialization and the use of fossil fuels over the past century. Due to the high price tag associated with O_3 plant damage, a large body of literature on the topic has been assembled by Jimenez-Montenegro et al. (2021). These writers reviewed studies from all around the world that studied the effects of ozone on plants.

Some cereal grains are consumed on a global scale more than others (*Zea mays* L). One defining characteristic of these plants is their extreme vulnerability to oxygen deprivation. Wheat grain quality is degraded and yields are reduced by an average of 7.1% due to O_3, with losses of up to 15% conceivable in areas such as China and India (Mills et al., 2011). It is estimated that O_3 causes yield losses of between 4.4% and 12.5% in rice fields in places like China, India, Indonesia, and Bangladesh, and between 6% and 15% in maize fields in China and the United States. Yield losses from O_3 in beans were reported at 19.0 percent a decade ago, and by 2030, in Europe's forecasted concentration range, that amount is expected to grow to 20 percent (Jimenez-Montenegro et al., 2021). In countries like the UK, where there is evidence that beans' nutritional content is dramatically diminished by O_3 exposure, the market value of beans has fallen (Mills et al., 2011).

The findings of the study by Yi et al. (2020), indicating that surface ozone concentration is adversely related to maize production, is consistent with previous ecological and economic research in China. Nonlinear regression analysis also revealed that maize is most susceptible to ozone damage during the second month of the growing season. While ozone's impact on immature plants may be negligible at first, research has revealed that this changes substantially as plants mature. Ozone pollution caused a 4.234% annual decrease in maize production between 2013 and 2015, or a 1.9% decline in overall domestic production, according to the authors' analysis of the available data (Yi et al., 2020). Although Shandong is the most productive province for growing maize, it loses more of its annual harvest—about 1 million tons—than any other region.

Changes in plant genetics, metabolism, and physiology brought on by O_3 lead to severe decreases in crop quality and output. Most O_3 studies concentrate on high-value crops such as horticultural crops (tomatoes and lettuce), beans, and species in the genus Brassicaceae (soybean, maize, wheat, and rice). Since the former are essential to human nutrition, a decrease in them could imperil global food supplies. The impact of O_3 on ecosystems increases with time, and it can be amplified by other pressures (such as protracted drought or high nitrogen deposition) (Grulke and Heath, 2019). Because of this shift in the competitive dynamics among species, sensitive species may become extinct as O_3-resistant types are favored through selection.

4.10 POLLUTION EFFECTS ON PLANT DEVELOPMENT

Pollutants can also make holes in the trunk or stem. Before a pollutant can reach the vascular system, it must first pass through the boundary layer and the layer of unperturbed air immediately adjacent to the leaf. Most gaseous contaminants, including carbon dioxide and oxygen, enter plants through stomata on the surfaces of leaves. The foliar surfaces serve as an initial collection point for pollutants in the 1–10 μm size range. Examining the phenotypic expression of air pollution's effects on plant development and yield is one way to learn more about the effects of air pollution on plants. Particles may raise leaf temperature and decrease available light for photosynthesis by altering the leaf's optical characteristics. The possibility of occlusion or damage to stomata, which can have consequences for photosynthesis and respiration, is just one of the numerous negative effects dust has on plants.

Gaseous pollutants take in air in the same way plants take in carbon dioxide and oxygen through their leaf stomata (Kozio and Whatley, 1984). However, the cuticle can effectively absorb a wide variety of organic contaminants due to the high lipid concentration found there. Some will be blocked by the leaf's cell walls, but others will be able to diffuse into the apoplast and react with the symplast there. When plants breathe out, ozone gets sucked into their tissues and causes harm. Yellowing, cell injury, tiny light-tan irregular patches, bronzing, and reddening are some indications of ozone damage in plants. This causes delays in crop maturity and lower overall yields. Soybeans, peanuts, and cotton are just some of the agricultural products that could be especially at risk from ozone damage. Air pollution is predicted to have a much greater impact on China's agricultural output than India's since China's soot and ozone precursor emission rates are significantly higher than India's.

4.11 EFFECTS OF SURFACE OZONE ON CROP YIELDS AND THEIR CAUSES

Combustion sources such as cars, power plants, biomass burning, and more all emit chemicals that can be further broken down into ozone. Ozone enters plants through the stomata in their leaves, which allow oxygen to enter and leave the plant during regular daytime gas exchange. As a result, photosynthesis is slowed, the plant's metabolism is thrown off, and the plant grows more slowly, produces lower-quality crops, and yields fewer fruits and vegetables. Ozone pollution reduces crop yield, endangering food security and costing the economy money. Ozone precursor emissions have increased as a result of economic growth, which is especially worrisome in southern and eastern Asia.

Since the beginning of the Industrial Revolution, ground-level ozone (O_3) concentrations have more than doubled across much of the globe. It is challenging to determine typical agronomic and economic consequences and to breed crops for O_3 tolerance due to the air pollutant's significant geographical and temporal variability. Because of the phytotoxicity it produces, increasing levels of tropospheric ozone pose a serious threat to the world's food supply. The globe over, this is devastating to C3 plants (Debaje, 2014). Scientists from the United States, Japan, and Europe have concluded that tropospheric ozone (O_3) is the most dangerous kind of air pollution. Oxygen usually has a detrimental effect on crop production and quality (Saxena et al., 2020).

Radiation, temperature, and humidity all have a role in determining how long O_3 lasts in the atmosphere and how much of its precursor molecules come from nearby vs distant sources. In the troposphere, O_3 has a mean lifetime of around 25 days (Young et al., 2013), whereas in the summertime surface boundary layer, it has a mean lifetime of fewer than 5 days. Ozone's 0.4 W m^2 of radiation sent into space has contributed to surface warming during the Industrial Revolution, indirectly affecting photosynthesis, transpiration, and respiration rates.

In addition to accelerating the ageing process, ozone harms plants by entering the leaf's intercellular air spaces through stomata and reacting with chemicals on the exposed, moist cell-wall surfaces to form damaging radicals and signaling. At high concentrations, O_3 causes programmed cell death and triggers a chain reaction that produces even more reactive oxygen species. Damage from O_3 depends on how easily it can pass through the stomata, which is affected by environmental conditions such as humidity. Therefore, O_3 damage is highly dependent on environmental and physiological factors that affect stomatal conductance in plants. Visual damage, decreased photosynthesis, changed carbon allocation, and lower yields are only some of how ozone affects agricultural crops. The ozone effect on different leaves is shown in Figures 4.3-4.5.

The most noticeable outward symptom of foliar injury is the appearance of tiny pale yellow, cream, or bronze patches (also called stipples) appearing between leaf veins (Harmens et al., 2020). From the results of two recent experiments conducted in northern India, it appears that farmers can benefit from learning about how various cultivars respond to O_3 pollution (Pandey et al., 2019). Examples of ozone-visible foliar injury (O_3 VFI) from Monitoring Ozone injury have been shown by Moura et al. (2022).

Using flux-based case studies, scientists in India and South Africa looked into how O_3 affects crop output. While 2015 estimations of O_3 flux across India revealed substantial variation, the highest amounts were regularly observed in the country's eastern and central regions. It is anticipated that India's wheat crop will fall short of projections by 11.8 million metric tons, or 11.3%. The two main wheat-growing regions in India are Uttar Pradesh and Madhya Pradesh, and they would be hit particularly severely by O_3 depletion, according to predictions (Pandey et al., 2019). Models of O_3 uptake by wheat and beans were utilized by South African researchers to show that ground-level ozone reduced agricultural output that growing season. In South Africa, O_3 caused a 4.2% average decrease in wheat yields and a 13.3% decrease in bean yields. Throughout the typical growth season, ambient O_3 levels below 50 ppb resulted in lower biomass and yield (December through March). Plant-1 grain weight declined drastically across all cultivars without EDU treatment, by as much as 33% for three types investigated in Varanasi during the 2015/2016 season, 43% for eleven variations tested in Lucknow, and 22% for eleven varieties tested in Banthra (Pandey et al., 2019). Several studies have used a flux-based technique to predict environmental conditions that affect stomatal uptake of O_3. To more accurately analyze the spatial distribution of yield losses due to O_3, O_3 fluxes should be employed instead of O_3 concentrations.

While other locations in South and East Asia have been backed by evidence, the vast majority of that evidence is restricted to a select few nations (such as China, India, and Pakistan). In some cases, ozone depletion can cause a 40 percent drop in harvests. Low oxygen levels are particularly harmful to certain types of beans and soybeans, wheat, maize (corn), and rice (in that order of least investigated) (Harmens et al., 2020).

FIGURE 4.3 Ozone-induced visible foliar injury index. (a) No O_3 injuries but with some insect injury; (b) less than 5% of ozone injuries on the trifoliates; (c) 5% to 25% of bronze patches indicating ozone injury on the leaves; (d) greater than 25% of ozone injuries; and (e) senesced leaf that is yellow instead of green, also showing previous ozone injuries. (Leung et al., 2020. Open Access.)

4.12 OZONE-RELATED PRODUCTIVITY LOSSES

The Iowa Department of Natural Resources showed that dicot species, such as soybeans, cotton, and peanuts, are more susceptible to yield loss from ozone exposure than monocot species such as cereals like sorghum, corn, and wheat (Figure 4.6a-d). Seasonal variations in daily mean surface ozone concentrations showed an effect on the yield of crops (www.iowadnr.gov/Environmental-Protection/Air-Quality/Air-Pollutants/Effects-Ozone.html). Negative effects on plant life and ecosystems have been connected to ozone because it prevents plants from opening the tiny pores on their leaves to breathe. This lowers the efficiency with which plants convert carbon dioxide into oxygen. Increased ozone levels reduce crop and commercial forest yields, decrease seedling growth and increase mortality, and make plants more vulnerable to viruses, pests, and environmental challenges like drought. Reduced global grain yields of 15% have been attributed to ozone's toxicity to plants (Bill Wellock, Florida State University, 2020; https://phys.org/news/2020-01-ozone-pollution-benefit.html). The annual loss to farmers is estimated to be between $10 billion and $25 billion. Seasonal variations in daily mean surface ozone concentrations and the effect of ozone on the yield of Crops are given in graphs (Figure 4.6a-d).

O_3 exposure has the greatest and usually unfavorable influence within the span of predicted environmental changes. However, the repercussions connected with a changing climate may be either negative or favorable, depending on crop, region, and climatic conditions. Human actions, like carbon fertilization, have an impact on yields in addition to natural factors, such as weather and precipitation. Reduced agricultural production is just one of O_3's many dangers to the world's food

FIGURE 4.4 Examples of ozone visible foliar injury (O_3 VFI) from Monitoring Ozone injury for setting new critical levels (MOTTLES) sites and the Free-air O_3 exposure (FO3X), A and G) Homogeneously distributed interveinal reddish (Rd); B, C, F, H and I) Reddish interveinal stippling (RdSt); D, K and L) Dark brownish interveinal stippling (BrwSt); E and J) Chlorosis (Cl). K and L) Homogeneously distributed interveinal brownish (Brw); C, F) Bronzing (Br) (Moura et al. (2022). Ecosystem Health and Sustainability, 8(1). https://doi.org/10.1080/20964129.2022.2144466; http://creativecommons.org/licenses/by/4.0.)

supply. Plants absorb O_3 during gas exchange through their stomata, making it a potential hazard for food crops. Foliar infections, decreased photosynthesis, and reduced yields are only some of the potential adverse results (such as necrosis and chlorosis).

High O_3 levels have devastating consequences on the global food supply across most crop-growing seasons. The impact of current and future O_3 on agricultural output has been modelled in recent studies. Some research has attempted to predict past and future crop yields by using Earth

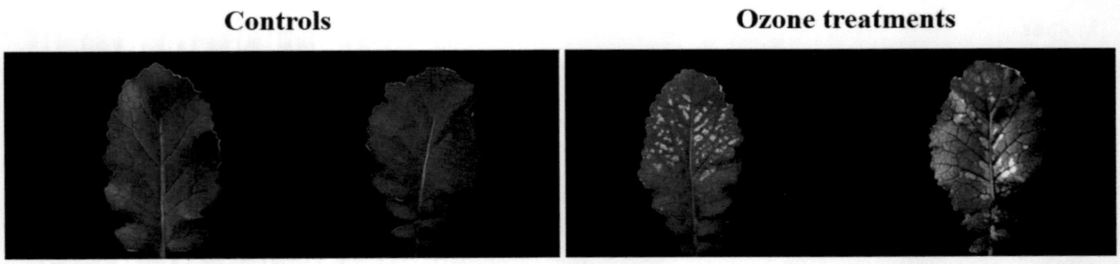

FIGURE 4.5 The visible symptoms of O_3 damage under an O_3 concentration of 100 nL L–1 at 14 days after exposure (Lee et al. (2022). https://doi.org/10.3390/horticulturae8030211.)

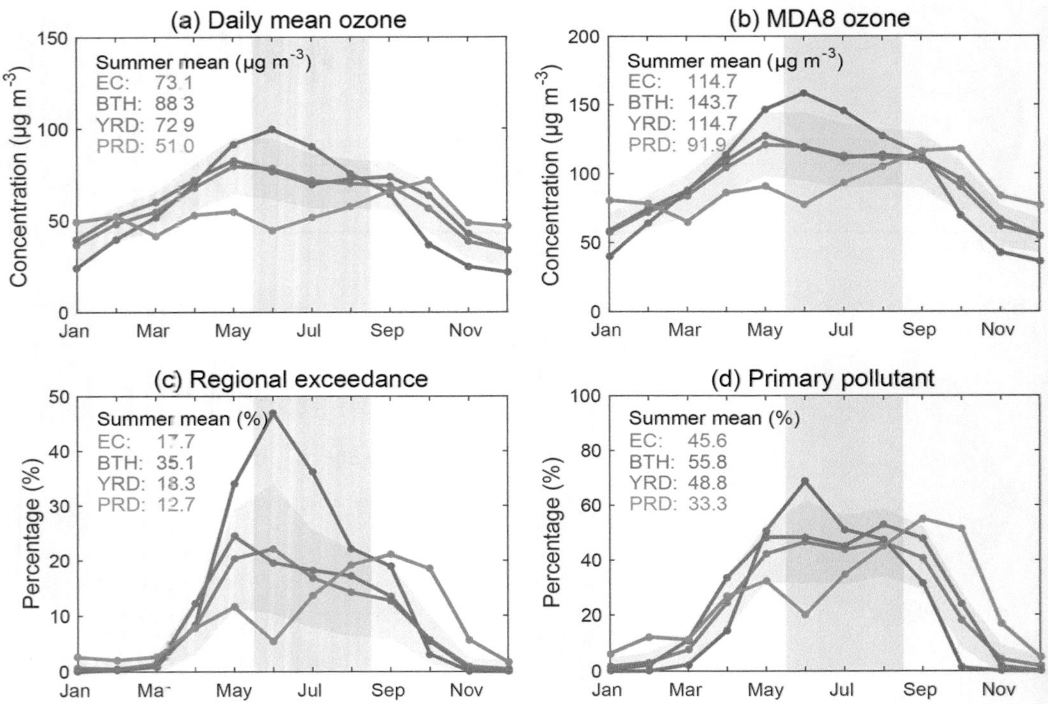

FIGURE 4.6 a-d Seasonal variations in daily mean surface ozone concentrations and the Effect of Ozone on the yield of Crops (Han et al. (2020), https://doi.org/10.5194/acp-20-203-2020.This work is distributed under the Creative Commons Attribution 4.0 License.

system models, atmospheric chemical transport models, and statistical relationships between crop yield and O_3 (Burney and Ramanathan, 2014).

4.13 OZONE'S EFFECT ON CROP YIELDS AND SENSITIVITIES

To damage plants, O_3 must first enter through the stomata and then bind to the plasmalemma and/or mesophyll. Research undertaken in the United States, Europe, and Japan showed that prolonged exposure to concentrations of O_3 above 50-70 ppbv throughout the growing season reduces crop yields. Winter wheat is an important part of China's agricultural output, and the country's high levels

TABLE 4.1
Production losses and economic damage (year 2005) for four Crop types

Commodities	Production (million tons)	Crop Production Loss (million tons)	Fractional Loss (%)	Damage (billion ISD)
Cotton	3.3	0.17(±0.10)	5.3(±3.1)	0.07(±0.12)
Soybean	8.6	0.23(±0.16)	2.7(±1.9)	0.06(±0.04)
Rice	95.1	2.1(±0.8)	2.13(±0.9)	0.54(±0.23)
Wheat	71	3.5(±0.8)	5.0(±1.2)	0.62(±0.15)
Total				1.29(±0.47)

Source: Ghude et al. (2014).

of O_3 pollution could have a devastating effect on this crop. During the next decade, China's haze is predicted to worsen (Chameides et al., 1999). The impact of surface ozone on crop yield has far-reaching implications for India's economy and the country's ability to feed its population of more than one billion. Research shows that rising levels of surface ozone significantly reduce crop yields. The importance of agriculture to the economic and well-being of people in the Indian subcontinent makes this a critical problem that demands immediate attention (Sharma et al., 2019). Avnery et al. (2011) indicated that the year 2000 O_3-induced global yield ranged, depending on the metric used, from 8.5–14% for soybean, 3.9–15% for wheat, and 2.2–5.5% for maize. Global crop production losses totaled 79–121 million metric tons, worth $11–18 billion annually (USD2,000). Avnery et al. (2011) calculated that yield reductions agree well with previous estimates. According to research by Ghude et al. (2014), global ozone pollution has a significant impact on the yields of crops critical to India's food security under the current emission scenario. More than 94 million people in India are malnourished, but if the country as a whole had a yield loss, there would not be enough food to feed everyone. Given the link between increased precursor emissions and growing O_3 levels at the surface, their findings suggest that the Indian government should give significant consideration to the new National Food Security Bill.

Four Crops and their Production losses and economic damage (year 2005) as reported by Ghude et al. (2014) are displayed in Table 4.1.

Sharma et al. (2019) estimated relative yield loss (in percent) for wheat and rice, and these are depicted in Table 4.2.

Yield losses from ozone were estimated to be 537 million tons in a study by Cai et al. (2022). The majority of these losses occurred at lower ozone concentrations (between 30 and 40 ppb) (325 million tons) (Figure 4.7).

Wheat output losses in India, due to the Ozone effect, are expected to total more than 22 million metric tonnes, with Uttar Pradesh (UP) and Madhya Pradesh losses alone accounting for roughly 5 and 5.5 million metric tonnes (Sharma et al., 2019). According to estimates, UP and Punjab lost a combined total of 6.5 million metric tonnes of rice, with UP losing 5 million and Punjab losing 1.5 million. The IGP (Indo-Gangetic Plains) region accounts for more than half of the country's rice and wheat crop production losses (CPLs) (consisting of the states of Punjab, Haryana, UP, Bihar, and WB). In light of these results, it is clear that the regional model was extremely conservative in its predictions for India (only 5 percent for wheat and only 2 percent overall). Tables 4.3 and 4.4 explain the sensitivity and vulnerability of agricultural yields to ozone pollution. Emberson and Büker (2011) presented sensitive, moderately sensitive, and tolerant values for different Crops (Table 4.3).

TABLE 4.2
Estimated relative yield loss (in percent) for wheat and rice based on a review of the relevant literature

Study	Method	Wheat (%)	Rice (%)	Remarks
Ghude et al. (2014)	WRF-Chem model	5	2	0.50 × 0.50 horizontal resolution
Sharma et al. (2019) and crop growing seasons taken same as in Ghude et al. (2014)	WRF-Chem model	21	6	0.30 × 0.30 horizontal resolution
Debaje (2014)	Observations	21	-	
Lal et al. (2017)	Observations	15	6	
Sharma et al. (2019) same as in Lal et al. (2017) and Debaje (2014)	WRF-Chem model	26	8	Crop growing seasons assumed
Van Dingenen et al. (2009)	TM5 global model	28	8	
Avnery et al. (2011)	MOZART-2 global model	30	-	2.80 × 2.80 horizontal resolution
Sinha et al. (2015) For states of Punjab and Haryana and using same AOT40 relationship as used in other studies here but for 4–4.5 month growing period	Observations	18–27%	12–14%	
Sharma et al. (2019)	WRF-Chem model	16	11	For states of Punjab and Haryana

Source: Sharma et al. (2019).

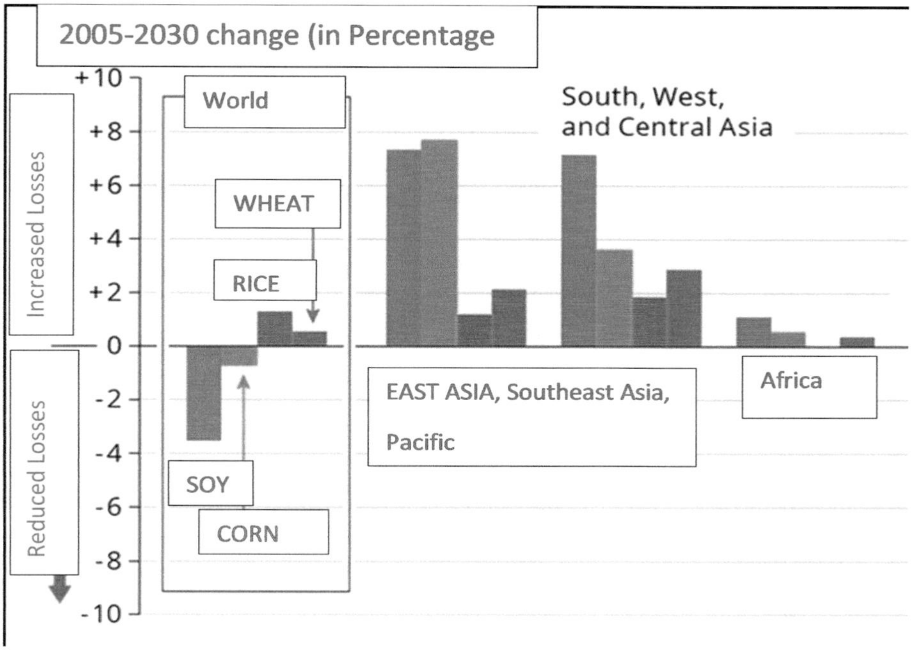

FIGURE 4.7 Projected yield losses due to elevated ozone concentrations (www.eea.europa.eu/legal/copyri ght .EEA holds the rights of use are public and may be reused without prior permission, free of charge, for commercial or non-commercial purposes, provided that the EEA is always acknowledged as the original source of the material and that the original meaning or message of the content is not distorted.)

TABLE 4.3
At 60 ppb of ozone, the yield drops by 30% according to the regression equation

Sensitive	Moderately sensitive	Tolerant
Peas and beans	Alfalfa (14)	Strawberry (1)
(including peanut) (30)	Watermelon (14)	Oat (-3)
Sweet potato (28)	Tomato (13)	Broccoli (-5)
Orange (27)	Olive (13)	
Onion (23)	Field mustard (12)	
Turnip (22)	Sugar beet (11)	
Plum (22)	Oilseed rape (11)	
Lettuce (19)	Maize (10)	
Wheat (18)	Rice (9)	
Soybean (18)	Potato (9)	
	Barley (6)	
	Grape (5)	

Source: Emberson and Büker. (2011).

Annual wheat, maize, and rice production lost under ozone 30–40 ppb and above 40 ppb scenarios are given by Feng et al. (2018) (Table 4.5).

Yield losses from ozone were estimated to be 537 million tons in a study by Cai et al. (2022). The majority of these losses occurred at lower ozone concentrations (between 30 and 40 ppb)

TABLE 4.4
Comparison between the models with different (Equation 4) and common O_3 sensitivity (Equation 5) of O3-FACE and OTC experiments. S_0 is the sensitivity in OTC, Sf is the sensitivity in FACE, and S_c^h is the common sensitivity in both OTC and FACE

	Different sensitivity		Common sensitivity	
Crop species	**S_0**	**Sf**	**S_c^b**	**ρValue**
Soybean	0.0102	0.0090	0.0097	0.048
Wheat	0.0215	0.0259	0.0222	0.023
Rice	0.0111	0.0203	0.0118	0.002

Source: Feng et al. (2018).

(325 million tons). Cai et al. (2022) mentioned that each country's annual loss rate of crop yield in the case of ozone above is 30, 40, 50, 60, 70, 80, and 90 ppb, and given the total yield for wheat, maize, rice, and soybean (Figure 4.8).

Debaje (2014) used M7 and AOT40 exposure-based indices and their CR relationships to calculate RYL of winter wheat and rabi rice (Table 4.6).

4.14 OZONE LEVELS AND THEIR VARIATIONS ACROSS TIME AND SPACE IN VARIOUS COUNTRIES

UK-based air pollution monitoring: The closeness to large sources of ozone precursors, geography, and weather patterns all play a role in the seasonal fluctuations in ozone concentration at any given spatial scale, as stated in A Miscellany of Monks et al. (2015). A number of factors, including generation, destruction, transfer from upwind sources, and removal at the surface, contribute to the total amount of ozone in the atmosphere at any time. Ozone exposure in the troposphere is harmful both immediately and over the long term. This highly reactive gas causes damage to plants and trees, decreasing food output. Damage to exposed structures and materials, especially rubbers and plastics, is possible due to their reactivity. Radiative forcing from ozone pollution is comparable to that from industrialization-related increases in atmospheric methane. Ozone pollution is a primary source of high tropospheric human greenhouse gas emissions (4-10 km). Due to the harm it causes to humans, agricultural products, and natural environments, ozone is classified as a pollutant in the air.

Ozone (O_3) is a gas that, when inhaled, can trigger asthma attacks and other respiratory problems in humans. Oxidative damage to plants and crops may be another way in which ozone wreaks havoc on ecosystems. Ozone can be harmful to human health and the environment; therefore, monitoring its concentration is crucial. The United Kingdom has no significant sources of ozone, unlike many other forms of air pollution. Ozone and the pollutants that contribute to its creation are challenging to monitor together because of their dependency. For instance, if the synthesis of ozone and its precursor pollutants, such as NO_x, is not drastically decreased, then a decrease in NO_x levels may not prevent an increase in ozone levels in metropolitan areas. It's hard to tell if someone has been exposed to ozone because the gas can travel long distances after being created. Because of these factors, it is impossible to gauge the severity of ozone's threat in the UK by relying just on emission estimates or quantitative measurements of ozone precursor pollutants. In order to assess the unintended consequences of measures made to reduce the emissions of other pollutants, including nitrogen oxides, it is essential to conduct regular monitoring of ozone levels in the United Kingdom. Clapp et al. (2001) reported that Monitoring data from the UK Automatic Urban and Rural Network are used to investigate the relationships between ambient levels of ozone (O_3), nitric oxide (NO) and

TABLE 4.5
Annual wheat, maize, and rice production lost under ozone 30–40 ppb and above 40 ppb scenarios (data from the world and the top five locations in grain production)

Wheat			Maize			Rice			Soybean		
Area	30-40 ppb	>40 ppb	Area	30-40 ppb	>40 ppb	Area	30-40 ppb	>40 ppb	Area	30-40 ppb	>40 ppb
Global	112.44	64.94	Global	99.51	82.76	Global	48.08	33.49	Global	64.87	30.93
China	28.17	23.42	America	42.09	27.44	China	22.59	18.97	America	29.93	13.46
Europe	25.37	12.62	China	23.40	32.77	India	6.15	3.59	Brazil	21.74	9.35
America	11.74	5.97	Europe	6.86	4.39	Vietnam	2.00	1.23	China	3.20	3.11
Russia	10.48	5.93	Brazil	6.43	3.75	Indonesia	1.68	1.77	Argentina	2.99	1.75
India	8.27	6.15	Argentina	3.08	2.39	Bangladesh	1.37	0.75	India	0.31	0.18
Other	28.40	10.86	Other	17.64	12.02	Other	14.29	7.19	Other	6.70	3.08

Source: Cai et al. (2022), Creative Commons Attribution 4.0 International.

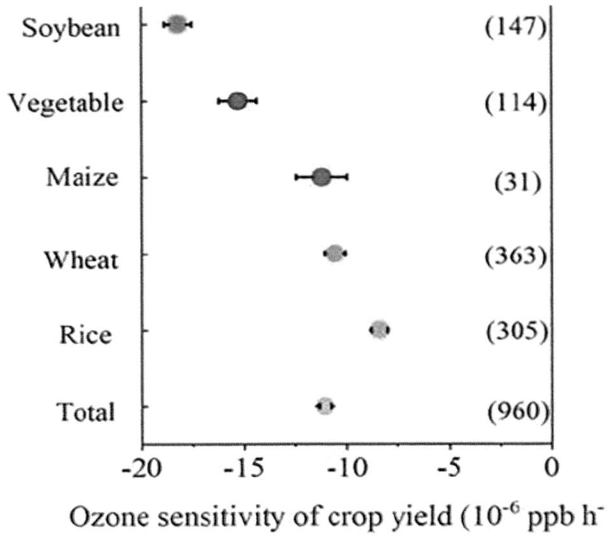

FIGURE 4.8 Ozone sensitivity of crop yield (ppb h-1). (Cai et al. (2022). Creative Commons Attribution 4.0.)

TABLE 4.6
Air quality indices used to evaluate the winter wheat and rabi rice crop yield losses over India. All surface ozone (O_3) concentration (Parts per billion by volume) refers to hourly values

Index	Unit	Definition	Exposure/doses response function: relative yield loss (RYL)	Wheat		Rice	
				a	b	a	b
M7	ppbv	7-h seasonal O_3 mean	1-exp[-(M7/a)^b]/exp[-(25/a)^b]	137	2.34	202	2.47
AOT40	Ppm h	3 months, 09-16 h (M7-29)/1.59 3 months	aAOT40	0.0161	0.99	0.00415	0.94

Source: Debaje (2014). Open Access. Creative Commons Attribution License.

nitrogen dioxide (NO_2) as a function of NOx, for levels ranging from those typical of UK rural sites to those observed at polluted urban curbside sites.

British ozone (O_3) measurements (1987 to 2023): Since 1992, annual maximum eight-hour mean concentrations of ozone have been increasing at urban background sites. The series value peaked in 2020 and declined by 7.2 g/m^3, or 3 percent, the following year, 2023 (Figure 4.8). Climate change is a major factor in the annual variance of ozone levels. Beginning in 1992, when the time series began, and continuing until the middle of the 2000s, ozone concentrations in the urban backdrop increased. This might have happened because the United Kingdom and Europe have reduced their emissions of nitrogen oxides, which contribute to the formation of ozone in urban areas. Annual mean concentrations of O_3 in the UK, 1987 to 2023 is depicted in (Figure 4.9).

In 2023, the annual average daily maximum eight-hour mean concentration of ozone reached 66.8 μg/m^3 in urban areas, which is the highest value in the time series. This was an increase of 3 percent compared to 2022.

Ozone concentrations fluctuate by around a factor of two throughout the United Kingdom each year (Figure 4.9). In most of Scotland, Northern Ireland, Wales, south-west England, and the Pennines (outside of urban areas), the average concentration is greater than 60 g m^{-3}, and it approaches 70 g

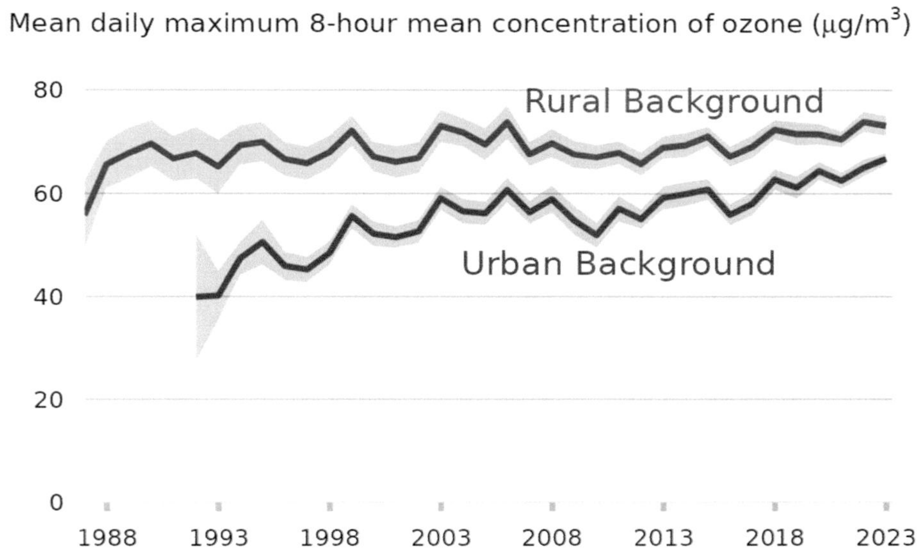

FIGURE 4.9 Average annual O_3 concentration in the United Kingdom from 1987 to 2023. (www.gov.uk/government/statistics/air-quality-statistics/concentrations-of-ozone. www.nationalarchives.gov.uk/doc/open-government-licence/version/3/. You are encouraged to use and re-use the Information that is available under this license freely and flexibly.)

m^{-3} on the highest ground. Average ozone levels are typically greater in upland sites, countering the ozone lost by dry deposition at the surface, because more ozone is mixed vertically with increasing height. The effect is stronger at night when there is less vertical mixing. Ozone levels in other rural areas of England average between 50 and 60 g m^{-3}. The annual average ozone concentrations in large urban/industrial districts and transport corridors are lower than the concentrations in the surrounding rural areas because ozone can be swiftly chemically removed by interaction with NO near the sources of NO emissions. There are strong intra-urban spatial gradients in ozone between the suburban backdrop and intensively travelled roadside areas, and average ozone concentrations in the center of the greatest conurbations are 40 g m^{-3} or less due to the normal urban decrement of roughly 10-15 g m^{-3}.

Ozone levels have increased annually on average in the United Kingdom over the previous 30 years. Researchers Diaz et al. (2020) found that while rural areas saw a rise of 0.13 ppb (0.5 percent per year), urban areas saw a rise of 0.20 ppb (1.1 percent per year) (Figure 4.8). Increases in annual average ozone have been connected to decreases in NO_X mixing ratios, which have been occurring at rates of 0.22 parts per billion per year (3.7 percent per year) for rural sites and 1.2 parts per billion per year (3.3 percent per year) for urban sites. The extent of the rising trend in rural sites is less than that seen in urban regions due to the strong hemispheric dependence of ozone concentrations. Peak and background ozone levels in Europe rise and fall at the same time (1995–2014) (Jenkin, 2008; Derwent et al., 2018). Due to the strong hemispheric dependence of ozone concentrations, the rising trend is less pronounced at rural sites than in urban regions.

European law has led to a reduction in emissions of nitrogen oxides and volatile organic compounds, both of which contribute to the production of ozone. The average ozone level in cities has been rising over the past two decades due to less NO being used to deplete O_3 locally (www.apis.ac.uk/overview/pollutants/overview_O3.html). These regulations have helped reduce peak ozone levels in the United Kingdom by around 30 parts per billion. Concentrations of O_3 in rural areas are rising because background O_3 levels have been rising at a rate of around 0.2 ppb each year, or about 5 ppb in total. Figure 4.9 depicts the ozone pattern in the UK over the past 25 years (1992–2018).

Changes in Britain's Ozone Levels are shown (Figure 4.10) by Diaz et al. (2020). The magnitude of the increasing trend at rural sites is smaller than that obtained at urban sites, due to the strong dependency of the concentrations of ozone on the northern hemispheric ozone baseline in rural areas.

Diaz et al. (2020) mentioned that the high exceedances in 2008 were mainly a result of the influence of the site Strathvaich, which reached 1000 h of exceedance in 2008 (Figure 4.11a, b).

The total exceedance is the total number of hours the ozone concentration exceeded the 50 ppbv limit. Due to lower NO_X emissions, Diaz et al. (2020) showed that ozone exceedances in rural areas were higher than in urban centers. According to the available data, ozone events only take place in the late spring and summer, when there is a lot of sunshine and the atmosphere makes it difficult for pollutants to disperse. Overshoot ozone levels tend to peak in the middle of the afternoon in both urban and suburban settings. In addition, weekday NO_X removal activity is lower, making weekend ozone events more likely (emissions are lower on the weekends). As NO_X levels have decreased, the researchers have noticed a shift in the day of the week when ozone outbreaks occur.

There is also an evaluation of the source locations for the three necessary O_3 readings. Model simulations are shown to exhibit a mean bias of -3.7 ppbv when attempting to estimate the level and position of surface O_3 over a domain (Romero-Alvarez et al., 2022). Although the United Kingdom and Europe's NOx emissions are responsible for about 13% and 16% of the monthly mean surface O_3, respectively, the majority of the O_3 in the United Kingdom is brought in from elsewhere in the hemisphere. Maximum amounts of ozone (O_3) from the northern and western hemispheres reach the United Kingdom in May, while those from the southern hemisphere reach their greatest levels in June and July. Romero-Alvarez et al. (2022)'s health-focused research shows that stricter emission regulations are needed over continental Europe, notably in western Europe, for levels of MDA8 O_3 over 50 and 60 ppbv. Reduce the ecological damage caused by the AOT40 by enforcing stricter emission limitations over a wider area, especially the Northern Hemisphere.

4.15 SURFACE NOX, CO_2, AND O_3 RELATIONS IN INDIA

The economies and infrastructures of India are developing rapidly. Increases in tropospheric ozone mixing ratios over the Indian subcontinent can be attributed to the increased emissions caused by the region's increasing industrialization, urbanization, and population in recent decades. More mixing is expected in the ozone layer in the future, according to forecasts. Too much ozone in the air is bad for business for farmers. The first time that tropospheric ozone was responsible for significant harm to vegetation was during heavy smog occurrences in Los Angeles. Arie Haagen-Smit and colleagues discovered in the early 1950s that nitrogen oxides (NOx) and organic trace gases (such as car exhaust) might cause damage to plants in the presence of sunlight.

The effects of ozone on plants are nuanced and dependent on both the concentration of the gas and the species of plant under study. Exposure to midday ozone through the stomata in the leaves of plants causes metabolic disruption and decreases agricultural productivity.

According to the 2011 census, India's population during the 2001–2011 decade was approximately 1210 million people. Its economy is expanding rapidly. Over the past decade, the number of registered automobiles has increased by around 10% annually (2001–2011). Delhi's population increased by 21.2% between the 2001 and 2011 censuses. There were 109.86 lakh registered automobiles in Delhi's NCT as of March 31st, 2018, up 5.81 percent from the previous year. There was a dramatic increase from 317 cars per 1,000 people in 2005–2006 to 598 cars per 1,000 people in 2017–2018. There were around 246 registered vehicles for every thousand people across India in the fiscal year 2020. Delhi had the highest number of registered vehicles, at over 11 million, across India at the end of the fiscal year 2020. It was calculated that the average hourly concentrations of

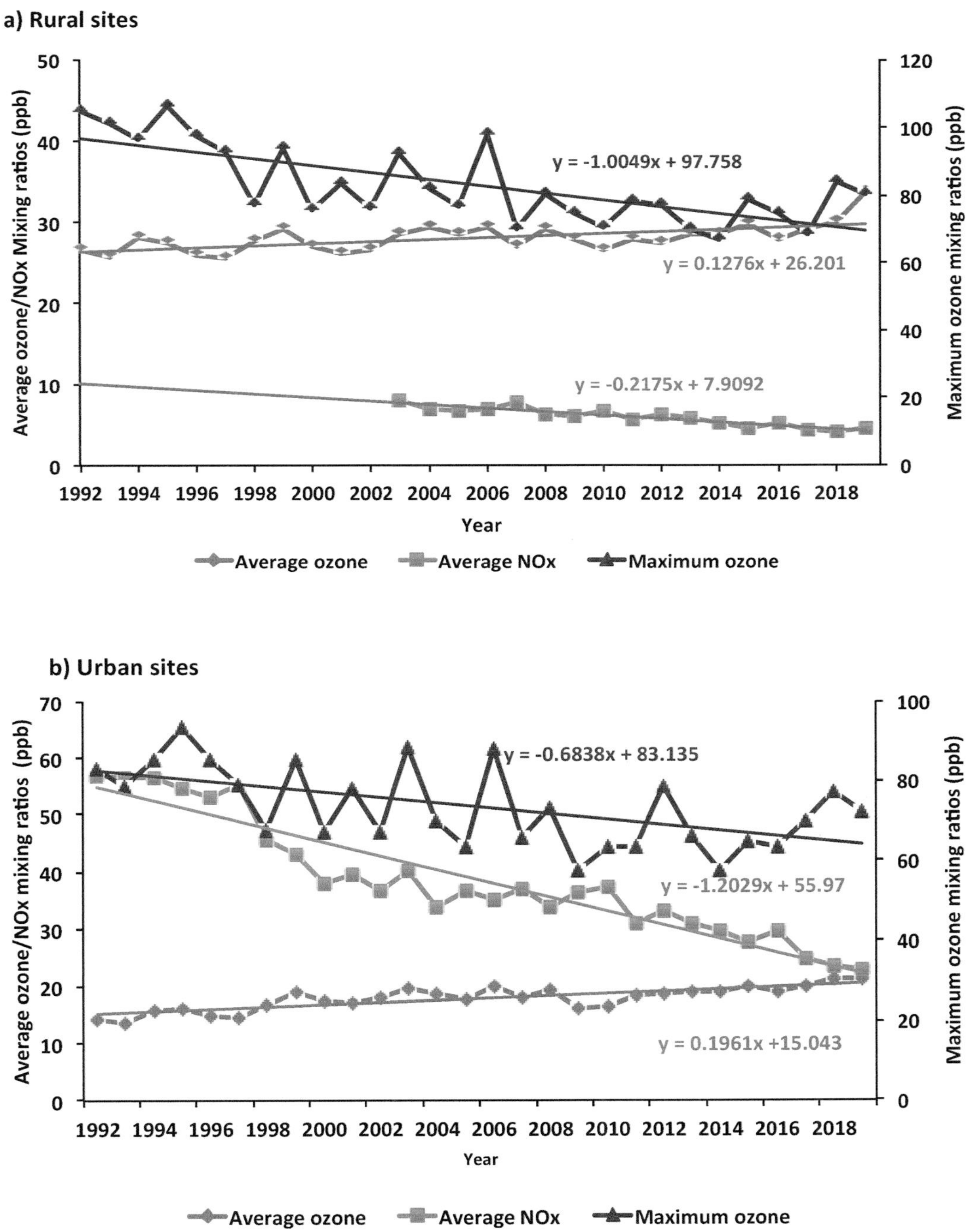

FIGURE 4.10 Changes in Britain's Ozone Levels (1992–2018) (Diaz et al. (2020).Open access.)

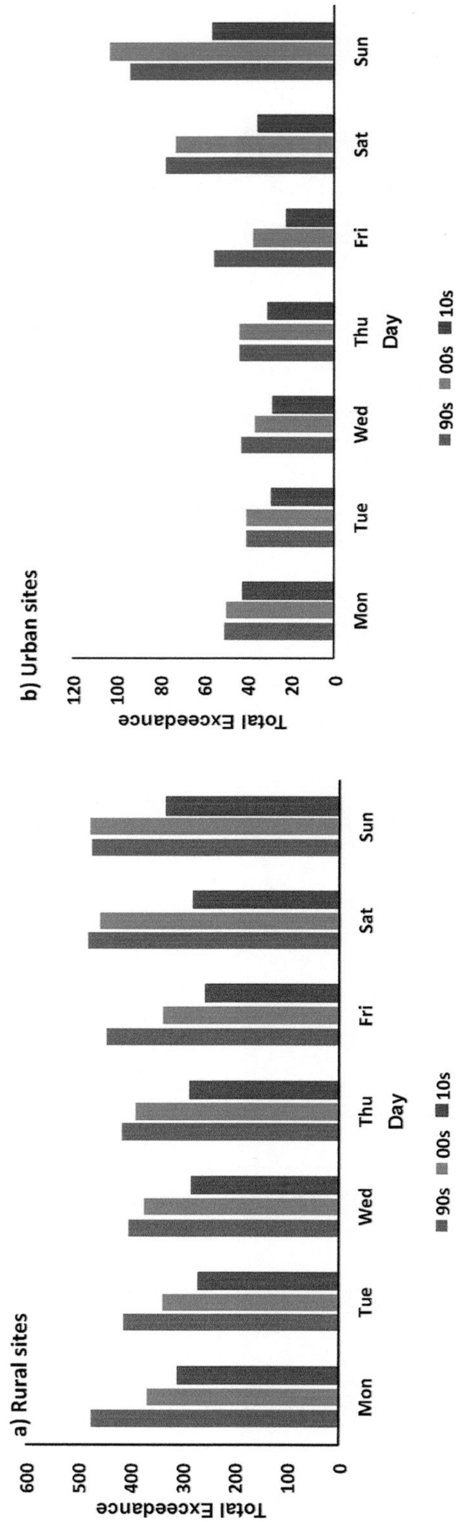

FIGURE 4.11 a, b Yearly total exceedances of ozone in (a) rural and (b) urban sites over the last 30 years. (Diaz et al. (2020). Open Access.)

O_3, NO, NO_2, CO, NOx (NO + NO_2), and OX were 32.44–36.57 ppb, 19.46–28.09 ppb, 20.83–26.89 ppb, 1.67–1.89 ppm, 43.04–54.99 ppb, and 1.67–1.89 ppm, respectively (Sinha et al., 2022). Changes in NOx and CO levels vary from day to night. The ozone levels are highest in the afternoon, but they are also higher in the morning and late at night (1900-2400hrs). Research into the molecular connection of the three species (NO, NO_2, and O_3) has also been linked to NOx levels during the day (0500hrs IST to 1900hrs IST). According to the study, lower ozone levels at night are associated with greater NOx levels during the day. To determine which areas produced the most oxidant pollution, Sinha et al. (2022) analyzed the correlation between [NOx] and oxidant concentrations ($NO_2 + O_3$). For each of the five locations, Sinha et al. (2022) determined the average yearly rate of change in O_3 concentrations (dO_3/dt). Temperature has been found to have a positive link with monthly and daily changes in O_X, while humidity has been found to have a negative correlation.

Roozitalab et al. (2022) studied the impacts of inter-annual variability in meteorological and emissions (dust, biogenic, and biomass burning) on air quality in northern India to better understand the chemistry underlying the fluctuations in ozone concentrations. They were able to do this by employing the IRR function of the WRF-Chem model version 4.0. The model indicated that the increase in air pollution will happen in both 2019 and 2020. Air pollution levels were predicted for April 2019 through April 2020, taking both weather and lockdown emissions into account (Roozitalab et al., 2022). Across the Indo-Gangetic Plain (IGP) as a whole, PM2.5 concentrations dropped by 12% in April 2020 due to weather, while they increased in central India due to higher emissions from biomass burning. However, because of the reduced emissions, the PM2.5 concentrations in the Indian parts of the domain dropped by 9 percent. Ozone concentrations were found to be higher over central India and lower over the IGP, just as was the case with PM2.5, due to prevailing weather patterns. Emissions controls have led to significant reductions in volatile organic compounds and nitrogen oxides, although ozone levels have not decreased uniformly over the region. While ozone levels went down across most of the domain, they found that they increased near large power plants and densely populated urban centers like Delhi. Understanding what caused the contaminants to react as they did during the COVID-19 lockdown could inform future efforts to improve air quality.

The amount by which pollutant concentrations changed before and after the severe lockdown was determined by Singh et al. (2020) (25th March to 3rd May 2020). India's improvements in air quality were the most noticeable and widespread with respect to the drop in particulate matter levels (PM2.5 and PM10). The lockdown resulted in a median reduction of 50-60% in PM2.5 and PM10 readings in the northwestern and IGP regions. There should be a 40% drop in PM2.5 levels and a 25% drop in PM10 levels in the south, and a 40% drop in both in the middle. The Indian Greater Peninsula (IGP) has the highest population density in the world. Dust storms are a major contributor to the high surface dust load and also have a major impact on the air quality in the IGP and northwest regions. There may be a correlation between PM levels and the amount of traffic, industry, and construction going on, as well as the amount of exhaust and non-exhaust emissions, especially resuspended dust.

The weather in April 2020 lowered PM2.5 concentrations in the IGP by 12%, but central India saw an increase in emissions from burning biomass. However, as a result of decreasing emissions throughout the entire Indian domain, PM2.5 concentration declined by 9%. Singh et al. (2020) found that due to meteorological patterns, ozone concentrations, like PM2.5 levels, were lower over the IGP and higher over central India. While NO_2 and VOC precursors were greatly reduced as a result of the lockdown emissions, ozone concentrations reacted in a uniform fashion across the domain. While NOx emissions decreased generally, they went up in heavily populated regions like Delhi and locations near several power stations.

It is vital to understand why the pollutants reacted as they did during the COVID-19 lockdown if we are to apply the lessons learned to future efforts to manage air quality. It was also shown that the concentration of O_3 decreases throughout the day. At night, NO_2 levels drop much more precipitously than they do during the day. Reduced nighttime O_3 removal causes a 30 percent increase in

O_3 concentration in the IGP and the northwest. Direct O_3 loss may also be caused by heterogeneous chemistry involving aerosols near the surface (Singh et al., 2020). While variations were high, it was shown that Southern regions generally had lower O_3 levels. The shutdown significantly reduced pollution levels in Delhi. Levels of NO_2, PM10, and PM2.5 were reduced by 40-60%, but CO and SO_2 were affected less. The average O_3 level in Delhi has increased by more than 20% over the previous few years due to a combination of factors, including decreased particulate matter over the city, increased photolysis production, precursor availability, and station chemistry.

Sharma et al. (2021a) reported a +0.158 Mann-Kendall test result, suggesting a significant increase in nocturnal ozone levels between 1990 and 2012. The Autoregressive Integrated Moving Average (ARIMA) model predicts a rise in nighttime ozone levels between 2013 and 2025 if the current trend continues. This is the first research to look at nighttime ozone levels in Delhi NCT for a full 30 years. This study highlights the significance of taking coordinated steps to reduce the generation of anthropogenic precursors of surface ozone because of the detrimental effects of high ozone levels on people's health, agricultural output, and urban air quality. Average nighttime ozone levels were above 18 ppb beginning in 2010, and they remained there throughout the study period. The models predicted an increase in both nighttime and daytime ozone concentrations between 2013 and 2025. According to Sharma et al. (2021a), this work is a major advancement in our understanding of the effects of ozone at the Earth's surface on agriculture, climate, and human health since it elucidates the significance of night-time chemistry and meteorology. As reported in Hindustan Times (www.hindustantimes.com/mumbai-news/pollution-destroys-21-wheat-6-rice-crop-every-year-iit-m-study/story-FfsjiUstkx62FL7ALe5uxI.html), the annual cost of pollution-related losses to agriculture is estimated to be greater than USD 5 billion for wheat and USD 1.5 billion for rice.

According to Ghude et al. (2014), global ozone pollution has a significant impact on the yields of crops critical to India's food security under the current emission scenario. Ozone-induced damage to wheat (3.5 ± 0.8 Mt) and rice (2.1 ± 0.8 Mt) is sufficient to feed roughly 35% (94 million poor people) of the 270 million below the poverty line population in India. Agricultural losses in India's north, east, west, and south were predicted using the mean ozone over 7 hours (M7) and the accumulated ozone beyond a threshold of 40 ppbv by Lal et al. (2017). Northern India, one of the country's most polluted regions, loses an estimated 8.9 million metric tonnes of wheat annually, while the Eastern region loses an estimated 1 billion kg of rice annually due to pollution (around 2.6 million tonnes). India loses an estimated 4.0-14.2 MMT of wheat every year (or 4.2-15.0 percent), and anything from 0.3 MMT to 6.7 MMT of rice (3 percent to 15 percent).

Maximum concentrations of PM1, PM2.5, PM10, and surface ozone were measured and reported by Sharma et al. (2021b) to be surface ozone of 159 ± 77 μg m^{-3}, 172 ± 79 μg m^{-3}, 280 ± 108 μg m^{-3}, and 335 ± 18 μg m^{-3}, respectively, throughout the crop cycle. Ascorbic acid was administered to the plants under a number of controlled laboratory conditions. During the vegetative, reproductive, and grain-forming stages of the crop cycle, a wide range of growth, biochemical, and yield characteristics were measured. When compared to crops grown in a controlled setting, those grown in natural conditions had a harvest yield that was 23% lower and a chlorophyll content that was 14% lower. Comparing crops exposed to only surface ozone during the vegetative stage with those exposed to only fine particulate matter revealed that root length, plant height, tiller count, and leaf count were all reduced by 13%, 5%, 15%, and 10%, respectively (Sharma et al., 2021a).

Two large-scale case studies of O_3's impact on crops in South Africa and India were undertaken by Harmens and associates (2020). While 2015 estimations of O_3 flux across India revealed substantial variation, the highest amounts were regularly observed in the country's eastern and central regions. India lost over 11.8 million metric tonnes of wheat, or 11.3% of their usual harvest. The regions of India most reliant on wheat production, such as Uttar Pradesh and Madhya Pradesh, were expected to be hit the worst by the decline in oxygen levels.

4.16 CONCLUSION

An effort was made to synthesize our existing knowledge of the ways in which ozone pollution harms agricultural products. This chapter presents a broad overview of ozone, from its generation and destruction in the stratosphere to its effects on crop yield in the troposphere, with the goal of increasing public understanding of the issue. In this chapter, we examined the research on how ozone damages plants and how breathing polluted air affects plant life. Significant crop losses and economic damage may lead to food security and wealth challenges in the affected regions. This chapter explains why stomata and cuticles are so crucial to plant life. There is also information on the differences in ozone levels between urban and rural locations throughout time, as well as data for various countries (including the United Kingdom, China, and India). Rice crops destroyed by ozone pollution in India in 2005 could have fed 94 million people, which is about a third of the country's poor. Recent reports expressed that the crop production loss (CPL) of about 7.39 million tons (Mt) of rice in 2005 increased to 11.46 Mt in 2020. While estimates of O_3 flux across India in 2015 varied widely, the central and eastern parts of the country consistently showed the highest values. Recent reports indicate that India's wheat harvest dropped by an average of 11.3%, or 11.8% in total, due to a variety of factors. Farmers who inject CO_2 into their soils, for instance, can make up for lost crop yields due to surface ozone pollution. Numerous studies have shown that a mixture of concentration-based and flux-based methods is required to assess the effects of CO_2 rises on O_3-induced losses in agricultural output. Soybean yields have dropped 6.7% worldwide over the past few decades, while wheat yields have dropped by 7.3%, rice yields have dropped by 1.1%, and maize yields have dropped by 3.6%, according to studies. The scientific community as a whole may have come a long way in its understanding of and response to air pollution's ecological repercussions, but there's still a long way to go. Institutional measures should be developed to increase emission regulation by regulatory agencies and encourage farmers to apply tactics for increased plant tolerance in order to lessen the harmful consequences of O_3 pollution on plants and guarantee the security of the food supply.

REFERENCES

Avnery, S., D.L. Mauzerall, J. Liu and L.W. Horowitz. 2011. Global crop yield reductions due to surface ozone exposure: 1. Year 2000 crop production losses and economic damage, *Atmospheric Environ.* 45(13): 2284–2296.

Baldwin, M. and M. Dameris. 2007. Climate ozone connections. In: *Scientific Assessment of Ozone Depletion: 2006. Global Ozone Research and Monitoring Project—Report No. 50;* 2002 Edition: Report No.47. Chapter 4: Global Ozone: Past and Future. WMO.

Bikle, D.D. 2014. Vitamin D metabolism, mechanism of action, and clinical applications. *Chem. Biol.* 21: 319–329.

Brasseur, G. 1984. Coupling between the thermosphere and the stratosphere: The role of nitric oxide. In: J. Taubenheim (ed.) *Handbook for MAP* (Vol. 10, pp. 116–121). Scientific Committee on Solar-Terrestrial Physics (SCOSTEP), University of Illinois.

Brasseur, G.P. 2020. The importance of fundamental science for society: The success story of ozone research. *Perspect. Earth Space Sci.* 2: e2020CN000136. https://doi.org/10.1029/2020CN000136

Brasseur, G. and M. Nicolet. 1973. Chemospheric processes of nitric oxide in the mesosphere and stratosphere. *Planetary Space Sci.* 21(6): 939–961. https://doi.org/10.1016/0032-0633(73)90141-4

Burney, J. and V. Ramanathan. 2014 Recent climate and air pollution impacts on Indian agriculture. *Proc. Natl Acad. Sci. USA* 111: 16319–16324.

Cai, A., B. Wang, T. Ren, W. Zhang, X. Wang, H. Dong and Y. Li. 2022. Large crop production losses induced by global ozone stress based on interval evaluation. *Res. Square.* https://doi.org/10.21203/rs.3.rs-1231026/v1

Chameides, W. and J.C.G. Walker. 1973. A photochemical theory of tropospheric ozone. *J. Geophys. Res.* 78(36): 8751–8760. https://doi.org/10.1029/JC078i036p08751

Chameides, W.L., X. Li, X. Tang, X. Zhou, C. Luo, C.S. Kiang, J. St. John, R.D. Saylor, S.C. Liu, K.S. Lam, T. Wang and F. Giorgi. 1999. Is ozone pollution affecting crop yields in China? *Geophys. Res. Letts*. 26(7): 867–870.

Chapman, S. 1930. A theory of upper atmospheric ozone. *Memoirs R Meteorol. Soc.* 3: 103–125.

Clapp, L.J. and M.E. Jenkin. 2001. Analysis of the relationship between ambient levels of O_3, NO_2 and NO as a function of NOx in the UK. *Atmos. Environ*. 35: 6391–6405.

Crutzen, P.J. 1970. The influence of nitrogen oxides on the atmospheric ozone content. *Quarterly J. R. Meteorol. Soc.* 96(408): 320–325. https://doi.org/10.1002/qj.49709640815

Debaje, S.B. 2014. Estimated crop yield losses due to surface ozone exposure and economic damage in India. *Environ. Sci. Pollution Res.* 21: 7329–7338. https://doi.org/10.1007/s11356-014-2657-6

Derwent, R.G., A.J. Manning, P.G. Simmonds, T.G. Spain, and S. O'Doherty. 2018. Long-term trends in ozone in baseline and European regionally-polluted air at Mace Head, Ireland over a 30-year period. *Atmos. Environ*. 179: 279–287.

Diaz, F.M.R., M.A.H. Khan, B.M.A. Shallcross, E.D.G. Shallcross, U. Vogt, and D.E. Shallcross. 2020. Ozone trends in the United Kingdom over the last 30 Years. *Atmosphere*. 11: 534. https://doi.org/10.3390/atmos11050534

Edwards, G. and D. Walker. 1983. C3, C4: *Mechanisms and Cellular and Environmental Regulation of Photosynthesis*. University of California Press.

Emberson, L. and P. Büker. 2011. Current knowledge of the impacts of ozone on food crops in South Asia. In: Mills, G. and H. Harmens (eds.) *Ozone Pollution: A Hidden Threat to Food Security*. ICP Vegetation Programme Coordination Centre, Centre for Ecology and Hydrology.

Feng, Z., J. Uddling, H. Tang, J. Zhu, and K. Kobayashi. 2018. Comparison of crop yield sensitivity to ozone between open top chamber and free-air experiments. *Glob. Change Biol.* 24: 2231–2238. https://doi.org/10.1111/gcb.14077

Ghude, S.D., C. Jena, D.M. Chate, G. Beig, G.G. Pfister, R. Kumar, and V. Ramanathan. 2014. Reductions in India's crop yield due to ozone. *Geophys. Res. Lett.* 41: 5685–5691. https://doi.org/10.1002/2014GL060930

Grulke, N.E., and R.L. Heath. 2019. Ozone effects on plants in natural ecosystems. *Plant Biol.*: 1–26, DOI :10.1111/plb.12971

Haagen-Smit, A.J. 1952. Chemistry and physiology of Los Angeles smog. *Ind. Eng. Chem.* 44(6): 1342–1346. https://doi.org/10.1021/ie50510a045

Han, H., J. Liu, L. Shu, T. Wang, and H. Yuan. 2020. Local and synoptic meteorological influences on daily variability in summer time surface ozone in eastern China. *Atmos. Chem. Phys*. 20: 203–222. https://doi.org/10.5194/acp-20-203-2020

Harmens, H., A. Holder, K. Sharps, F. Hayes, M. Vieno, and R. Beck. 2020. *Ozone impacts on crops in ODA countries: Evidence, risk and policy implications*. Contract AQ0846, ICP Vegetation, Issue Number 1. UK Centre for Ecology & Hydrology, Environment Centre Wales Bangor.

Houzeau, A . 1858. The ozone layer. www.albany.edu/faculty/rgk/atm101/ozmeas.htm

Jenkin, M. 2008. Trends in ozone concentration distributions in the UK, since 1990: Local, regional and global influences. *Atmos. Environ*. 42: 5434–5445.

Jimenez-Montenegro, L., M. Lopez-Fernandez, and E. Gimenez.2021. Worldwide research on the ozone influence in plants. *Agronomy.* 11: 1504. https://doi.org/10.3390/agronomy11081504

Juráň, S.J. Grace, and O. Urban. 2021. Temporal changes in ozone concentrations and their impact on vegetation. *Atmosphere.* 12: 82. https://doi.org/10.3390/atmos12010082

Kozioł, M.J. and F.R. Whatley. 1984. *Gaseous air pollutants and plant metabolism.* Butterworth-Heinemann. https://doi.org/10.1016/C2013-0-04164-3

Lal, L., S. Venkataramani, M. Naja, J.C. Kuniyal, T.K. Mandal, P.K. Bhuyan, K.M. Kumari, S.N. Tripathi, U. Sarkar, T. Das, Y.V. Swamy, K. Rama Gopal, H. Gadhavi and M.K.S. Kumar. 2017. Loss of crop yields in India due to surface ozone: An estimation based on a network of observations. *Environ. Sci. Pollution Res.* 24(2):20972–20981. https://doi.org/10.1007/s11356-017-9729-3

Lee, J.K., M.J. Kwak, S.G. Jeong, and S.Y. Woo. Individual and interactive effects of elevated ozone and temperature on plant responses. *Horticulture.* 8: 211. https://doi.org/10.3390/horticulturae8030211

Leung, F., J.Y.S. Pang, A.P.K. Tai, T. Lam, D.K.C. Tao, and K. Sharps. 2020. Evidence of ozone-induced visible foliar injury in Hong Kong using *Phaseolus vulgaris* as a bioindicator. *Atmosphere.* 11: 266. https://doi.org/10.3390/atmos11030266

Mills, G., F. Hayes, D. Norris, J. Hall, M. Coyle, H. Cambridge, S. Cinderby, J. Abbott, S. Cooke, and T. Murrells. 2011. Impacts of ozone pollution on food security in the UK: A case study for two contrasting years, 2006 and 2008. *Report for Defra contract AQ08610*. http://icpvegetation.ceh.ac.uk/; http://uk-air.defra.gov.uk/library/

Molina, M.J. and F.S. Rowland. 1974. Stratospheric sink for chloro fluoromethanes: Chlorine atom-catalyzed destruction of ozone. *Nature*. 249: 810–812. https://doi.org/10.1038/249810a0

Monks, P.S., A.T. Archibald, A. Colette, O. Cooper, M. Coyle, R. Derwent, D. Fowler, C. Granier, K.S. Law, G.E. Mills, D.S. Stevenson, O. Tarasova, V. Thouret, E. von Schneidemesser, R. Sommariva, O. Wild, and M.L. Williams. 2015. Tropospheric ozone and its precursors from the urban to the global scale from air quality to short-lived climate forcer. *Atmos. Chem. Phys.* 15: 8889–8973, www.atmos-chem-phys.net/15/8889/2015/

Morales-Gonzalez, O.M., M. Escribà-Gelonch, and V. Hessel. 2019. Life cycle assessment of vitamin D3 synthesis: From batch to photo-high p, T. *Int. J. Life Cycle Assess.* 24: 2111–2127. https://doi.org/10.1007/s11367-019-01634-6

Moura, B.B, E. Carrari, L. Dalstein-Richier, P. Sicard, S. Leca, O. Badea, D. Pitar-Silaghi, A. Shashikumar, M-L. Ciriani, E. Paoletti and Y. Hoshika. 2022. Bridging experimental and monitoring research for visible foliar injury as bio-indicator of ozone impacts on forests. *Ecosystem Health Sustainability*. 8(1). https://doi.org/10.1080/20964129.2022.2144466

Nicolet, M. 1971. Agronomic reactions of hydrogen and ozone. In: G. Fiocco (ed.) *Mesospheric Models and Related Experiments* (Chap. 1, pp. 1–51). D. Reidel.

Pandey, A.K., B. Majumder, S. Keski-Saari, S. Kontunen-Soppela, V. Pandey, and E. Oksanen. 2019. High variation in resource allocation strategies among 11 Indian wheats (*Triticum aestivum*) cultivars growing in high ozone environment. *Climate.* 7: 23.

Perdomo, J.A., S. Capó-Bauçà, E. Carmo-Silva, and J. Galmés. 2017. Rubisco and Rubisco activase play an important role in the biochemical limitations of photosynthesis in rice, wheat, and maize under high temperature and water deficit. *Front. Plant Sci., Sec. Plant Physiol.* https://doi.org/10.3389/fpls.2017.0

Portis Jr, A.R. and M.A.J. Parry, 2007, Discoveries in Rubisco (Ribulose 1,5-bisphosphate carboxylase/oxygenase): A historical perspective. *Photosynth. Res.* 94: 121–143. https://doi.org/10.1007/s11120-007-9225-6

Romero-Alvarez, J., A. Lupaşcu, D. Lowe, A. Badia, S. Acher-Nicholls, S.R. Dorling, C.E. Reeves, and T. Butler. 2022. Sources of Surface O_3 in the UK: Tagging O_3 within WRF-Chem. *Atmos. Chem. Phys. Discuss.* 22(20): 13797–13815.

Roozitalab, B., G.R. Carmichael, S.K. Guttikunda and M. Abdi-Oskouei. 2022. Elucidating the impacts of COVID-19 lockdown on air quality and ozone chemical characteristics in India. *Environ. Sci. Atmos.* 2: 1183–1207.

Saxena, P., M. Chakraborty and S. Sonwani. 2020. Phytotoxic effects of surface ozone exposure on rice crop—A case study of tropical megacity of India. *J. Geoscience Environ. Protect.* 8: 322–334. https://doi.org/10.4236/gep.2020.85020

Schönbein, C.F. 1840. Recherché sur la nature de l'odeur qui se manifeste dans certaines actions chimiques. *Comptes Rendus de l'Académie des Sciences*. 10: 706–710.

Sharma, A., N. Ojha, A. Pozzer, G. Beig, and S.S. Gunthe. 2019. Revisiting the crop yield loss in India attributable to ozone. *Atmos. Environ. X* 1: 100008. https://doi.org/10.1016/j.aeaoa.2019.100008

Sharma, R., K. Sachdeva, and A.R. Sharma. 2021a. Assessment of synergistic impact of ambient surface ozone and fine particulate matter on experimentally grown wheat. *Asian J. Atmos. Environ.* 15(1): 2020080. https://doi.org/10.5572/ajae.2020.080

Sharma, R., K. Sachdeva, and A.R. Sharma. 2021b. Variations of surface ozone levels in urban area of India: A focus on night-time residual concentrations. *EQA –Intern. J. Environ. Quality.* 42: 22–35.

Singh, V., S. Singh, A. Biswal, A.P. Kesarkar, S. Mor, and K. Ravindra. 2020. Diurnal and temporal changes in air pollution during COVID-19 strict lockdown over different regions of India. *Environ. Pollut.* 266(Pt 3): 115368. https://doi.org/10.1016/j.envpol.115368

Sinha, B., K. Singh Sangwan, Y.Maurya, V. Kumar, C. Sarkar, B.P. Chandra and V. Sinha. 2015. Assessment of crop yield losses in Punjab and Haryana using 2 years of continuous in situ ozone measurements. *Atmos. Chem. Phys.* 15: 9555–9576. https://doi.org/10.5194/acp-15-9555-2015

Sinha, P., Singh, and P. Saroj. 2022. Relationship of Surface Ozone (O_3) with its precursors and meteorological parameters over New Delhi, India. *Mausam*. 73(4): 829–842.

Solomon, S. 1999. Stratospheric ozone depletion: A review of concepts and history. *Rev. Geophys*. 37(3): 275–316. https://doi.org/10.1029/1999RG900008

Solomon, S., R.R. Garcia, F.S. Rowland, and D.J. Wuebbles.1986. On the depletion of Antarctic ozone. *Nature*, 321(6072): 755–758. https://doi.org/10.1038/321755a0/

Soret, J.L. 1865. Recherche sur la densité de l'ozone. In: *Compte rendus hebdomadaires des séances de l'Académie des Sciences, Série C* (Vol. 61, p. 941). Académie des Sciences. Jacques Louis Soret, 1827–1890, Editor-in-chief of Archives des sciences physiques et naturelles.

Van Dingenen, R., F. Raes, M.C. Krol, L. Emberson, J. and Cofala, J. 2009. The global impactof O_3on agricultural crop yields under current and future air quality legislation. *Atmos. Environ*. 43: 604–618.

Vingarzan, R. 2004. A review of surface ozone background levels and trends. *Atmos. Environ*. 38: 3431–3442.

www.albany.edu/faculty/rgk/atm101/ozone.shtml

www.apis.ac.uk/overview/pollutants

www.apis.ac.uk/overview/pollutants/overview o3.html

www.differencebetween.com/distinguish-between-stratospheric-ozone-and-tropospheric-ozone/#Key

www.eea.europa.eu/legal/copyright

www.freethink.com/science/rubisco

www.gov.uk/government/statistics/air-quality-statistics/concentrations-of-ozon

www.hindustantimes.com/mumbai-news/pollution-destroys-21-wheat-6-rice-crop-every-year-iit-m-study/story-FfsjiUstkx62FL7ALe5uxI.html

www.iowadnr.gov/Environmental-Protection/Air-Quality/Air-Pollutants/Effects-Ozone

www.nationalarchives.gov.uk/doc/open-government-licence/version/3/

www.suzannegazdamd.com/blog/vitamin-d-deficiency-is-it

www.theozonehole.org/ozonedestruction.htm

www.theozonehole.org/ozonelayer.htm

www.whatarethe7continents.com/the-ozone-layer-what-how-formed-important

Yi, F.-J., J. Feng, Y.J. Wang, and F. Jiang. 2020. Influence of surface ozone on crop yield of maize in China. *J. Integrative Agric*. 19(2): 578–589.

Young, P.J., A.T. Archibald, K.W. Bowman, J.F. Lamarque, V. Naik, D.S. Stevenson, S. Tilmes, A. Voulgarakis, O. Wild, D. Bergmann, P. Cameron-Smith, I. Cionni, W.J. Collins, S.B. Dalsøren, R.M. Doherty, V. Eyring, G. Faluvegi, L.W. Horowitz, B. Josse, Y.H. Lee, I.A. MacKenzie, T. Nagashima, D.A. Plummer, M. Righi, S.T. Rumbold, R.B. Skeie, D.T. Shindell, S.A. Strode, K. Sudo, S. Szopa, and G. Zeng. 2013. Pre-industrial to end 21st century projections of tropospheric ozone from the Atmospheric Chemistry and Climate Model Inter comparison Project (ACCMIP). *Atmos. Chem. Phys*. 13: 2063–2090. https://doi.org/10.5194/acp-13-2063-2013

Wagg, S., G. Mills, F. Hayes, S. Wilkinson, D. Cooper, and W.J. Davies. 2012. Reduced soil water availability did not protect two competing grassland species from the negative effects of increasing background ozone. *Environ. Pollut*. 165: 91–99.

5 Biogenic Volatile Organic Compounds Effects on Health and the Environment

5.1 INTRODUCTION

Heart ailments, as well as several forms of cancer, have all been linked to pollution in the environment. Low-income people are more prone to reside in polluted locations with unsanitary water supplies. Additionally, pregnant women and young children are especially vulnerable to pollution-related health issues. Plants' ability to grow, develop, reproduce, and defend themselves is dependent on the biogenic volatile organic chemicals they produce. Within plant communities, between plants, and between plants and insects, they serve as a means of communication. Many of these compounds have huge mass emission rates from vegetation into the atmosphere, and their high chemical reactivity has a profound effect on the chemical composition and physical properties of the atmosphere. The connection between the biosphere and the atmosphere is therefore mediated by biogenic volatile organic molecules. Anthropogenic environmental changes, such as global climate change, have the potential to alter this link, which in turn could have unintended and difficult-to-predict effects on the Earth system. The generation of secondary pollutants such as peroxyacetyl nitrate (PANs), secondary organic aerosols (SOA), particulate matter (PM), aldehydes, and ketones is facilitated by BVOCs, which are the primary precursors to generate tropospheric ozone and atmospheric aerosols. In terms of overall annual emissions, isoprene accounts for around half of all BVOC emissions and is roughly equivalent to global methane emissions. Pollutant absorption capacity and BVOC production vary by species and habitat. Different studies have shown that coniferous species, and generally species with a complex foliar system, can absorb more pollutants than other types of plants.

Emissions of BVOCs into the urban atmosphere are significant since they are not proportional to other chemicals generated by human activities, particularly nitrogen oxides (NOx). The photochemical interaction between NOx and BVOCs can produce ozone, peroxyacetyl nitrates, aldehydes, ketones, hydrogen peroxide, secondary organic aerosol, and particulate matter (Fehsenfeld et al., 1992; Guenther et al., 1995). Especially in regions subjected to in situ release or long-range transport of anthropogenically polluted air masses, the climate and air quality may be affected by biologically volatile organic compounds (BVOC), which play a crucial role in biosphere-atmosphere interactions and are key constraints on the physical and chemical properties of the atmosphere. Ozone and particle formation may be aided by reactive BVOCs in certain settings. Methanol and isoprenoids are the most abundant BVOCs released by plants, accounting for the vast majority of the estimated 1'000 Tg C released annually by vegetation. Forest species are primarily responsible for this. This high emission rate presumably indicates crucial involvement in plant activity, as it is associated with a substantial metabolic expenditure. Both constitutive and induced BVOC emissions by biotic and abiotic stimuli have been seen in plants (Calfapietra et al., 2013). Isoprenoids, both constitutive and induced, play important roles as protective chemicals for plants in hostile conditions.

DOI: 10.1201/9781003618164-5

The impact of BVOC emissions on the composition of the atmosphere and the chemistry-climate system is of interest to atmospheric chemists. The roles that BVOCs play in plant biology and ecology have piqued the curiosity of plant biologists. Both local and global atmospheric chemistry are affected by biological emissions. The volatile organic molecules isoprene and monoterpenes (BVOCs) released by living organisms can have an indirect impact on climate change by contributing to the creation of ozone and secondary particles in the troposphere. Fine particulate matter (PM2.5) and ground-level ozone pollution are harmful to human health, especially in metropolitan areas where they are caused by the emission of biologically volatile organic compounds (BVOCs). However, most estimates of BVOCs failed to account for emissions from urban green spaces, leading to an incomplete picture of regional BVOC emissions and their impact on the environment and human health. To a large extent, the oxidants and aerosols that regulate air quality and climate are determined by biological volatile organic compounds (BVOC) released into the atmosphere by terrestrial ecosystems. To effectively manage air quality and climate, it is necessary to have precise quantitative estimates of BVOC emissions (Baghi et al., 2012).

According to numerous studies (Komenda et al., 2001), BVOCs emitted from plants are the most common. Many biogenic volatile organic compounds (BVOCs), such as aerosols and oxidants, are produced and released into the atmosphere by terrestrial ecosystems. Many components of terrestrial ecosystems (including flowers, stems, trunks, roots, leaf litter, soil microbes, insects, and animals) contribute to the worldwide total emission of these BVOCs, but foliage accounts for the vast majority of this total. Numerical models of regional air quality and global climate now commonly include BVOC emissions, estimated as a function of land cover and environmental driving variables, due to the growing recognition of the importance of these emissions for Earth system modelling. Some BVOCs have a comparatively large impact due to their reaction rates, products, ozone production potentials, organic aerosol yields, and other features, which can affect atmospheric composition if emitted at high rates. Therefore, determining the chemical variety of BVOC emissions is crucial. Studies like these have been conducted because of the relevance of BVOCs.

5.2 THE CHARACTERISTICS OF BIOGENIC VOLATILE ORGANIC COMPOUNDS/VOCS

In atmospheric chemical processes in the troposphere, biogenic volatile organic compounds (BVOCs) are crucial. Low-boiling-point substances known as biogenic volatile organic compounds (BVOCs) are frequently produced by plants' secondary metabolic processes. BVOCs can be released into the atmosphere by a variety of vascular plants. Volatile organic compound (VOC) content is a commonly used and significant indicator of environmental contamination. Industrial, municipal, and even biological activity release volatile organic compounds (VOCs), which are responsible for the emissions of many types of pollutants in ambient air. Around the world, ecosystems contribute far more to VOC emissions than human activity does. Significant amounts of BVOCs are also released by flora, such as street trees and urban forests, which are planted to enhance the urban environment and give residents a calmer ambiance. Because BVOCs may effectively react with NO_X generated by neighboring autos to form O_3, they are receiving increasing attention from metropolitan areas. June, July, and August were the summer months when the emissions of BVOCs surged. Due to their high emission and reactivity, biogenic volatile organic compounds (BVOCs) can significantly affect photochemical pollution and haze.

5.3 VALUE OF BVOC EMISSIONS IN RELATION TO THEIR EFFECTS ON CLIMATE CHANGE

It is now widely acknowledged that biological processes in terrestrial ecosystems broadly affect the atmosphere and climate system of the Earth, which consequently implies potentially significant feedback effects on current global changes in atmosphere and climate arising from human activities. Of these biological processes, those related to the carbon cycle (i.e., the assimilation and release of CO_2 and the carbon sequestration in organic material) have been the focus of much attention. However, gases other than CO_2 are exchanged between the biosphere and the atmosphere. Vegetation covering the landmasses releases biogenic volatile organic compounds (BVOCs), which comprise a large variety of molecules that differ in size and physicochemical properties as well as in their metabolic origin. The emission rates of BVOCs are determined by their synthesis rates and by their physicochemical characteristics, mainly their solubility, volatility, and diffusivity. Their emission rates are therefore heavily modulated by internal and external factors. There is a broader range of production capacity for BVOCs than for photosynthetic carbon assimilation between plant species. As a result, changes in species and community structure of ecosystems, land uses, climate, and resource availability can lead to major changes in regional BVOC fluxes, even if the standing biomass and net primary productivity remain unchanged.

5.4 THE BIOLOGICAL PROCESS – PRODUCTION OF VOLATILE ORGANIC COMPOUNDS

The lower border of the atmosphere is given by the contact between the terrestrial biosphere and the atmosphere, and this boundary has a significant impact on the chemical, biological, and physical features of the atmosphere. By affecting soil moisture, light environment, and temperature, vegetation shifts the biogeophysical boundaries of the system. As much as 55% of Earth's surface area is covered by leaves during the boreal summer, increasing the total area available for emissions. Vegetation has multiple effects on the biogeochemical dynamics of the system, including the release of a wide range of biogenic volatile organic compound (BVOC) emissions and climate-relevant primary biological aerosol particles (PBAP). Surface-level meteorological and climatic parameters, like temperature, light exposure, soil moisture, and wind speed and direction affect many of these emissions. When released into the troposphere, they undergo a sequence of chemical processes that yield beneficial byproducts. Both PBAP and BVOC have the potential to degrade into climate-relevant tiny particles, with the latter also contributing to the formation of ozone and secondary organic aerosols (SOA).

These emissions and the aerosols they produce can affect the weather on the ground through processes like reducing the amount of sunlight reaching the ground, increasing the concentration of greenhouse gases in the atmosphere (which can have a positive feedback effect on the temperature of the air near the ground), and altering cloud formation and, in turn, precipitation. These atmospheric changes may reflect back to the land surface and the emissions themselves, creating positive or negative feedback loops, which may either reduce the emission response or increase it. The most common BVOC, isoprene, has a temperature-dependent feedback reaction that can be either positive or negative. A positive, negative, or uncoupled reaction to soil moisture and precipitation might occur depending on the soil moisture content and the total atmospheric aerosol loading.

Whether or not the isoprene molecule reacts positively to light depends on how diffuse the illumination is. These feedbacks demonstrate the dynamical response of the biosphere to shifting atmospheric conditions throughout a wide range of periods, from minutes for trace gases and aerosols to months for phonological changes to years for land cover and land use changes. The complex feedbacks between the biosphere and the atmosphere, and how they affect climate and global change, are vital to study, model, and anticipate because of the dynamic character of the system. Predictive modelling tools can shed light on how these feedbacks can change in response to a warming climate, which is important because it is difficult to identify them in observations.

Not all pollution is caused by humans. New research from the University of California, Riverside, estimates that a rise in global temperature of 4 degrees Celsius will result in a 14 percent increase in toxic plant emissions and dust. According to Sánchez-Gómez et al. (2023) and Castorina et al.

(2020), plants are expected to be responsible for almost two-thirds of future pollution. Air pollution from natural sources, like plants, will increase in average global temperatures (Figure 5.1). *Cannabis,* belonging to the Flora species, it is obvious that there will be a release of BVOCs.

Tropical rainforests are the largest source of biogenic volatile organic compounds (BVOCs) in the world. Plants emit a broad number of Biogenic Volatile Organic Compounds (BVOCs) that can impact urban ozone (O_3) production. Conversely, the O_3 is a phytotoxic pollutant that causes unknown alterations in BVOC emissions from native plants. BVOCs are highly reactive in the atmosphere, reacting, for example, with ozone. Plants collectively produce thousands of biogenic volatile organic compounds (BVOCs).

FIGURE 5.1 The VOC emissions from flowers and foliage of *Cannabis sativa* (Kumeroa et al., 2022. https://doi.org/10.3390/agronomy12112651.)

According to Moura et al. (2022), three native species from Brazil were exposed to constitutive and O_3-induced BVOCs for two ($2dO_3$) and four ($4dO_3$) days at an O_3 concentration of 80 ppb. The O_3 response was assessed using histochemical techniques to identify hydrogen peroxide (H_2O_2) and programmed cell death (PCD). *Piptadenia gonoacantha* (Mart) J.F. Macbr, *Croton floribundus* Spreng, and *Astronium graveolens* Jacq are from distinct ecological succession groups (acquisitive and conservative), have various carbon-saving defense mechanisms, and emit particular BVOCs. Green leaf volatiles (GLV), monoterpenes (MON), sesquiterpenes (SEQ), and other chemicals (OTC) comprised the BVOC composition that was emitted by the three species. Compared to *A. graveolens*, *C. floribundus* is more acquisitive.

Wang et al. (2019) measured biogenic volatile organic compounds (BVOCs) emitted by *Cannabis* plants grown under conditions mimicking greenhouse cultivation. Plants also release BVOCs in response to biotic interactions, including herbivory and pathogen attacks, as well as abiotic stress such as drought, increased salinity, temperature, windstorms, or other mechanical damage. The research used an ensemble of regional-scale regulatory modeling to predict the impacts of *Cannabis* terpenes (BVOCs) on ozone concentrations. BVOC emissions from indoor cultivated *Cannabis* in Colorado could contribute to ozone formation and particulate matter pollution. Methanol, ethylene, formaldehyde, ethanol, acetone, and acetaldehyde are all examples of low-molecular-weight (C5) BVOCs that are released by plants. Plants emit more than 30,000 different BVOCs, including terpenoids, green leaf volatiles, phenylpropanoids, benzenoids, and methyl esters of plant hormones (i.e., methyl jasmonate and methyl salicylate). BVOCs include the aroma of freshly cut grass and the flavor of a ripe strawberry. Sánchez-Gómez claimed that plants constantly release them. In and of themselves, BVOCs pose no harm. The soil-based emissions and biotic interactions of VOCs are shown in Figure 5.2. Soils may act as sources or sinks of volatile organic compounds (VOCs). Many of the formed VOCs are produced by microorganisms, and it would be a challenge to investigate soil microbial communities by studying their VOC profile. Such "volatilomics" would have the advantage of avoiding extraction steps that are often a limit in genomic or proteomic approaches. As the vapor pressures of the different VOCs vary and correspond with temperature, soil may function either as a sink or as a source of BVOCs. Agricultural BVOC emissions can significantly impact atmospheric chemistry, affecting air quality and human health, especially in regions with a high proportion of agricultural fields surrounding populated areas (Kevin et al., 2017). Soil BVOC emissions are influenced by factors like temperature, moisture conditions, microbial metabolism, and soil properties. Figure 5.2 shows the soil-based emissions and biotic interactions of VOCs (volatile organic compounds).

Organic aerosols are created when these substances combine with oxygen (Knudsen and Gershenzon, 2006). Infant mortality, asthma in children, and adult cardiovascular disease and lung cancer are all linked to inhalation of these aerosols. Plants' BVOC production rises in response to two factors: rising temperatures and atmospheric carbon dioxide levels. Both of them are expected to keep growing in the future. To be absolutely clear, plant cultivation has good environmental effects. They cut down on atmospheric carbon dioxide, which aids in the regulation of global warming. The BVOCs produced by home gardeners are safe to breathe. BVOC emissions released (Ma et al., 2019) from the urban landscape had significant effects on the production of ozone concentration (Figure 5.3) (Gu et al., 2021). Vehicular exhaust and evaporative emissions are also major emitters of VOCs in urban areas.

The greatest mass emission rates are found in leaves, but the greatest diversity of BVOCs is released by flowers and fruits, with emission rates peaking at maturity. Terpenoids, such as isoprene, monoterpenes, sesquiterpenes, and some diterpenes, are more commonly released from the vegetative sections of woody plants. Oxygenated BVOCs and some monoterpenes are released at rather high levels by grass species. These chemicals, along with other so-called green leaf volatiles (C6 aldehydes and ketones), may be released at higher rates from injured plants. Some BVOCs, like terpenes, methyl jasmonate (MeJA), and methyl salicylate (MeSA), can be induced from leaves by

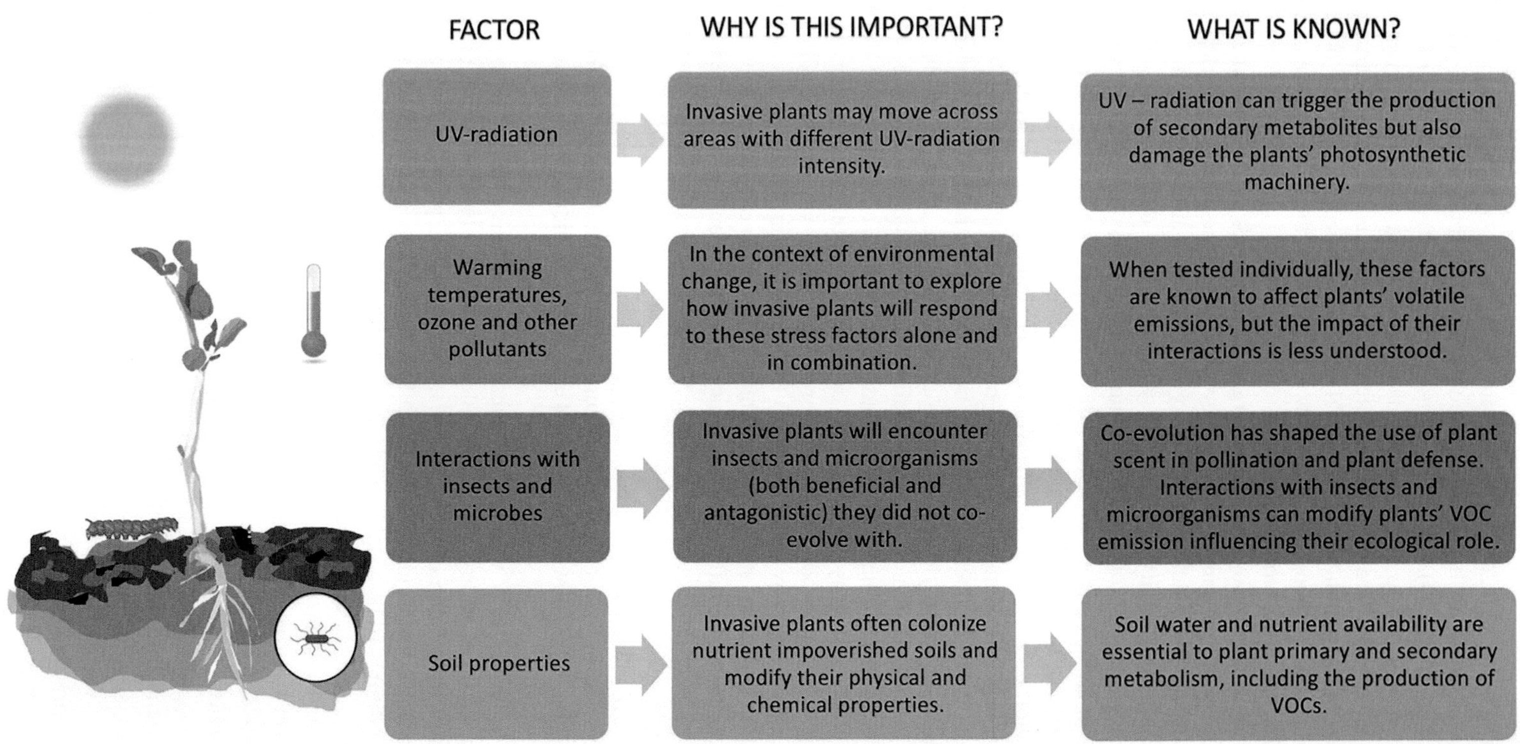

FIGURE 5.2 The soil-based emissions and biotic interactions of VOCs (volatile organic compounds). Soil may function either as a sink or as a source of BVOCs. (McCormick et al., 2023. This work is distributed under the Creative Commons Attribution 4.0 License.)

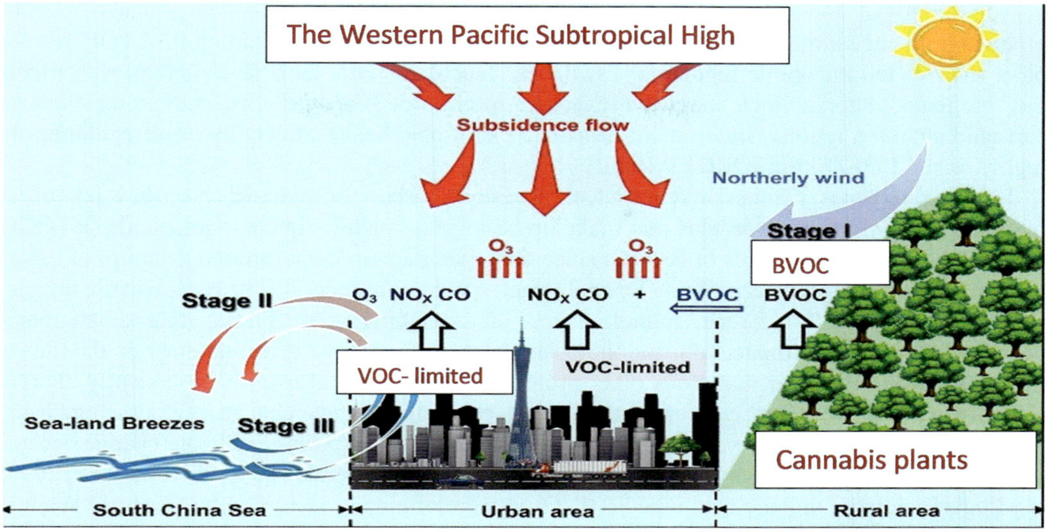

FIGURE 5.3 The characteristics of BVOC emissions released from the urban and rural areas. (Ma et al. (2019). Under Creative Commons Attribution License (CC BY 4.0).)

biotic and abiotic stressors, with the extent and quality of production depending on the nature of the damage. Laothawornkitkul et al. (2009) describe the main categories of BVOCs, the main genera of BVOC-emitting plants, and projections of current and future fluxes into the atmosphere.

For instance, as Sánchez-Gómez pointed out that your lawn won't produce enough BVOCs to make you sick (Gómez et al., 2020). Large-scale increases in carbon dioxide cause an uptick in biosphere-produced volatile organic compounds (BVOCs) and, ultimately, organic aerosols. Airborne pollutants with a diameter of 2.5 micrometers or less are grouped together under the umbrella term "PM2.5," which includes organic aerosols, dust, sea salt, black carbon, and sulphate. This research found that as CO_2 levels rose, so did naturally occurring PM2.5 pollution. More and more PM2.5 pollution is being produced. This research suggests that a significant temperature rise, specifically a 4°C increase, would likely lead to a substantial increase in PM2.5 pollution levels due to enhanced emissions of Biogenic Volatile Organic Compounds (BVOCs) from vegetation.

As CO_2 emissions rise, so does the amount of PM2.5 released into the air, and vice versa. The air quality will improve as we continue to cut back. For instance, the study indicated that even with a temperature rise of 2 degrees Celsius, PM2.5 levels would rise by only 7 percent. Since the focus of the research was on the effects on human health, these findings only relate to variations in air quality over land. Sánchez-Gómez warns that drastic reductions in carbon dioxide emissions are necessary to improve air quality in the future.

However, the mechanisms through which exposure to rural and coastal environments improves health are not fully understood. It has been argued that escaping to one of these settings can have positive effects on one's mental and physical well-being by acting as a buffer against the stressors of daily life. Inhaling and ingesting airborne bacteria and particles expose people living in natural habitats to a wide variety of biogenic substances. By suppressing the activity of interconnected cell signaling systems, especially PI3K/Akt/mTORC1 (Moore, 2015), the "biogenics" theory proposes that chronic exposure to low concentrations of combinations of natural substances and toxins in natural environments provides pleiotropic health advantages. Many pathogenic processes, such as cancer, diabetes, inflammation, immunosuppression, and neurological disorders, are associated with overactive Akt and mTOR (mTORC1). The activity of these protein kinases is inhibited by a wide variety of natural products (i.e., from bacteria, algae, fungi, and higher plants). Natural products influence autophagy and other cytoprotective mechanisms in addition to mTOR and the other interrelated metabolic regulatory "switches" (such as PTEN and NF-B). Here, we propose the "biogenic" theory, which states that exposure to airborne biogenic substances in their natural surroundings on a regular, intermittent basis has pleiotropic health effects by downregulating the highly linked PI3K/Akt/mTORC1 system.

Isoprene, terpenes, alkanes, alkenes, alcohols, esters, carbonyls, and acids are just a few of the many different types of molecules that make up biological volatile organic compounds (BVOCs). Plants are the primary sources of BVOCs since they manufacture them through a variety of tissue-specific physiological mechanisms. Several plant species produce and release volatile organic compounds (BVOCs) isoprene, monoterpenes, and sesquiterpenes. These volatile isoprenoid chemicals have major impacts on the photochemistry and radiative characteristics of the atmosphere, in addition to their numerous other protective and ecological roles. Consequently, there is a lot of curiosity about how environmental factors, especially climate change, affect the emissions of volatile isoprenoids from various species. In addition, BVOCs promote climate change because they alter the chemical and physical features of the atmosphere (Bema, 1997; Courtois et al., 2009).

Biogenic volatile organic compounds (BVOCs) were originally identified by Georgian scientist Sanadze (1956) in the early 1950s. However, much of the early data was published in Russian literature and not understood widely by scientists throughout the world. Went (1955, 1960) postulated that biogenic volatile organic compounds (VOCs) played a role in the formation of the blue haze found in the southeastern US Blue Ridge mountains, at the same time as Rasmussen and Went (1965) analyzed emissions from individual plants. The first regional-scale BVOC emissions model linked these emissions with the atmospheric community and models of tropospheric chemistry, allowing researchers to better understand the role of BVOC in the formation of ozone and secondary organic aerosol (SOA).

Vegetation releases volatile organic molecules into the atmosphere, as shown by Fehsenfeld et al. (1992). At least three major research communities are interested in understanding the mechanisms by which plants generate BVOCs and the effects or functions these compounds have. The impact of BVOC emissions on the composition of the atmosphere and the chemistry-climate system is of interest to atmospheric chemists. Researchers in plant biology are curious about the ecological and biological effects of BVOCs. Entomologists are curious about them because of the messages they send. There has been little coordination between these fields of study so far, despite the fact that numerous lines of evidence have shown that the roles of BVOCs in the biosphere and the atmosphere are intricately intertwined. Therefore, the purpose of this review is to provide a synopsis of our current understanding of BVOCs in the Earth system, with an emphasis on the roles they play, and to highlight any gaps in that understanding (Choomanee and Blauert, 2022). Further, it shows how an integration of knowledge and resources between the research domains of atmospheric chemistry is important to increase our understanding of the Earth system by highlighting the significant interlinkages between the roles of BVOCs in the biosphere and the atmosphere.

Plants, like all other forms of life, are vulnerable to the numerous consequences of climate change. When plants are under stress, they release biogenic volatile organic compounds (BVOC) to protect themselves. The goal of the research conducted by Janyasuthiwong et al. (2022) was to determine how temperature affected the pace at which the tropical plants *Syzygium cumini* and *Peltophorum pterocarpum* released their BVOC (isoprene and monoterpenes). The identified species were both plentiful in the urban green space and known to generate significant levels of BVOC. This investigation centered on three temperature ranges: the low end (16-18 °C), the middle (36-38 °C), and the high end (40-42 °C). A VOCs analyzer was used to determine the levels of BVOC. According to their findings, the rate of BVOC emission increased up to 38–40 °C and then declined. *S. cumini* had a BVOC maximum of 44.04 g/m^2 h, while *P. pterocarpum* peaked at 13.12 g/m^2 h. Because of differences in leaf architecture, namely the number and width of stomata, *S. cumini* emitted lighter than *P. pterocarpum*. The findings can be utilized to create a model investigation of plant BVOC emissions in urban green space, with a focus on megacities prone to the heat island effect.

5.5 IMPACTS OF EMISSIONS CAUSED BY BIOLOGY

VOCs such as isoprene, terpenes, alkanes, alkenes, alcohols, esters, carbonyls, and acids are released into the atmosphere as a result of biological processes (www.sciencedirect.com/topics/earth-and-planetary-sciences/biogenic-emission). Both local and global atmospheric chemistry are affected by biological emissions. The most common volatile organic compound (VOC) released by plants is isoprene (C_5H_8). Contributing to the creation of O_3 and influencing the longevity of other species are two of its most important roles. Ambient environmental variables, especially temperature and light, regulate biological outputs. Isoprene emissions tend to rise as temperatures rise (Janyasuthiwong et al., 2022). Biogenic emissions, which are linked to climate change, are predicted to rise in the future due to factors such as expected temperature increases and may have an impact on air quality in the region. The Sulphur cycle in the atmosphere depends significantly on biological emissions. Dimethyl sulfide (DMS) is oxidized by oxygen in the remote, unpolluted marine atmosphere after being emitted by marine organisms. Emissions of volatile organic compounds (VOCs) into the atmosphere can be traced back to both natural biogenic sources and human activities such as vehicle use, solvent disposal, and industrial processes. Numerous modelling studies and analyses of primary data have been conducted to determine the efficacy of pollution removal. The research relies on the accumulation of air pollutants and the resulting temperature shifts. Ozone at ground level is formed when there is a high concentration of nitrogen oxides and biogenic volatile organic emissions. Reacting with NOx, the naturally emitted VOCs such as monoterpenes, isoprene, etc., produce O_3. The rate of VOC emission is influenced by the plant species used in its production. Researchers found that biogenic VOC production and subsequent conversion to O_3 and PM had a considerable impact. Pollutant gases are taken in through the stomata and the leaf surface of green plants. Some gases can diffuse into the intercellular space, where they can remain for much longer. Pollutants absorbed by the leaf surface and other sections are washed away by rain and other precipitation. Each plant species has a unique leaf structure, texture, morphology, and length that affects the effectiveness with which it absorbs pollutants. The broad, rough surface of the leaves is excellent for storing contaminants for extended periods.

The volatile organic molecules isoprene and monoterpenes (BVOCs) released by living organisms can have an indirect impact on climate change by contributing to the creation of ozone and secondary particles in the troposphere. Reactivity with chemicals generated from anthropogenic sources, particularly nitrogen oxides (NOx), gives BVOCs in the urban atmosphere a specific significance. The photochemical reaction between NOx and BVOCs can produce ozone, peroxyacetyl nitrates, aldehydes and ketones, hydrogen peroxide, secondary organic aerosol, and particulate matter. Fine particulate matter (PM2.5) and ground-level ozone pollution are dangerous to human health, especially in metropolitan areas where they are caused by the emission of biologically volatile organic compounds (BVOCs). However, most estimates of BVOCs failed to account for emissions from urban green spaces, leading to an incomplete picture of regional BVOC emissions and their impact on the environment and human health. In particular, compared to AVOCs, BVOCs derived from plants are thought to play more significant roles due to their greater source strengths and chemical reactivity. Few quantitative studies have looked into the effects of BVOC on the environment and human health. Furthermore, BVOCs from green spaces can have significantly bigger health effects when they exist in metropolitan regions with high human population densities than when they occur in wild forests (Ren et al., 2017). The health losses associated with BVOC emissions in urban areas will also be considerably underestimated if urban emissions are ignored. This could have a big impact on regional and global emissions.

5.6 THE PRODUCTION OF BIOGENIC VOCS AND THEIR REACTIVITY FROM VARIOUS ACTIVITIES

Other than carbon dioxide and monoxide, organic trace gases found in the atmosphere are referred to as biogenic volatile organic compounds (biogenic VOCs). Nonmethane hydrocarbons (NMHCs) refer specifically to hydrocarbons other than methane, whereas nonmethane volatile organic compounds (NMVOCs) refer to VOCs other than methane. The isoprenoids (iso-prene and monoterpenes), alkanes, alkenes, carbonyls, alcohols, esters, ethers, and acids are all examples of biological VOCs. Isoprene and monoterpenes stand out as the most abundant chemicals in emission inventories. Next to isoprenoids, the most common classes of compounds include alcohols and carbonyls (https://iopscience.iop.org/article/10.1088/1748-9326/abcee9).

Biogenic volatile organic molecules have enormous mass emission rates from plants into the atmosphere and a high chemical reactivity, both of which have an important influence on the chemical composition and physical features of the atmosphere. Plants' ability to grow, develop, reproduce, and defend themselves is dependent on the biogenic volatile organic chemicals they produce. Within plant communities, between plants, and between plants and insects, they serve as a means of communication. Many of these compounds have huge mass emission rates from vegetation into the atmosphere, and their high chemical reactivity has a profound effect on the chemical composition and physical properties of the atmosphere. Thus, volatile organic molecules produced by living organisms serve as a go-between for the biosphere and the atmosphere (Laothawornkitkul et al., 2009).

Biogenic VOCs, such as terpenoids (e.g., isoprene and monoterpenes), hexenal family chemicals (hexenals, hexenols, and hexenyl esters), methanol, and acetone, originate mostly from plant matter. The reactivity of volatile organic compounds (VOCs) contributes to the creation of ozone when UV radiation and nitrogen oxides (NOx) are present. The effect of BVOC on O_3 generation is amplified during the summer, when both photochemical activity and BVOC emissions are at their highest. Recent investigations of Komenda et al. (2001) and Tambunan et al. (2006) have shown that plant-based emissions are the most prevalent BVOCs. Important for secondary organic aerosol generation and atmospheric oxidative capacity, BVOCs include isoprene, terpenes (monoterpenes, sesquiterpenes), and oxygenated VOCs (alcohols, aldehydes, ketones, and acetates). The effects of BVOCs on global warming were investigated by Pacfico et al. (2009) and Peuelas and Staudt (2010). If the average world temperature is anticipated to rise by another 2 to 3 degrees Celsius by the turn of the century, this might lead to a rise in global emissions of BVOCs, which in turn would raise tropospheric ozone and methane and alter the atmosphere's oxidation capacity. The amount of isoprene released by deciduous trees changes with the season. Heat and light trigger an increase in isoprene production in trees. This means that peak levels of biogenic VOCs coincide with peak levels of other main pollutants due to human activity. When temperatures go from 77 to 95 degrees Fahrenheit, for instance, four times as much isoprene is released.

The largest contribution to global VOCs is biogenic VOCs (BVOCs), which are gases given out by living creatures or biological processes vital to the upkeep of life cycles. The anticipated worldwide emission range for these sources is between 760 and 1,150 TgC per year-1, which is around ten times higher than the equivalent anthropogenic emissions (Figure 5.4). Flowers, stems, trunks, roots, leaf litter, soil microbes, insects, and mammals are all potential producers of BVOCs in terrestrial ecosystems, but foliage accounts for the vast majority of the world's total emissions. VOCs are also emitted into the atmosphere from a range of sources, including anthropogenic activities like industrial processes, transportation, petroleum refineries, domestic processes, fossil fuel-burning power plants, textile cleaning, construction activities, and pharmaceutical industries, as well as from biogenic or natural sources like vegetation, volcanoes, anaerobic moorland processes, and natural forest fires.

The physical and chemical features of the atmosphere and the climate are heavily influenced by a class of compounds called biologically volatile organic compounds (BVOCs). Ozone levels are

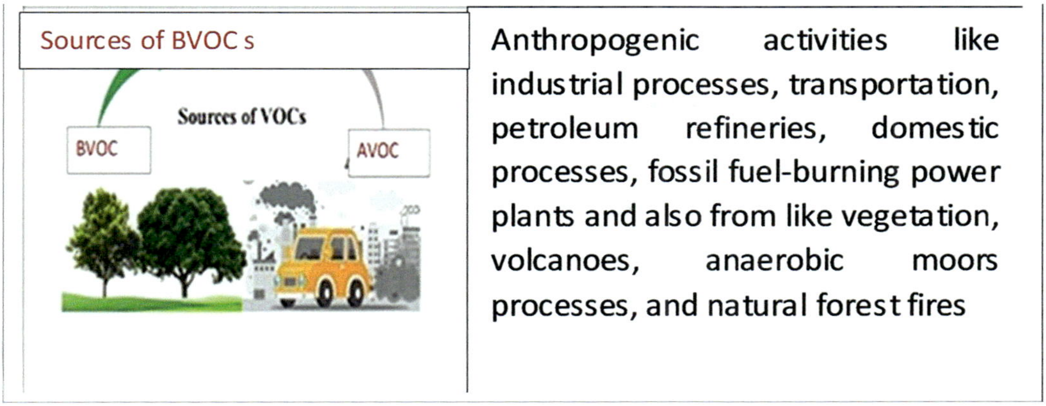

FIGURE 5.4 Anthropogenic emissions of BVOC. (Mula et al., 2024. https://doi.org/10.4209/aaqr.230170. CC BY 4.0 License.)

affected by BVOCs in the air in several ways, which in turn depend on the levels of NOx (nitrogen oxides) in the air, which are caused mostly by human activities such as the combustion of fossil fuels, and the specifics of the surrounding environment. Ozone is produced from BVOCs by the set of reactions: nitrogen oxides + BVOCs + sunlight > O_3. Reactions between BVOCs and NOx often cause ozone when NOx levels are high enough. However, their reaction can lead to ozone degradation under specific situations.

Anthropogenic volatile organic compounds (AVOCs) come from sources such as cars, solvents, fossil fuel combustion, etc., while biological volatile organic compounds (BVOCs) come mostly from plants, animals, and microbes. The annual emissions of BVOCs are significantly higher than those of AVOCs, and BVOCs are also more reactive. As a result, BVOCs play a crucial role in the production of O_3 in the troposphere. Additionally, by 2052, surface O_3 will increase by 19%-20% and the emission of isoprene, the dominating BVOCs accounting for nearly 50% of the world flux, may increase by 20%-92% due to climate change. To summarize, it is important and worthwhile to focus on BVOC emissions. The sources of BVOCs emissions, methods of detection, impacts of land use and climate change, emission estimates, forest ecosystem roles, and ozone depletion have all been examined in certain research.

Many biogenic volatile organic compounds (BVOCs) are produced by terrestrial ecosystems and released into the atmosphere, where they interact with other atmospheric components such as aerosols and oxidants (Guenther, 2013). Many components of terrestrial ecosystems (including flowers, stems, trunks, roots, leaf litter, soil microbes, insects, and animals) contribute to the production of these BVOCs, although leaves account for the vast majority of the global total emission (Baghi et al., 2012; Barney et al., 2009). Numerical models of regional air quality and global climate now commonly include BVOC emissions, estimated as a function of land cover and environmental driving variables, due to the growing recognition of the importance of these emissions for Earth system modelling. Hundreds of distinct BVOC chemical species are released into the atmosphere, and the ability of different plant species to synthesize and emit terpenoids and other BVOCs varies greatly. In addition, several activities in an ecosystem might influence the emission of a given molecule from a range of distinct sources. Some molecules are kept in open structures that are constantly being emitted into the environment, while others are stored in plant tissues that are insulated from the atmosphere and are only emitted if these tissues are damaged. Further chemicals are produced, but may only be released in reaction to stress or under very precise environmental conditions, and hence are not kept in tissues.

Went's (1960) pioneering study began quantifying BVOC emissions to the atmosphere by extrapolating data of a single class of compounds, monoterpenes, from a single plant species, *Artemisia tridentata*, to the entire planet. To measure regional emissions, Rasmussen (1972) proposed a method for categorizing the biosphere into different vegetation groups based on species' BVOC emission capacity. Atmospheric processes and biological interactions are both influenced by BVOCs. Biogenic volatile organic compound (BVOC) emissions at the ecosystem level are typically attributed to plant leaves (Laothawornkitkul et al., 2009). Natural ecosystems emit a complex mixture of molecules, and research has shown that stems, including bark (Amin et al., 2012), plant roots and the rhizosphere, decomposing litter, and even microbes all generate BVOCs. However, little is known about the emissions that come from the soil and underground plant parts (such as roots and rhizomes).

When compared to respiratory CO_2 effluxes, the role of BVOCs in soil carbon loss is negligible. The oxidation products of BVOCs, however, contribute to the generation of secondary organic aerosols and may even be implicated in the formation of new particles in the atmosphere, highlighting the importance of BVOCs as reactive atmospheric trace gases (Riipinen et al., 2012). Biologically active compounds (BVOCs) are crucial to many cellular processes and also contribute to air chemistry. Some soil microbes use BVOCs in the soil's atmosphere as a carbon source, but these compounds can disrupt biogeochemical cycles and affect microbial activity, which can have far-reaching consequences on ecosystem processes. Biogenic volatile organic compounds (BVOCs) are crucial to the chemical reactions that take place in the atmosphere because of their high reactivity. The hydroxyl radical (OH) in the atmosphere quickly oxidizes these pollutants, and their subsequent reactions with other atmospheric components produce ozone or secondary organic aerosols. Emissions of biological VOCs have less effect on the radiative budget than those of greenhouse gases. To provide a complete picture of chemical and climatic processes and their evolutions, including all possible feedback loops, knowledge is no less crucial.

Knowing how BVOCs are produced is critical to any effort to manage air quality, because hydrocarbons are a primary component in the production of ozone in the atmosphere. Researchers and those responsible for air quality are both hampered by a lack of data on these emissions at the level of individual plant species. There is an urgent need to collect more information, especially about the greenhouse gas emissions caused by fynbos, urban trees, commercial forestry, and agriculture. Extending models to a landscape and, eventually, a regional scale requires an understanding of the elements that govern emissions at the plant level. In response, urbanization, agricultural practices, and climate change are modifying these landscapes. Currently, BVOC emissions are not considered in any air quality control plan in southern Africa. Understanding BVOC emissions at the plant species level and landscape scales allows better predictions of future BVOC emissions as land-use changes take place across the region, and is valuable input to air quality management at all spatial scales.

It is anticipated that Europe will soon become a global leader in the manufacture of bioenergy plants. The environment and the availability of resources could both be impacted. This novel mixture of highly reactive molecules may alter the chemical makeup of the atmosphere, as the biogenic volatile organic compounds (BVOCs) emitted by bioenergy plants are not the same as those emitted by typical agricultural crops. The rapid interactions of BVOCs with OH have the potential to accelerate photochemical O_3 generation, boost the creation of secondary organic aerosols (SOA), and extend the lifetime of CH_4. However, accurate field observations of important crops are lacking, making it challenging to quantify the environmental impacts of bioenergy plants on air quality and the regional climate. The air quality and local climate are being impacted by BVOCs.

5.7 ANTHROPOGENIC SOURCES OF VOLATILE ORGANIC COMPOUNDS (VOCS) AND CLIMATE CHANGE

Plants release volatile organic compounds (VOCs) for a variety of reasons, including photosynthesis, fertilization, and defense. Bio volatile organic compound (BVOC) reactions with hydroxyl (OH), nitrate (NO_3), and ozone (O_3) radicals increase the atmosphere's oxidation capacity (Figure 5.5), generate fewer volatile compounds that can form and grow atmospheric clusters, and influence cloud condensation (Müller et al., 2017) and solar radiation scattering, thus influencing biosphere-atmosphere interactions and local/regional climate change (Liu et al., 2021).

The emissions and oxidation of BVOCs are influenced by the changing climate (Figure 5.5; red arrows). As a result of global warming, it has been determined that isoprene emissions are the primary cause of rising BVOC levels around the world (Guenther et al., 1991). Isoprene is generally emitted through the leaf pores (stomata). Although stomatal closure may affect the isoprene production rate because it affects the internal concentration of CO_2, it does not affect the isoprene emission rate, presumably because isoprene continues to be emitted through the leaf cuticle (Pacífico et al.,

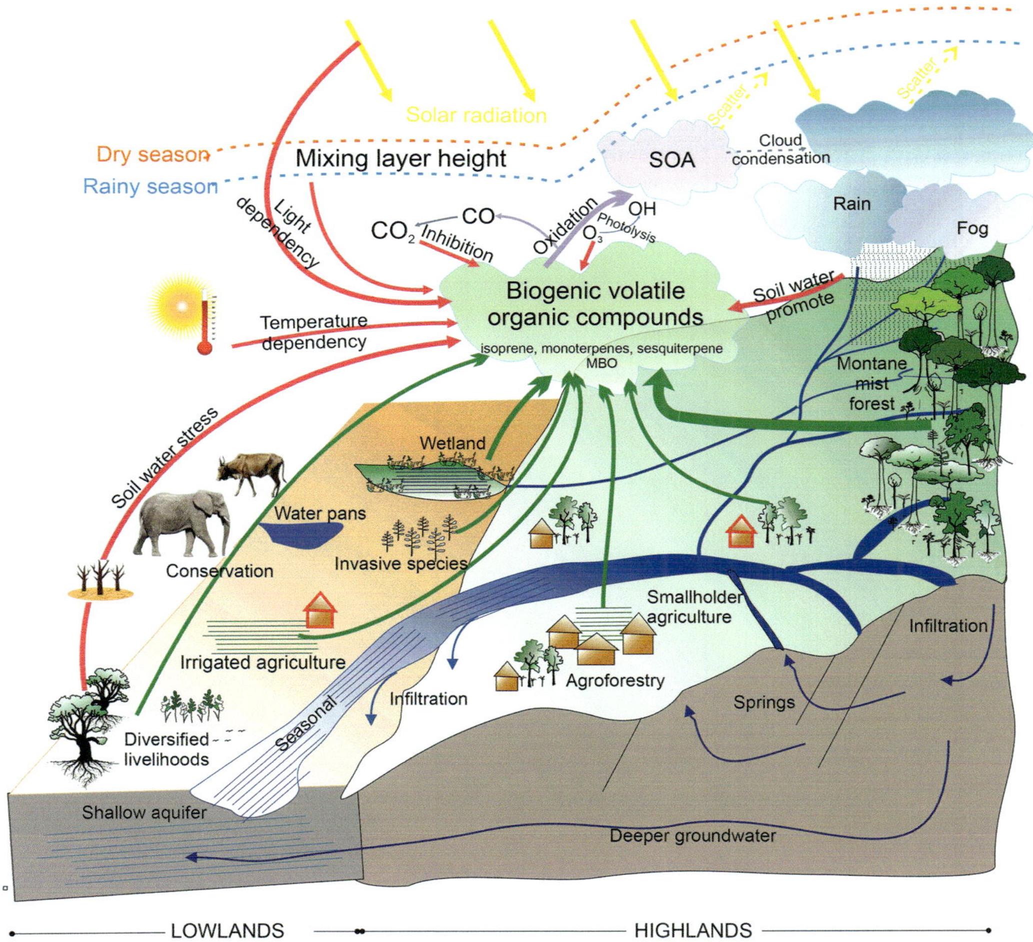

FIGURE 5.5 Atmospheric oxidation and abiotic effects on biogenic volatile organic compounds (BVOCs) in the African highland (Liu et al., 2021. This work is distributed under the Creative Commons Attribution 4.0 License.)

2009). Isoprene emission is influenced by soil water availability in addition to temperature and light, and thus reacts to soil water stress (Guenther et al., 2012). While it's common knowledge that temperature has a major role in regulating monoterpene emission, it's less common knowledge that the availability of light plays a significant role in the emission of some monoterpenes (such as ocimene). With a 1 °C rise in air temperature owing to climate change, Mochizuki et al. (2020) predicted a 15% rise in monoterpene emissions.

Some ecosystems are more sensitive and vulnerable to climate change than others. These include drier ecosystems and human-modified systems, including savannas, bushland, grassland, and agroforestry. It is estimated that grass, bushes, and crops account for about 18% of global BVOC emissions (Guenther, 2013). Since BVOC data from these habitats are uncommon, this estimate comes with a high degree of uncertainty (e.g., Guenther, 2013). Widespread and covering 55.2% of tropical Africa, these climate-sensitive ecosystems have a great deal of influence on native ecosystem changes, such as the expansion of human-modified systems at the expense of grassland and savannas, which can reduce the global BVOC levels (Unger, 2014b). It is predicted that in the future climate, extreme weather events, including heat waves, hot nights, droughts, and flooding, will become more often in the aforementioned climate-sensitive ecosystems.

Duan et al. (2023) found that isoprene emission (Figure 5.6) was directly affected by photosynthetic rates, and indirectly affected by photosynthetically active radiation and air temperature (Tair). Monoterpene emissions were directly affected by atmospheric pressure, relative air humidity, and specific leaf weight, and indirectly affected by Tair.

Feldner et al. (2022) found that cities around the world are reacting to adjust to responsibilities associated with climate change, frequently by significantly extending the urban green infrastructure. Biogenic volatile organic compounds (BVOCs), primarily isoprene and terpenes, may be emitted

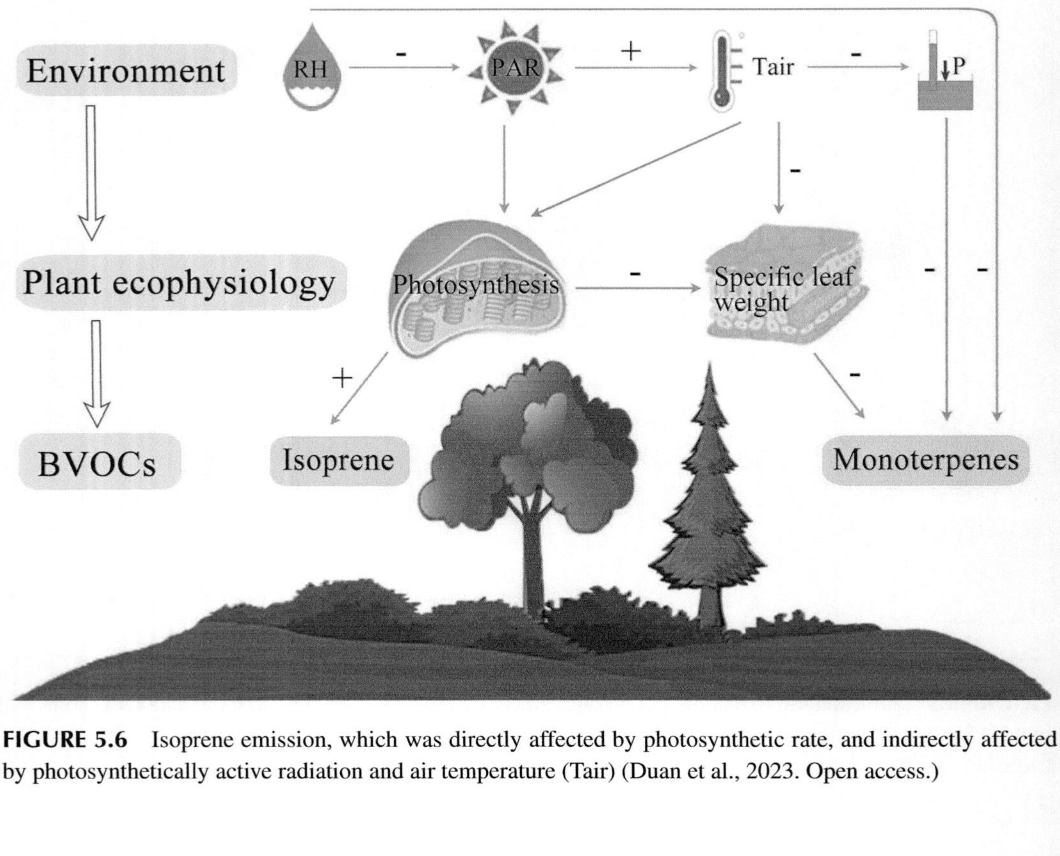

FIGURE 5.6 Isoprene emission, which was directly affected by photosynthetic rate, and indirectly affected by photosynthetically active radiation and air temperature (Tair) (Duan et al., 2023. Open access.)

by trees and other plants, contributing to the pollution of urban air. These authors noted that the current state of urban air quality is rather poor and that BVOC emissions are affected by the stress of rising heat and dryness. The effect of abiotic stresses on BVOC emissions and the contribution of these emissions to ozone concentrations in a Northern European city was evaluated by Feldner et al. (2022). Both emissions and concentrations of isoprene were dramatically reduced due to drought stress. Future increases in BVOC emissions caused by expanded green infrastructure may increase photochemical generation of ozone, according to an analysis of the chemical regime. Plants maintain a baseline level of BVOC emissions as a bottom-up constitutive defense, emitting compounds that act as repellents or deterrents to feeding and/or egg deposition by herbivores (Niinemets and Monson, 2013).

The pace of the reaction depends on the BVOC and the oxidizing partner, but once released into the atmosphere, ozone, hydroxyl (OH), and nitrate (NO_3) radicals swiftly oxidize BVOCs. High quantities of nitrogen oxides (NOx) can be found in metropolitan areas, and the photochemical ozone formation in these areas is connected to the NOx cycle. Nitric oxide (NO) reacts with ozone to form nitrogen dioxide (NO_2) and oxygen (O_2), and this in turn can be regenerated by photolyzing NO_2 to recover NO and generate ground-state oxygen, which in turn can regenerate ozone. In the intervals between these reactions, a photo-stationary state is attained, which is disrupted by BVOC. When BVOCs are oxidized, organic peroxyl radicals are produced, which combine with NO to create NO_2 in place of ozone. Additionally, the hydroperoxyl (HO_2) radical is produced during the chemical breakdown of BVOCs, which converts NO to NO_2 (Atkinson and Arey, 2003).

In addition to converting NO to NO_2, Feldner et al. (2022) found that the chemical breakdown of BVOCs results in the formation of the hydroperoxyl (HO_2) radical. Thus, in high NOx conditions, oxidation of BVOCs will raise ozone levels at ground level. Reduced agricultural crop and commercial forest yields, stunted development and decreased survival of tree seedlings, and heightened vulnerability to diseases, pests, and other pressures like harsh weather are all results of higher ozone levels. On hot, bright days, ozone levels in the air can rise to dangerously high levels, posing a threat to human health. Exposure to even moderate levels of ozone is harmful. Human health is negatively impacted by ground-level ozone because it reduces lung and heart function (https://climate-adapt.eea.europa.eu/en/observatory/evidence/health-effects/ground-level-ozone). Long-term ozone exposure has been linked to worsened asthma and an increased risk of stroke. In contrast, short-term exposure is connected with respiratory symptoms, decreased lung function, and airway inflammation. Acute ozone exposure was a contributing factor in the hospitalization of 12.253 persons in 23 European countries in 2019. Northern European countries often have lower rates of death and morbidity due to ozone exposure compared to the rest of Europe. Reducing BVOC emissions by 50% overall and isoprene, limonene, and SOA emissions by up to 60% lowers Germany's average ozone concentration by 2.5%.

5.8 THE MULTIFACETED RESPONSIBILITIES THAT BVOCS PLAY

BVOCs, or biological volatile organic compounds, are released mostly by plant life. Although a wide variety of BVOCs exist, isoprenoids are by far the most common. Isoprene (small carbon chain), monoterpenes (twice as long), and sesquiterpenes (very long carbon chain) are all examples. Trees are the most influential emitters because of their size and longevity. Furthermore, around 80% of all isoprenoids are released into the atmosphere by tropical trees alone (www.attoproject.org/bvoc-emissions-across-the-seasons/).

BVOCs, especially isoprenoids, perform a wide variety of tasks and are responsible for the aromas typically associated with plants. However, plants also employ BVOCs as a means of communication with other plants and even insects, such as pollinators and predators of herbivores. In addition to acting as a shield for plant cells, these molecules also react with other chemicals in the air. Plants actively aid in the creation of clouds by producing these reactive gases. Simply put, they generate their own precipitation. Eliane Gomes-Alves (www.attoproject.org/bvoc-emissions-across-the-seasons/) and her team's conclusion was the decrease in isoprene emissions seen as the wet season ended and the dry season began. The emissions of heavier monoterpenes and sesquiterpenes also rose throughout this time.

Throughout the 20th century, the average global temperature rose by 0.76 °C; during the 21st century, it is expected to rise by another 1.8–4.0 °C. Both direct and indirect impacts of temperature on biochemical events in the BVOC-producing metabolic pathways impact the amount of BVOCs released from plants during the growth season. Although BVOC emissions are sometimes inevitable due to their volatility, the vast majority of these compounds have evolved to serve vital physiological and ecological purposes. Variations in temperature and global changes in tissue BVOC concentrations and emission rates will greatly impact such physiological and ecological functions, since these factors determine these functions. The physiological and ecological functions and environmental significance of BVOCs will be impacted by the changing emissions, and the most likely overall consequence is an increase in BVOC emissions, primarily driven by present warming.

Plants' ability to grow, develop, reproduce, and defend themselves is dependent on the biogenic volatile organic chemicals they produce. Within plant communities, between plants, and between plants and insects, they serve as a means of communication. Many of these compounds have huge mass emission rates from vegetation into the atmosphere, and their high chemical reactivity has a profound effect on the chemical composition and physical properties of the atmosphere. Biogenic volatile organic compounds (BVOCs) are released by plants in large quantities, and they have profound impacts on the chemistry, physics, and biology of the surrounding environment. As a result of their huge emissions and high reactivity, biogenic volatile organic compounds (BVOCs) have recently received a great deal more attention from researchers interested in their effects on human health and climate change.

5.9 PLANTS CAN COMBAT BVOCS AND GLOBAL WARMING

Plants emit a wide variety of volatile organic compounds (VOCs) into the atmosphere. These BVOCs can provide as much as 10% of the carbon fixed by photosynthesis under stressful conditions and as much as 100% of the carbon fixed yearly in some tropical habitats. There are several recognized and unverified BVOC functions in plants and involvement in atmospheric processes, and new research links emission of these compounds to climate (Peuelas and Llusià, 2003). The research shows that BVOCs can protect plants from high temperatures. A rise in BVOC emissions is probably caused by warming and other variables associated with global change, such as changes in land cover (Sonwani et al., 2016). Increases in the emissions of volatile organic compounds (BVOCs) could have serious consequences for the intricate processes associated with global warming (through both negative and positive feedback).

BVOCs are a type of organic compounds produced by plants, and they have crucial functions in plant growth, reproduction, and defense. Since they are produced by plants and released into the air, they have a significant effect on the chemistry and physics of the atmosphere as well as the life of other species. Peuelas and Staudt (2010) outlined the conditions that can alter the emission rates of BVOCs, including warming, drought, changes in land use, elevated levels of atmospheric CO_2, ozone, and increased UV radiation. These changes have species- and dose-specific effects on BVOC emissions, with a general trend toward increased BVOC emissions. These emission changes have the potential to influence the direction and severity of the biosphere's feedback on atmospheric chemistry and temperature; therefore, they warrant additional investigation.

Plant communities are shaped in large part by plant competition, which affects the availability of resources like light, nutrients, and water. Plants regulate the release of several biogenic volatile organic molecules in response to these conditions (BVOCs). Because of their ability to stimulate responses in nearby plants, BVOCs serve as a stage for communicative exchanges between plants.

In 2010, Kegge and Pierik analyzed the connection between BVOC emissions and rivalry between nearby plants for resources. New research suggests that BVOCs can function as allelochemicals as well as neighbor-detecting signals. It is hypothesized that BVOCs reveal details about nearby rivals, such as their identity or development rate, that are unavailable from traditional neighbor detection signals.

According to Sonwani et al. (2016), BVOCs not only contribute to air chemistry but also participate in several plant physiological functions, such as plant signaling. Emission of BVOC from plants is caused by any kind of stress, whether biotic or abiotic. Positive and negative feedback mechanisms will determine the net impact of rising BVOC emissions. Plants emit these gases to aid in their own defensive processes. There has been surprisingly little focus on this process, even though BVOCs were highlighted by Sonwani et al. (2016) as an essential part of atmospheric sciences and plant physiology. BVOCs also have an impact on the critical environmental issue of global warming. The creation of tropospheric ozone, the development of aerosols, and the subsequent generation of SOA all had significant contributions from BVOCs. The precursors, like BVOCs, contribute to climate change issues, leading them to their conclusion that precursors play a crucial role in the creation of secondary pollutants.

5.10 REACTIONS OF BIOGENIC VOLATILE ORGANIC COMPOUND EMISSIONS TO CLIMATE CHANGE

Nitric oxide, carbon monoxide, and other so-called biogenic volatile organic compounds are just some of the volatile molecules that all plants give out. In the recent two decades, researchers have paid more attention to BVOC emissions because these compounds are highly reactive and may significantly alter the chemical and physical characteristics of the atmosphere. Carbon emissions from BVOCs are estimated to be roughly 1.1 Pg per year, which is on par with methane emissions. Many forest and agroforest species, the primary BVOC emitters, have a close relationship between BVOC emissions and carbon absorption, although BVOCs play a different function in biosphere-atmosphere interactions. Forest and agroforest vegetation absorbs a significant amount of atmospheric CO_2 and accounts for over 90% of global biomass carbon. As a result, it is crucial to slow down climate change through photosynthesis (www.cnr.it/en/focus/046-4/bvoc-biogenic-volatile-organic-compound-emission-responses-to-climate-change).

The creation of tropospheric ozone and aerosols, on the other hand, is a key contributor to the deleterious influence that BVOCs have on atmospheric chemistry. Tropospheric ozone and photochemical smog are formed when BVOCs react quickly with other molecules in the atmosphere, both manmade and natural. The generation of secondary aerosols, a component of PM10, may also be influenced by BVOCs, which can alter the residence duration of other greenhouse gases (such as methane). This highlights the significance of biogenic emissions, making inventories of BVOC emissions an important topic in atmospheric sciences. Furthermore, climate change (increasing [CO_2] and temperature) has the potential for extraordinarily severe effects on the biogeochemical carbon cycle and on atmospheric chemistry because of the importance of both BVOC emissions and carbon assimilation for the biosphere-atmosphere interactions. Predicting future emissions accurately, then, necessitates an in-depth knowledge of BVOC synthesis pathways and plant and ecosystem responses to climate change. According to research conducted by the CNR-Institute for Atmospheric Pollution (CNR-IIA), climate change may have divergent effects on the emission of isoprene from forest and agroforest vegetation. Isoprene is the most important BVOC, with a lifetime of less than an hour and possibly accounting for up to 50 percent of the atmospheric loading of non-methane hydrocarbons (Rasmussen, 1972). According to research conducted on poplar plants subjected to free-air CO_2 enrichment, increasing CO_2 levels may result in less isoprene emission. However, it is reasonable to assume that the rate of isoprene emission will accelerate as temperatures rise. Primary physiological processes like breathing may mediate the trade-off between these opposing effects, although this is not certain. Sun et al. (2013) mentioned that isoprene emissions importantly protect plants from heat stress, but the emissions become inhibited by the instantaneous increase of CO_2, and it is currently unclear how isoprene-emitting plants cope with future, more frequent and severe heat episodes under high CO_2.

5.11 BVOCS' EFFECT ON THE ATMOSPHERE IN BOREAL AND SUBARCTIC FORESTS

Insect damage, as studied by Yu et al. (2021), can rupture resin canals in needles, bark, and xylem, leading to separate outbreaks of BVOCs. There are periodic epidemics of Geometrid moths in subarctic mountain birch forests. The assessment of plant stress responses and the atmospheric chemistry of BVOC emissions have recently received increased attention in the BVOC literature (Duan et al., 2021). Many times, during epidemics, trees lose all their leaves, thus there is no vegetation around to release BVOCs. In the years after an outbreak, however, shoot growth is prolonged, leaf count is increased, and the density of glandular trichomes that store BVOCs increases. As a result, the subarctic forest may emit the most BVOCs in the years following an insect outbreak, as the chemical defense response may be slowed. Warmer seasons and the introduction of invasive herbivore species are projected to enhance insect outbreaks in high latitudes as a result of climate change (Kleist et al., 2012).

Yu et al. (2021) found that elevated BVOC emissions influence the generation of tropospheric Ozone and the oxidation of BVOCs by O_3. The rate of SOA production and particle development in the atmosphere is anticipated to be accelerated by herbivore-induced BVOC emissions from deciduous and coniferous trees. Camphene, a monoterpene, was shown to be emitted at higher rates at the start of the next growing season due to warming and herbivory from the previous growing season, indicating the possible long-term implications of herbivory under climate warming. Herbivores can experience thermal stress from a temperature rise, and one study found that a 1 °C rise in temperature throughout three growing seasons reduced feeding by *A. posticalis* and herbivory-induced total MT emissions by 77% (Cao et al., 2022; Kleist et al., 2012).

Vegetation can have an impact on the abiotic environment at the ecosystem scale due to processes such as the release of BVOCs and water, as well as the development and condensation of SOA. According to the available BVOC literature, two significant factors influencing BVOC emissions from woody species are temperature and herbivory. The ability of conifer trees to retain terpene-rich resins in resin canals above and below ground makes them a potential major source of biogenic mono- and sesquiterpene emissions. The boreal conifer forest is the biggest terrestrial biome. Saukko et al. (2012) found that when temperatures are raised, BVOCs in resin are more likely to diffuse outward. In addition, insect damage can create characteristic bursts of BVOCs during outbreaks by breaking resin canals in needles, bark, and xylem. There are periodic epidemics of Geometrid moths in subarctic mountain birch forests.

Yu et al. (2021) discussed secondary organic aerosol (SOA) generation from oxidation processes of BVOCs in the atmosphere as well as the impact of a warming environment on herbivore pressure. It is yet unclear how alterations in ecosystems can affect the chemistry of the atmosphere. Ecology and atmospheric chemistry studies working together may provide light on BVOC-controlled biosphere-atmosphere interactions and lead to more effective techniques for reducing the biological impacts of climate change (Ylivinkka et al., 2020). When unrelated events, such as tree falls or windstorms, occur, BVOC-mediated plant defense is ineffective at attracting herbivore foes. Because this technique can change BVOC emission patterns from both intact and damaged leaves, these researchers also advise avoiding ex situ BVOC sampling of cut plants.

5.12 THE EFFECT OF BVOC'S METEOROLOGICAL SOURCE STRENGTH

There has been a rise in concern over BVOC emissions. Research activity is highest in the fields of meteorology and atmospheric sciences, environmental science, and ecology. The United States, which has worked closely with Germany and China, has published the most articles in this field by a wide margin. In the United States, the most well-known researcher is Guenther, who, among other things, models BVOC emissions and works in the field of meteorological atmospheric science (Guenther et al., 2012).

Samples are typically processed using dynamic enclosure techniques for field measurements, and BVOCs are then detected using GC, GC-MS, or PTR-MS. Several models, including G95, BEIS, and MEGAN, are capable of estimating BVOC emissions. Most studies, however, approximated the BVOC emissions based on an unobservable basal emission rate and a less-detailed distribution of vegetation, leading to considerable model uncertainty (Duan et al., 2021). Therefore, more field observations are needed to strengthen the reliability of models. Additionally, theoretical guidance for air pollution control may be gleaned from models that appropriately predict BVOC emissions. The biosphere and the atmosphere rely heavily on BVOCs. When plants experience abiotic challenges like strong light, heat, drought, or high ozone, they may release these chemicals as a means of mediating intra- and interspecific interactions within the ecosystem.

For air quality authorities to successfully design emission control measures, a more precise quantification of the contribution of biogenic VOC (BVOC) emissions to 4 regional ozone production is required. The estimation of photosynthetically active radiation (PAR) 7 reaching the canopy is one of the most significant sources of uncertainty in BVOC emission models (6, 7). When it comes to depicting the regional and temporal fluctuations of PAR, satellite insolation retrieval data offer an alternative to predictive 8 meteorological models. This work (https://scholarship.rice.edu/bitstream/handle/1911/98795/ATMENV-D-17-00094R2.pdf?sequence=4) uses biogenic emission estimates developed with the MEGAN and BEIS biogenic 10 emissions models utilizing satellite or prognostic PAR to investigate the contribution 11 of BVOC to ozone in the USA.

Since estimates of BVOC emissions are very model-dependent and might differ depending on the amount of solar radiation available, two popular BVOC emission models (BEIS and MEGAN) are employed using data derived from both models and satellites. By incorporating satellite insolation/PAR data into the forecasting weather model (i.e., WRF), we can decrease the under-prediction of clouds and increase agreement with surface radiation sensors. For modelling ground-level isoprene and ozone concentrations, the best results were found using BEIS coupled with the GOES insolation retrieval (https://scholarship.rice.edu/bitstream/handle/1911/98795/ATMENV-D-17-00094R2.pdf?sequence=4). Overestimating isoprene and ozone concentrations, MEGAN yielded greater estimates of BVOC emissions.

According to Tsui et al. (2009), ozone and secondary organic aerosols are formed when BVOCs in the atmosphere react, leading to poor air quality, negative effects on human health, and a possible role in climate change. The purpose of this research is to make an initial evaluation of BVOC emissions in the Hong Kong Special Administrative Region. Thirteen tree species were investigated to determine their isoprene emission potential. Based on field survey information, an estimate of tree distribution was made for country park areas. Hong Kong's annual BVOC emissions were estimated to be 8.6 × 109g C based on data from measurements and the literature on plant emissions, estimates of tree distribution, land use, and weather (Tsui et al., 2009). About 30%, 40%, and 30% of the anticipated total annual emissions were accounted for by isoprene, monoterpenes, and other VOCs, respectively. There are hundreds of different plant species in Hong Kong's parks, yet the model results show that only ten tree species account for around 76 percent of the annual VOC emissions. The model also accurately anticipated the noticeable seasonal and diurnal fluctuations in emissions. This work provides a strong groundwork for future studies in the area, and its findings can be used for the investigation of BVOC emissions in southern China and other Asian locations with similar climates and plant patterns.

Knowing what kinds of chemical compounds plants release into the air is useful for air quality assessments. The levels of these emissions are highly species- and cultivar-specific. Long-term field observations from relevant crops are uncommon; therefore, these changes have not been well accounted for (Havermann et al., 2022). As a result, they monitored and modelled the emission of maize, ryegrass, and oil-seed rape for whole crop rotations in northeastern Germany. They used the measurements as a basis for parametrization, then ran separate simulations of each detected chemical to assess their significance to air chemistry. Significant variations among the examined crops have been established. The annual emissions from oil-seed rape are 37 times higher than those from maize, yet its impact on air chemistry is only 6 times higher due to the lower reactivity of the emitted chemicals.

It is generally known that many plants release biogenic volatile organic molecules (BVOCs). Depending on the circumstances, BVOCs can significantly boost the photochemical production of ozone (O_3) and influence the concentrations of atmospheric photo-oxidants. Because abiotic stresses like heat and drought can drive BVOC emissions, urban areas impacted by climate change may be at an increased risk of enhanced ozone production potentials. The impact of heat waves and declining soil water potentials on urban air quality is unclear, however. During the vegetative period of 2018, which was marked by noteworthy drought and hot episodes, the impact of abiotic stress on BVOC emissions and urban O_3 production in the coastal metropolitan area of Hamburg, Germany, was evaluated. The third edition of the Estimate of Emissions of Gases and Aerosols from Nature (MEGAN), which takes into account many abiotic stressors, was used to model BVOC emissions. The majority of the BVOCs in the area of investigation come from isoprene (60 percent). The results of this study show that drought stress is the primary abiotic stressor influencing BVOC emissions in this region.

5.13 BIOGENIC VOLATILE ORGANIC COMPOUND EMISSIONS AND THEIR IMPACT ON AIR QUALITY AND HUMAN HEALTH

Fine particulate matter (PM2.5) and ground-level ozone pollution are dangerous to human health, especially in metropolitan areas where they are caused by the emission of biologically volatile organic compounds (BVOCs). However, most estimates of BVOCs failed to account for emissions from urban green spaces, leading to an incomplete picture of regional BVOC emissions and their impact on the environment and human health. Most recent field survey data on local vegetation were used by Yuan et al. (2017) to estimate geographical and temporal trends, air quality impacts, health harm, and solutions for controlling BVOC emissions in the Greater Beijing Area. Their research revealed that (1) the urban core was the epicenter of regional BVOCs emissions with the highest region-based emission intensity (3.0 g C m^2 yr^1) out of the 11 sub-regions; (2) urban green spaces played much more important roles (account for 62% of total health damage) than rural forests; and (3) BVOCs emissions from green spaces will more than triple by 2050 due to urban area expansion, tree growth, and environmental changes.

The local air quality can be negatively impacted by certain trees. Biogenic volatile organic compounds (BVOCs) (including isoprene and terpenes) are released by almost every type of tree. These are necessary for the production of ground-level ozone. In urban green spaces, where many people go to relax, the improper tree population is especially undesirable because it can result in excessive exposure to harmful ozone. Isoprene, a biogenic hydrocarbon, is one example. Due to its high reactivity, it is capable of producing ozone. In reality, isoprene could initiate a formation with less concentrated O_3 than anthropogenic hydrocarbons.

5.14 A FEW SIGNIFICANT FINDINGS ABOUT HOW BIOLOGICAL VOLATILE ORGANIC COMPOUNDS (BVOC) AFFECT THE CLIMATE

Biogenic volatile organic compounds (BVOC) include a large variety of volatile molecules emitted by all plants, including nitric oxide, carbon monoxide, and non-methane volatile organic compounds. Carbon emissions from BVOCs are estimated to be roughly 1.1 Pg per year, which is on par with methane emissions. Many forest and agroforest species, the primary BVOC emitters, have a close relationship between BVOC emissions and carbon absorption, although BVOCs play a different function in biosphere-atmosphere interactions. Because of their significant role in the generation of tropospheric ozone and aerosols, BVOCs have a significant, deleterious impact on atmospheric chemistry. Tropospheric ozone and photochemical smog are formed when BVOCs react quickly with other molecules in the atmosphere, both manmade and natural. The generation of secondary aerosols, a component of PM10, may also be influenced by BVOCs, which can alter the residence duration of other greenhouse gases (such as methane). This highlights the significance of biogenic emissions, making inventories of BVOC emissions an important topic in atmospheric sciences. Calfapietra et al. (2013) note that BVOCs have a role in air chemistry processes that determine the generation of ozone and particles (aerosols), and hence impact air quality and the climate system. Ground-level ozone (O_3) and secondary organic aerosols (SOA) are largely derived from BVOC, which are released primarily by plants (Kota et al., 2015). Ground-level ozone is dangerous to human health, and BVOCs contribute to its creation. Biogenic volatile organic compounds (BVOC) and anthropogenic volatile organic compounds (AVOC) refer to those that originate in natural and artificial environments, respectively. Up to 90% of all VOC emissions worldwide are BVOC emissions, with vegetation responsible for 98% of these emissions.

Air chemistry and climate modelling might benefit from a more detailed mechanistic description of BVOC emission; however, current approaches vastly oversimplify the processes involved. Aerosol production, interactions with nitrous oxide and radicals that affect ozone concentration, and biological functions, including protection and signaling, are the three main areas to consider when assessing the environmental significance of BVOCs. A major function of BVOCs is the transmission of signals between leaves, inside plants, and between plants and insects. In doing so, they contribute to many interactions within the ecosystem by discouraging herbivores and parasites while luring their natural predators. There are several biogenic volatile organic molecules released by the land's vegetation (BVOCs). Secondary organic aerosol generation and the oxidative capacity of the atmosphere are both aided by BVOCs like terpenes (isoprene, monoterpenes, and sesquiterpenes) and oxygenated chemicals (alcohols, aldehydes, ketones, and acetates). According to Laothawornkitkul et al. (2009), annual emissions of isoprene account for around half of the total BVOC emissions and are comparable to the total emission of methane from all sources. Furthermore, BVOCs influence atmospheric chemistry due to their high reactivity; they photochemically combine with nitrogen oxides (NOx) to produce tropospheric ozone. Isoprene, the most abundant BVOC, has a half-life of less than an hour and may account for up to 50 percent of the atmospheric loading of non-methane hydrocarbons, but its emission from forest and agroforest plants may be affected differently by climate change. Species and environmental factors, such as temperature and pollution levels in the air, affect the quality of pollutants absorbed and BVOCs released.

Since it is now generally accepted that biological processes in terrestrial ecosystems have far-reaching effects on Earth's atmosphere and climate system, this could imply significant feedback effects on the current global changes in atmosphere and climate resulting from human activities. Many researchers have focused on the biological processes involved in the carbon cycle (i.e., the intake and release of CO_2 and the storage of carbon in organic matter) (Guenther, 2013). However, the biosphere and the atmosphere exchange several gases besides CO_2. Biogenic volatile organic compounds (BVOCs) are emitted by the world's terrestrial plants, and they come in many different sizes, chemical compositions, and metabolic origins. Synthesis rates and physicochemical properties, especially solubility, volatility, and diffusivity, establish BVOC emission rates. Therefore, both internal and external influences greatly modulate their emission rates. The ability of different plant species to produce BVOCs varies more than their ability to assimilate carbon through photosynthesis. Therefore, even if the standing biomass and net primary production remain unchanged, changes in species and community structure of ecosystems, land uses, climate, and resource availability can lead to significant changes in regional BVOC fluxes (Unger, 2014a). Multiple warming and cooling climatic pollutants are affected by the rapid oxidation of biological volatile organic compounds (BVOC) emissions from terrestrial ecosystems. When accounting for the warming effect of human-caused VOC emissions from fossil fuel consumption and industry during the same period, the net chemical forcing of the global climate due to all known anthropogenic influences on BVOC emissions is $0.17Wm^2$ (cooling). Anthropogenic radiative forcing calculations should factor in BVOC emissions. Numerical models of regional air quality and global climate now frequently contain BVOC emissions that are estimated as a function of land cover and environmental driving variables, thanks to the growing recognition of the importance of BVOC emissions for Earth system modelling. Ozone generation will improve as a result of climatic shifts, especially when temperatures rise and BVOC emissions rise. Ozone levels are expected to rise annually as a result of rising temperatures. Precipitation, changes in land use, and the possible impact of pests on plant and BVOC emissions should be taken into account in future studies. To reduce BVOC emissions, Chang et al. (2012) recommend replanting low-emitting species and reestablishing broad-leaved forests in peri-urban and rural parts of the Greater Hangzhou Area in China.

The O_3 to BVOCs ratio is a strong predictor of the likelihood of serious O_3 pollution. High O_3 levels are likely generated by high levels of NOx in Intensively Used Areas and Transport Class, but can be explained by reduced O_3 absorption in Commercial/Industrial Class due to the lower proportion of tree cover (Calfapietra et al., 2013). Because of the significant NOx typically generated from anthropogenic sources, BVOC can play an even more essential role in urban environments. In such situations, the selection of a low BVOC emitter is critical for lowering the O_3-forming potential since the load of BVOC emission into the environment can be quite relevant.

Biogenic volatile organic compounds (BVOCs) such as isoprene, monoterpenes (MT), and sesquiterpenes (SQ) are released into the atmosphere by a large number of plants (SQT). These chemicals' photochemical reaction can lead to ozone and secondary organic aerosol production (Fehsenfeld et al., 1992). Depending on physiological and environmental conditions, BVOC emissions from trees can vary greatly between and even within species. Calfapietra et al. (2013) surveyed the research on BVOC in metropolitan areas. Planting one million low-BVOC tree species has the potential to release 19,000 kg of total VOC emissions per year, according to modelling research (Curtis et al., 2014). Choosing tree species with low emission rates over those with high emission rates for volatile organic compounds (VOCs) is the same as removing roughly 500,000 cars from the city's inner traffic.

Simpson et al. (2011) describe a Tree BVOC index that takes into account reduced emissions from low-emitting trees. They demonstrated how the index may be used in practice, as well as how to keep tabs on its progress. This measure may prove valuable in incorporating urban trees into strategies for reducing air pollution. They proposed a method that allows for extensive customization. As long as they maintain a TBI (The Tree BVOC Index) of one or less at the end of the year, they are allowed to choose whatever species they see appropriate for the local site conditions.

To rate potential tree species for urban plantation, urban planners currently consider their size, aerodynamic qualities, aesthetic value, and resistance to air pollution (API). In selecting species, the index does not take into account whether or not they contribute to air pollution by producing allergic pollen or highly reactive ozone. The new Air Quality Impact Index (AQII) developed by Datta et al. (2021) rates options in a more comprehensive way by considering their aerodynamic features, leaf structure, pollutant-absorbing capability, pollution tolerance, ozone and aerosol precursor emissions, and the impact on pollen allergies. They showed the value of the AQII rating by analyzing the effect of two species with similar APIs but very different positions on the AQII scale on urban air quality throughout the summer. Datta et al. (2021) identified 149 of the 280 tree species most frequently evaluated for urban cultivation as having detectable VOC emissions. Scientists have collated data on the allergy risk and air pollution tolerance of 98 species to determine the AQII.

A "Tree BVOC" refers to the Biogenic Volatile Organic Compounds emitted by trees, while a "Tree TBI" (Tree BVOC Index) is a numerical value used to compare the emission rates of different tree species, allowing planners to select trees that emit lower amounts of BVOCs, which can be beneficial for managing air quality, especially in urban environments by minimizing ozone precursor emissions; essentially, a lower TBI value indicates a tree that emits less BVOCs, making it a better choice for planting in areas concerned about air pollution. The TBI is calculated by comparing the BVOC emission rate of a specific tree species to a reference emission rate, with a value less than 1 indicating a lower-emitting tree. The BVOC emission rate of a tree species can vary depending on environmental factors like climate and soil conditions, so it's important to consider local data when selecting trees based on TBI values.

The Tree BVOC Index (TBI) is an alternative prescriptive approach that estimates emission reductions, provides users with a clearly defined target to reach and a method to continuously monitor progress, is completely transparent to users and regulators, and eliminates labelling of tree species to make verification and enforcement easier in a regulatory setting. If the TBI is less than or equal to 1, it means that the user's tree-planting efforts are on track to succeed. *Mangifera indica* and *Azadirachta indica*, both of which had computed TBI values less than 1, were found to be appropriate for planting (in India and other Tropical Asian Countries) and can be employed as in greenbelt development schemes by Saxena and Ghosh (2018). The tree BVOC index can be utilized in landscape improvement projects and as a long-term strategy to reduce ozone pollution. A TBI greater than one will force the city to reduce BVOCs from the population planted by adjusting the numbers planted of each species in a way that best suits their circumstances. Alternatively, a TBI less than one affords more flexibility in species selection. *Dalbergia sissoo* and *Butea monosperma* have high TBI values (2-3). This interpretation has been given by Saxena and Ghosh (2018). *Butea monosperma* is a species native to tropical and sub-tropical parts of South Asia and Southeast Asia. *Dalbergia sissoo* (sissoo) is an extensively planted tree species in India and Bangladesh.

Some common tree species, such as oak and pine, emit large quantities of BVOC into the atmosphere; consequently, planting such tree species in large numbers could potentially worsen inner-city air quality. These researchers developed a methodology based on taxonomic relationships to assign emission rates to trees in Southern California and ranked over 300 tree species on a scale of low-emitting to high-emitting species. The scale defined species emitting 0.01^{-1} µg g dw^{-1} h^{-1} VOC as low emitters, 1–10 µg g dw^{-1} h^{-1} as moderate emitters, and 10–1,000 µg g dw^{-1} h^{-1} as high emitters.

5.15 CONCLUSION

Human activities produce volatile organic molecules known as anthropogenic volatile organic compounds (AVOCs), while terrestrial ecosystems generate biogenic volatile organic compounds (BVOCs). Plants' growth, development, reproduction, and defense depend on biogenic volatile organic chemicals (BVOCs). These chemicals, including isoprene, terpenes, alkanes, alkenes, alcohols, esters, carbonyls, and acids, are crucial for communication within plant communities and between plants and insects. Human activities produce anthropogenic BVOCs, while terrestrial ecosystems generate BVOCs. BVOCs transmit signals between leaves, inside plants, and between plants and insects, facilitating interactions within ecosystems. BVOC production increases due to rising temperatures and atmospheric carbon dioxide levels, which are expected to continue growing. The negative health effects of BVOCs are especially pronounced in densely populated metropolitan areas.

Strategic tree planting can significantly reduce urban BVOC flux and atmospheric VOC burden, reducing emissions equivalent to removing approximately 500,000 cars from inner traffic. Low-emitting tree species are best for urban areas, as they can improve or worsen air quality. Ispir Oak, Oriental Sweetgum, and Cilician Fir are the highest BVOC emitters, while *Mangifera indica* and *Azadirachta indica* are suitable for greenbelt development programs. The Tree BVOC index can be used in landscape improvement projects and as a long-term strategy to reduce ozone pollution. Growing trees is an affordable and effective method of preserving and improving the physical environment.

Plants release biogenic volatile organic chemicals (BVOCs), which impact human and environmental health by causing climate and air pollution. Urban air contains a mix of gaseous and particulate species from both anthropogenic and biogenic sources, which are further processed in the atmosphere. Variations in terrestrial biosphere emissions can lead to feedback loops, worsening environmental effects. Understanding how BVOC emissions respond to environmental changes can help predict future consequences. This study aims to develop comprehensive prediction models of the Earth system, combining computer modeling, gradient studies, experimental data, and remote sensing. Forest species should not be used for greening due to their high BVOC emissions and negative impact on the atmosphere.

REFERENCES

Amin, H., P.T. Atkins, R.S., Russo, A.W. Brown, B. Sive, A.G. Hallar, and K.E. HuffHartz. 2012. Effect of bark beetle infestation on secondary organic aerosol precursor emissions. *Environ. Sci. Technol.* 46: 5696–5703. https://doi.org/10.1021/es204205m

Atkinson, R., and J. Arey 2003. Atmospheric degradation of volatile organic compounds, *Chem. Rev.,* 103: 4605–4658.

Baghi, R., D. Helmig, A. Guenther, T. Duhl, and R. Daly. 2012. Contribution of flowering trees to urban atmospheric biogenic volatile organic compound emissions. *Bio Geosciences*. 9: 3777–3785.

Barney, J.N., J.P. Sparks, J. Greenberg, T.H. Whitlow, and A. Guenther. 2009. Biogenic volatile organic compounds from an invasive species: Impacts on plant-plant interactions. *Plant Ecol.* 203(2): 195–205.

Bema. 1997. A European Commission project on biogenic emissions in the Mediterranean area. *Atmos. Environ.* 31: 1–256.

Calfapietraa, C., S. Faresc, F. Manesd, A. Morania, G. Sgrignaa,e, and F. Loreto. 2013. Role of Biogenic Volatile Organic Compounds (BVOC) emitted byurban trees on ozone concentration in cities: A review, *Environmental Pollution* 183: 71–80. https://doi.org/10.1016/j.envpol.2013.03.012

Cao, J., S. Situ, Y. Hao, S. Xie, and L. Li. 2022. Enhanced summertime ozone and SOA from biogenic volatile organic compound (BVOC) emissions due to vegetation biomass variability during 1981–2018 in China. *Atmos. Chem. Phys*. 22: 2351–2364. https://doi.org/10.5194/acp22-2351-2022

Castorina, G., F. Grassi, G. Consonni, S. Vitalini, R. Oberti, A. Calcante, E. Ferrari, M. Bononi, and M. Iriti. 2020. Characterization of the biogenic Volatile Organic Compounds (BVOCs) and analysis of the PR1 molecular marker in *Vitis vinifera* L. inoculated with the Nematode *Xiphinema index*. *Int. J. Mol. Sci.* 21: 4485. https://doi.org/10.3390/ijms21124485

Chang, J., Y. Ren, Y. Shi, Y. Zhu, Y. Ge, S. Hong, L. Jiao, F. Lin, C. Peng, T. Mochizuki, A. Tani, Y. Mu, and C. Fu. 2012. An inventory of biogenic volatile organic compounds for a subtropical urban–rural complex. *Atmos. Environ.* 56: 115–123.

Choomanee, J.P., and S. Blauert .2022, Biogenic volatile organic compound emission from tropical plants in relation to temperature changes. *Environ. Challenges.* 9: 100643. https://doi.org/10.1016/j.envc.2022.100643

Courtois, E.A., C.E. Paine, P.A. Blandinieres, D. Stien, J.M. Bessiere, E. Houel, C. Baraloto, and J. Chave. 2009. Diversity of the volatile organic compounds emitted by 55 species of tropical trees: A survey in French Guiana. *J. Chem. Ecol.* 35(11):1349–1362. https://doi.org/10.1007/s10886-009-9718-1

Curtis, A.J., D. Helmig, C. Baroch, R. Daly, and S. Davis. 2014. Biogenic volatile organic compound emissions from nine tree species used in an urban tree-planting program. *Atmos. Environ.* 95: 634–643. https://doi.org/10.1016/j.atmosenv.2014.06.035

Datta, S., A. Sharma, V. Parkar, H. Hakkim, A. Kumar, A. Chauhan, S. Singh Tomar, and B. Sinha. 2021. A new index to assess the air quality impact of urban tree plantation. *Urban Climate*. 40: 100995. https://doi.org/10.1016/j.uclim.2021.100995

Duan, C., S. Zuo, Z. Wu, Y. Qiu, J. Wang, Y. Lei, H. Liao and Y. Ren. 2021. Review of research hotspots and trends in biogenic volatile organic compounds (BVOCs) emissions combining bibliometrics with evolution tree methods. *Environ. Res. Lett.* 16: 013003. https://doi.org/10.1088/1748-9326/abcee9

Duan, C., Z. Wu, H. Liao, and Ren. 2023. Interaction processes of environment and plant ecophysiology with BVOC emissions from dominant greening trees. *Forests*, 14: 523. https://doi.org/10.3390/f14030523

Fehsenfeld, F.C., J. Calvert, R. Fall, P. Goldan, A.B. Guenther, C.N. Hewitt, B. Lamb, S. Liu, M. Trainer, H. Westberg, and P. Zimmerman. 1992. Emissions of volatile organic compounds from vegetation and the implications for atmospheric chemistry. *Global Biogeochem. Cycles*. 6: 389–430.

Feldner, J., M.O.P. Ramacher, M. Karl, M. Quante and M.L. Luttkus. 2022. Analysis of the effect of abiotic stressors on BVOC emissions from urban green infrastructure in northern Germany. *Environ. Sci.: Atmos.* 2: 1132.

Gómez, M.C., N. Durana, J.A. García, M. de Blas, E.S. de Cámara, E. García-Ruiz, G. Gangoiti, E. Torre-Pascual and J. Iza. 2020. Long-term measurement of biogenic volatile organic compounds in a rural background area: Contribution to ozone formation. *Atmos. Environ.* 224: 117315.

Gu, S., A. Guenther, and C. Faiola. 2021. Effects of anthropogenic and biogenic volatile Organic compounds on Los Angeles air quality. *Environ. Sci. Technol.* 55 (18): 12191–12201. https://doi.org/10.1021/acs.est.1c01481

Guenther, A. 2013. Biological and chemical diversity of biogenic volatile organic emissions into the atmosphere. *ISRN Atmos. Sci.*, 2013: 1–27. https://doi.org/10.1155/2013/786290

Guenther, A. B., Jiang, X., Heald, C. L., Sakulyanontvittaya, T., Duhl, T., Emmons, L. K., and Wang, X. 2012. The model of emissions of Gases and aerosols from nature version 2.1 (MEGAN2.1): An extended and updated framework for modeling biogenic emissions. *Geoscientific Model Development*, 5(6): 1471–1492.

Guenther, A., C.N. Hewitt, Erickson, R. Fall, C. Geron, T. Graedel, P. Harley, K. Lee, M. Lerdau, W.A. McKay, T. Pierce, B. Scholes, R. Steinbrecher, R. Tallamraju, J. Taylor, and P. Zimmerman. 1995. A global model of natural volatile organic compound emissions. *J. Geophys. Res.* 100: 8873–8892.

Guenther, A.B., R.K. Monson, and R. Fall. 1991. Isoprene and monoterpene emission rate variability: Observations with eucalyptus and emission rate algorithm developmen. *J. Geophys. Res.* 96: 10799–10808. https://doi.org/10.1029/91JD00960

Havermann, F.A., J.P. Ghirardo, C. Schnitzler, M. Nendel, D. Hoffmann, R. David Kraus, and R. Grote. 2022. Modeling intra- and interannual variability of BVOC emissions from maize, oil-seed rape, and ryegrass. *J. Adv. Modeling Earth Syst.* 14: e2021MS002683.

Janyasuthiwong, S., P. Choomanee, S. Bualert, S. Maneejantra, T. Charoenpun, W. Chommon and S. Jitjun. 2022. Biogenic volatile organic compound emission from tropical plants in relation to temperature changes. *Environ. Challenges*. 9: 100643. https://doi.org/10.1016/j.envc.2022.100643

Kegge, W., and R. Pierik. 2010. Biogenic volatile organic compounds and plant competition. *Trends Plant Sci.* 15(3): 126–132. https://doi.org/10.1016/j.tplants.11.007

Kevin, P., C. Monard, J.L. Le Garrec, N.L. Bris, and F. Binet. 2017. Organic amendment practices as possible drivers of biogenic Volatile Organic Compounds emitted by soils in Aerosystems. *Agricul. Ecosyst. Environ.* 250: 25–36. https://doi.org/10.1016/j.agee.2017.09.007

Kleist, E., T.F. Mentel, S. Andres, and A. Bohne. 2012. Irreversible impacts of heat on the emissions of monoterpenes, sesquiterpenes, phenolic BVOC and green leaf volatiles from several tree species. *Bio Geosci.* 9: 5111–5123.

Komenda, M., E. Parusel, A. Wedel, and R. Koopman. 2001. Measurements of biogenic VOC emissions: Sampling, analysis and calibration. *Atmos. Environ.* 35(12): 2069–2080.

Kota, S.H., G. Schade, M. Estes, D. Boyer, and Q. Ying. 2015. Evaluation of MEGAN predicted biogenic isoprene emissions at urban locations in Southeast Texas. *Atmos. Environ.* 110: 54–64. https://doi.org/10.1016/j.atmosenv.2015.03.02

Knudsen, J.T., and J. Gershenson. 2006. The chemistry diversity of floral scent. In: Dudareva, N., and E. Pichersky (eds.) *Biology of Floral Scent* (pp. 27–52). CRC Press.

Kumeroa, F., S. Komahan, S. Sofkova-Bobcheva, and A.C. McCormick. 2022. Characterization of the volatile profiles of six industrial Hemp (*Cannabis sativa* L.) Cultivars. *Agronomy*. 12: 2651. https://doi.org/10.3390/agronomy12112651

Laothawornkitkul, J., J.E. Taylor, N.D Paul, and C.N. Hewitt. 2009. Biogenic volatile organic compounds in the Earth system. *New Phytol.* 183: 27–51.

Liu, Y., S. Schallhart, D. Taipale . Tykkä, M. Räsänen, L. Mer bold, H. Hellen, and P. Pellikka. 2021. Seasonal and diurnal variations in biogenic volatile organic compounds in highland and lowland ecosystems in southern Kenya. *Atmos. Chem. Phys.* 21: 14761–14787, https://doi.org/10.5194/acp-21-14761-2021

Ma, M.Y., Y. Gao, S. Wang, and L.R. Zhang, C. Leung, S. Liu, B. Wang, X. Zhao, Chang, T.H. Su, T. Zhang, L. Sheng, X. Yao, and H. Gao. 2019. Substantial ozone enhancement over the North China Plain from 2 increased biogenic emissions due to heat waves and land cover in 3 summer. *Atmos. Chem. Phys. Discuss.* 1–27. https://doi.org/10.5194/acp-2019-362

McCormick, A.C., E.V. Effah, and A.N. Rodriguez. 2023. Ecological aspects of volatile organic compounds emitted by exotic invasive plants. *Front. Ecol. Evol.* 11: 1059125. www.frontiersin.org/articles/10.3389/fevo.2023.1059125

Mochizuki, T., F. Ikeda, and A. Tani. 2020. Effect of growth temperature on monoterpene emission rates of *Acer palmatum*. *Sci. Total Environ.* 745: 140886. https://doi.org/10.1016/j.scitotenv.140886

Moore, M.N. 2015. Do airborne biogenic chemicals interact with the PI3K/Akt/mTOR cell signaling pathway to benefit human health and wellbeing in rural and coastal environments? *Environ. Res.* 140: 65–75. https://doi.org/10.1016/j.envres.2015.03.015

Moura, B.B., V.P. Bolsoni, M.D. de Paula, G.M. Dias, and S.R. de Souza. 2022. Ozone impact on emission of biogenic volatile organic compounds in three tropical tree species from the Atlantic forest remnants in Southeast Brazil. *Front. Plant Sci., 23, Sec. Functional Plant Ecol.* 13. https://doi.org/10.3389/fpls.2022.879039

Mula, V., Bogdanov, J., Stanoeva, J.P., Zeneli, L., Zdravkovski, Z. (2024). Monitoring Volatile Organic Compounds in Air Using Passive Sampling: Regional Cross-Border Study between N. Macedonia and Kosovo. *Aerosol Air Qual. Res.* 24: 230170. https://doi.org/10.4209/aaqr.230170

Müller, A., Y. Miyazaki, E. Tachibana, K. Kawamura, and T. Hiura. 2017. Evidence of a reduction in cloud condensation nuclei activity of water-soluble aerosols caused by biogenic emissions in a cool-temperate forest. *Sci. Rep.* 7(1): 8452. https://doi.org/10.1038/s41598-017-08112-9

Niinemets, Ü., and K.R. Monson. 2013. *Biology, Controls and Models of Tree Volatile Organic Compound Emissions*. Springer.

Pacífico, F., S. Harrison, S. Jones, and C.S. Sitch. 2009. Isoprene emissions and climate. *Atmos. Environ.* 43: 6121–6135.

Penuelas, J., and J. Lluvia. 2003. BVOCs: Plant defense against climate warming? *Trends Plant Sci.* 8(3): 105–109. https://doi.org/10.1016/S1360-1385(03)00008-6

Penuelas, J., and M. Staudt. 2010. BVOCs and global change. *Trends Plant Sci.* 15(3):133–44. https://doi.org/10.1016/j.tplants.2009.12.005

Rasmussen R.A., and F.W. Went. 1965. Volatile organic material of plant origin in the atmosphere. *Proc. Natl. Acad. Sci. USA*. 53(1): 215–220.

Rasmussen, R.A. 1972. What do the hydrocarbons from trees contribute to air pollution? *J. Air Poll. Control Assoc.* 22(7): 537–543.

Ren, Y., Z. Qu, Y. Du, R. Xu, D. Ma, G. Yang, Y. Shi, X. Fan, A. Tani, P. Guo, Y. Ge, and J. Chang. 2017. Air quality and health effects of biogenic volatile organic compounds emissions from urban green spaces and the mitigation strategies. Environ. *Pollut.* 230: 849–861. https://doi.org/10.1016/j.envpol.2017.06.049

Riipinen, I., T. Yli-Juuti, J.R. Pierce, T. Petäjä, D.R. Worsnop, M. Kulmala, M. Neil, and N.M, Donahue. 2012. The contribution of organics to atmospheric nanoparticle growth. *Nat. Geosci.* 5: 453–458.

Sánchez-Gómez, A., S. Martínez-Pérez, S. Leduc, A. Sastre-Merlín, and E. Molina-Navarro. 2023. Streamflow components and climate change: Lessons learnt and energy implications after hydrological modeling experiences in catchments with a Mediterranean climate. *Energy Rep.* 9: 277–291.

Sanadze, G.A. 1956. Emission of gaseous organic substances from plants. *Rep. Akad. Nauk GruzSSR.* 17: 429–433.

Saukko, E., A.T. Lambe, P. Massoli, T. Koop, J.P. Wright, D.R. Croasdale, D.A. Pedernera, T.B. Onasch, A. Laaksonen, P. Davidovits, D.R. Worsnop, and A. Virtanen. 2012. Humidity-dependent phase state of

SOA particles from biogenic and anthropogenic precursors. *Atmos. Chem. Phys.* 12: 7517–7529. https://doi.org/10.5194/acp-12-7517-2012
Saxena, P., and C. Ghosh., 2018. A sustainable way to mitigate ozone pollution by reducing biogenic VOCs through landscape management programme. *Int. J. Eng. Trends Tech.* 56(2): 87–91.
Simpson, J.R., and E.G. McPherson. 2011. The tree BVOC index. *Environ. Poll.* 159(8–9): 2088–2093. https://doi.org/10.1016/j.envpol.2011.02.034
Sonwani, S., P. Saxena, and U. Kulshrestha. 2016. Role of global warming and plant signaling in BVOC Emissions. In: Kulshrestha, U., and P. Saxena, (eds.) *Plant Responses to Air Pollution*. Springer. https://doi.org/10.1007/978-981-10-1201-3_5
Sun, Z., K. Hüve, V. Vislap, and Ü. Niinemets. 2013. Elevated [CO2] magnifies isoprene emissions under heat and improves thermal resistance in hybrid aspen. *J. Exp. Bot.* 64(18): 5509–5523. https://doi.org/10.1093/jxb/ert318
Tambunan, P., S. Baba, A. Kuniyoshi, H. Iwasaki, T. Nakamura, H. Yamasaki, and H. Oku. 2006. Isoprene emission from tropical trees in Okinawa Island, Japan. *Chemosphere*. 65(11): 2138–2144.
Tsui, J.K.N., A. Guenther, W. Yipa, and F. Chen. 2009. A biogenic volatile organic compound emission inventory for Hong Kong. *Atmos. Environ.* 43: 6442–6448.
Unger, N. 2014a. Human land–use–driven reduction of forest volatiles cools global climate. *Nat. Clim. Change*. 4: 907–910. https://doi.org/10.1038/nclimate2347
Unger, N. 2014b. On the role of plant volatiles in anthropogenic global climate change. *Geophysics. Res. Lett.* 41: 8563–8569. https://doi.org/10.1002/2014GL061616
Wang, C.T., C. Wiedinmyer, K. Ashworth, P.C. Harley, J. Ortega, and W. Vizuete. 2019. Leaf enclosure measurements for determining volatile organic compound emission capacity from *Cannabis* spp. *Atmos. Environ*. 199:80–87.
Went, F.W. 1955. Air pollution. *Scient. Amer.* 192: 63–72.
Went, F.W. 1960. Blue hazes in the atmosphere. *Nature.* 187(4738): 641–643.
Ylivinkka, I., J. Itämies, T. Klemola, K. Ruohomäki, M. Kulmala, and D. Taipale. 2020. Investigating evidence of enhanced aerosol formation and growth due to autumnal moth larvae feeding on mountain birch at SMEAR I in northern Finland. *Boreal Environ. Res.* 25: 121–143.
Yu, H., Holopainen, J.K., Kivimäenpää, M., Virtanen, A., and Blande, J.D. 2021 . Potential of Climate Change and Herbivory to Affect the Release and Atmospheric Reactions of BVOCs from Boreal and Subarctic Forests. Molecules, 26, 2283. https://doi.org/10.3390/molecules26082283
Yuan, R., Z. Qu, Y. Du, R. Xu, D. Ma, G. Yang, Y. Shi, X. Fan, A. Tani, P. Guo, Y. Ge, and J. Chang. 2017. Air quality and health effects of biogenic volatile organic compounds emissions from urban green spaces and the mitigation strategies. *Environ. Poll.* 230: 849–861.

6 Pollutants in Wastewater Effluents

Impacts and Remediation Processes

6.1 INTRODUCTION

The health, food, energy, and economics of humans are just a few of the many areas where water plays a significant role. A lack of clean water threatens the health of children and the poor, in addition to having environmental, economic, and social consequences. It is believed that every year between 10 and 20 million people die from water-related illnesses, and that over 200 million people die from nonfatal infections. About 5,000 to 6,000 children under the age of five succumb to diarrhea-related illnesses every day (Amin et al., 2014). Major health issues are being caused by the fact that more than a billion people lack access to clean water. Rapid population growth, increasing industrialization, urbanization, and intensive agricultural practices have all contributed to the surge in wastewater production, which in turn has made the water both unsanitary and potentially lethal. There are annual deaths in the millions caused by diseases spread through drinking water tainted with harmful microorganisms.

Effluent is the term used to describe the water that remains after wastewater treatment, and the primary goal of this process is to remove as many solids as possible. Oxygen is used up during the breakdown of solid matter, which is detrimental to aquatic life. The environment and human health are at risk if wastewater is not handled adequately. Without access to clean water, life on Earth would not be sustainable. Although clean water is essential for human survival, many people overlook the fact that wastewater is also a valuable resource. Because it includes dangerous elements, wastewater must be treated before it can be reused or released back into the environment. Protecting public health is a primary goal of wastewater treatment, which entails transporting wastewater away from populous regions and repurposing it into a non-hazardous form. Bacteria, chemicals, and other poisons abound in wastewater. To make the water safe for discharge back into the environment, it is treated to reduce the toxins to tolerable levels.

Wastewater treatment is crucial because it helps replenish water supplies and keeps harmful chemicals out of the environment. Beach closures and other restrictions on recreational water usage, limits on fish and shellfish harvesting, and contamination of drinking water are only some of the potential consequences of these actions. Wastewater treatment is critical for two reasons: restoring the water supply and keeping harmful chemicals out of the environment. More than 80% of the world's wastewater is simply discharged back into waterways and oceans. The good news is that there is still plenty of time to remediate this area to make the planet more habitable.

Water science is fascinating since it encompasses so many subfields (chemistry, biology, and physics). Water is essential for life, but it also has many practical uses in industries, including agriculture, tourism, and transportation. Contaminated water poses serious risks to public health

DOI: 10.1201/9781003618164-6

and must be addressed immediately. The water's purity, scent, and flavor will all be enhanced by the filtration processes. The need for novel methods to enhance water quality is urgent. The malleability and versatility of functional materials are crucial to the development of these technologies. Innovative schemes that use garbage recycling to purify water supplies have been discussed. Wastewater treatment can be tailored to meet the water quality requirements of a planned reuse. In-depth discussion of functional materials and their applications in wastewater treatment constitutes this chapter. In this chapter, we'll talk about why it's important to treat wastewater and what kinds of materials can be used to do so efficiently. Functional materials are essential to most wastewater treatment methods. We briefly go over some of the methods used to treat agricultural wastewater, including desalination, biological aerated filters, and Biochar-based functional materials. Effectively prepared and used biochar in water treatment has become a method to improve the efficiency and economic benefits of thermochemical processes (Li et al., 2019). In addition to highlighting bioinspired materials for water purification, this chapter highlights the use of biomass waste turned into valuable materials for wastewater treatment. Wastewater analysis is just one of the methods presented in this chapter that can shed light on water use. Various sources of wastewater are provided in Figure 6.1.

One of the most pressing issues of the 21st century will be ensuring that all people, animals, and plants have access to clean water. This necessitates cutting down on pollution and making sure there's plenty of clean water available without breaking the bank.

More and more people are interested in finding efficient ways to turn trash into useful products. The "waste-to-value" concept is extremely important from both a financial and ecological standpoint. Water purification capability has recently been demonstrated by a wide variety of waste-based carbon- and transition-metal-based compounds. There has been and continues to be significant investment in the research and development of functional materials to support technical advancements in wastewater treatment. Due to the potentially catastrophic effects of water pollution on ecosystems and human life, the development of reliable wastewater treatment techniques has assumed paramount importance and urgency. Because of their malleable features and functionalities, functional materials are crucial to the development of modern technologies.

In the fight against environmental pollution, specially engineered functional materials are crucial. The use of a wide variety of functional materials has led to significant progress in wastewater

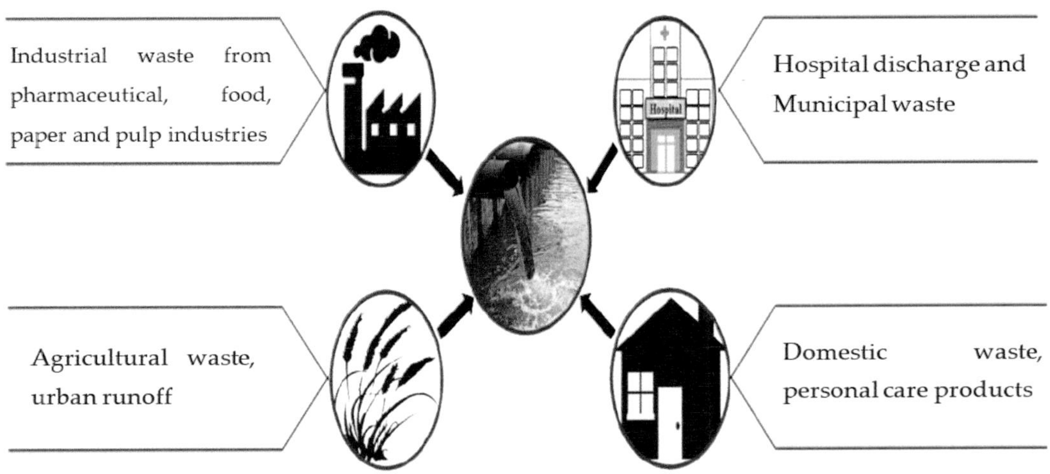

FIGURE 6.1 Various sources of wastewater. (Jain et al., 2021. Open access.)

treatment technology in recent years (Chen et al., 2018; Khan et al., 2022). Sustainable biomass and natural polymers, as well as traditional organic and inorganic materials, can be modified to create a wide variety of innovative functional materials. Functional nanoparticles, carbon nanofibers/nanotubes, and nanoporous materials are only a few examples of the nanotechnological products that have been developed as high-performance modifiers to radically alter the surface properties of adsorbent materials and separation devices (e.g., meshes, membranes).

Different approaches are needed to develop successful solutions for cleaning up oils that have been spilled on ocean surfaces or mixed throughout the water. Using various functional materials as effective adsorbents or employing separation devices made possible or improved by novel functional materials are two examples of how this concept might be put into practice. New functional adsorbent materials, such as biofilm carriers, sand, activated carbon, zeolite, clay, silica aerogel, and biopolymers and synthetic polymers, have been developed based on surface changes of classic adsorbent materials. These materials' applicability in treating various contaminants in wastewater can be substantially expanded by tuning their adsorptive performance by adjusting surface attributes such as roughness, porosity, hierarchical micro-/nano-structures, and wettability. When designing new functional materials and technologies for wastewater treatment, it is important to keep in mind production costs, ease of use, and environmental friendliness.

Industrial and agricultural wastewater, as well as home sewage, have been increasingly discharged into the environment as the economy and human activities have improved, resulting in severe environmental problems. Physical, chemical, and biological processes are now used in China for wastewater treatment. Physical approaches are typically less effective than other methods on their own and come with additional costs related to their implementation, energy consumption, and physical footprint. Large quantities of chemicals are required for chemical approaches; these are not only costly but also detrimental to environmental quality. Wastewater is typically treated using biological technologies due to their low consumption and expense.

The effectiveness of wastewater treatment using algal-bacterial symbioses needed to be improved, although it was a more promising method. Oxidative breakdown and nitrogen uptake by algal-bacterial symbioses are crucial for pollution clearance. Microalgae biomass was improved by the addition of functional materials that stimulated metabolism, regulated photosynthesis (material conversion spectrum, light capture, and carbon sequestration), increased pigment accumulation, and boosted lipid accumulation, all of which contributed to the indirect removal of pollutants. The list of water/wastewater contaminants shown in Figure 6.2 includes both organic and inorganic substances.

6.2 RELEVANCE AND IMPORTANCE OF WASTEWATER TREATMENT

The annual volume of wastewater generated worldwide is around 359 billion cubic meters, or enough to fill 144 million Olympic-sized swimming pools. In many parts of the world, it is standard practice to release massive amounts of sewage into waterways. This is a true waste of water and has devastating repercussions on wildlife, aquatic ecosystems, and the surrounding area. Destroying aquatic ecosystems and posing health risks to individuals, wastewater is defined as water that has been contaminated to the point where it cannot be used again. Wastewater treatment has been made mandatory by the government to prevent water pollution and lessen the impact of water scarcity around the world. Aerobic sewage treatment systems, like those used today, take advantage of ambient airflow. By creating an environment in which the organic pollutants may be completely oxidized into carbon dioxide, nitrogen, and water, the related odor is also eliminated. Thus, the treated effluent is safe for reuse and discharge after being thoroughly purified. To hasten the purification process, water treatment plants are built (Mathew et al., 2022). The natural mechanism is strained under the weight of billions of humans. The amount of effluent would create devastation without effective wastewater treatment, as it still does in underdeveloped nations today. More than 80% of wastewater worldwide

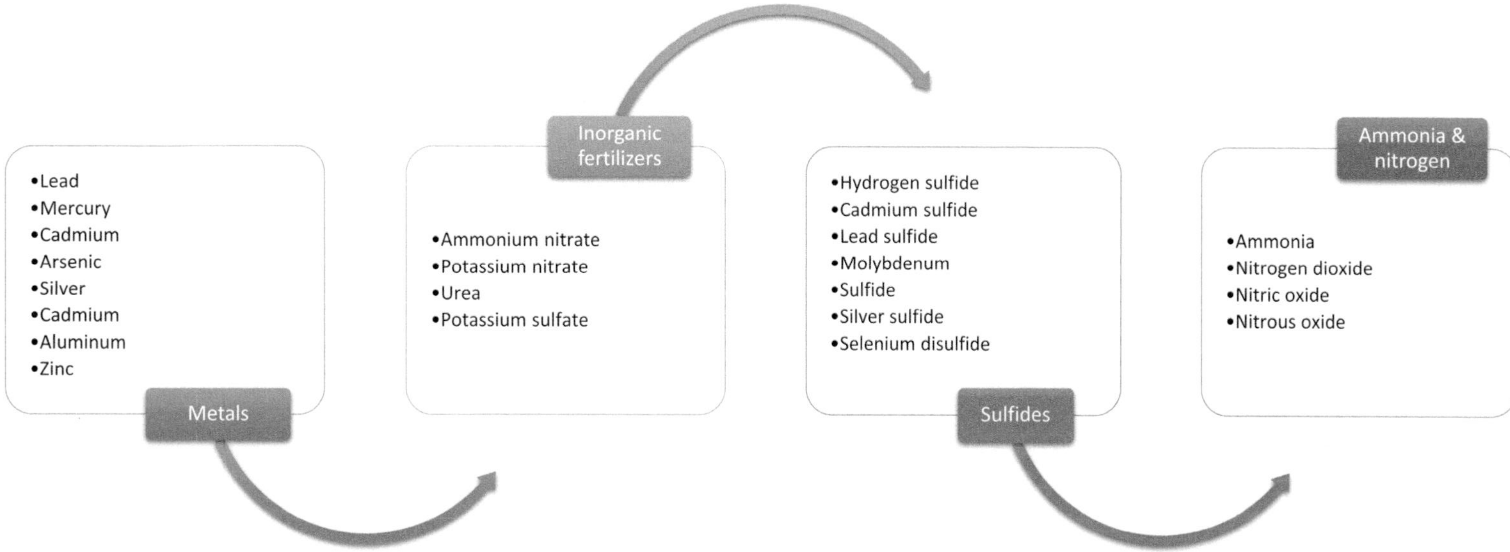

FIGURE 6.2 List of water/wastewater contaminants, including biological, organic, and inorganic contaminants (Kumar et al., 2023. https://creativecommons.org/licenses/by/4.0/. Open access.)

is released untreated. To keep individuals and the planet healthy and prospering, countries with water treatment facilities utilize a variety of techniques to purify wastewater before releasing it back into the environment (Molazadeh et al., 2019).

Humans and wildlife alike are put at risk by the presence of harmful substances in wastewater. The existing water purification and sanitation problems experienced in many developing countries could be alleviated with the installation of wastewater treatment plants (Desai et al., 2021). More than 90% of the world's 1.7 million annual deaths are attributable to the consumption of untreated water; these deaths primarily occur in underdeveloped nations. In many third-world nations, where only a tiny fraction of household and city wastewater is cleaned before being released into the environment, water-related diseases such as cholera and schistosomiasis persist. Heavy metals in drinking water are a leading cause of numerous health problems, such as heart disease, brain damage, kidney damage, and an increased chance of developing cancer and diabetes. Despite Nature's best efforts, the volume of wastewater that needs to be processed naturally is too great. With the rising global population comes a corresponding rise in wastewater production.

Half to eighty percent of wastewater in India and other developing nations is released untreated or only partially treated. As the world's urban areas and population grow, so does the volume of wastewater they produce. To slow the rate of eutrophication near sewage discharge sites, traditional sewage treatment technologies like activated sludge must reduce the nutrient concentration of wastewaters to acceptable levels. Microalgae have many uses, including (a) removing excess inorganic nitrogen, phosphorus, and heavy metals; (b) preventing the expansion of harmful bacteria; (c) sequestering carbon dioxide through the action of microalgae; and (d) producing biofuel. The heterotrophic bacteria have the potential to degrade the organic matter without aerated O_2, as O_2 is provided by the microalgal photosynthesis, and the need for CO_2 sparging is also eliminated, as this is produced by the bacterial respiration, making the integrated bacterial-microalgal approach even more promising. Thus, the integrated method of bacteria-microalgae in treating wastewater would call for single-stage treatment, simplifying the procedure.

6.3 SOME BENEFITS OF WASTEWATER TREATMENT

To guarantee high-quality water and sanitation, the wastewater treatment process must be broken down into several phases. Paper, plastic, and metal are just some of the inorganic materials that are filtered out in the first stage of the treatment process using massive filtration screens. Grit and silt, which can damage plant machinery, are then filtered out of the wastewater (https://blog.constructionmarketingassociation.org/7-benefits-of-wastewater-treatment/).

6.3.1 Possible Illnesses Spread by Rodents

The germs and bacteria that cause illness are removed and killed by wastewater treatment systems. These harmful substances are removed by the filter before the effluent is released from the tank. This filtration process helps to safeguard the environment by preventing the spread of illness to water sources, as well as to plants and farm animals (www.carlowtanks.ie/the-benefits-of-a-modern-wastewater-treatment-system/).

6.3.2 Environmental Advantages

Polluted water is a result of the release of wastewater into natural water sources like lakes and rivers. Fish, crops, and animals all drink this contaminated water. The filthy water is evaporating and will eventually end up in the rain as well. Wastewater treatment is critical for maintaining the health of a wide variety of ecosystems.

6.3.3 Water Reused

Water recycling is the ideal solution because companies and industries use so much water every day. Plants can purify even the filthiest wastewater and put it to good use again. Hundreds of liters of water are not wasted during this procedure.

6.3.4 Generation of Energy

Sludge is a byproduct of treating wastewater and is rich in biodegradable material. These components can be employed in energy production in various forms. The wastewater treatment plants may use the energy produced here, making the area self-sufficient.

6.3.5 Keep You From Paying Hefty Penalties

A special committee has been formed by the Government of Maharashtra and MPCB (Maharashtra Pollution Control Board, India) to monitor the development and compliance of sewage treatment facilities. Realizing the significance of wastewater treatment and putting it into practice is crucial. Here is an example of a British penalty. Following an inquiry into historic sewage discharges, Britain's water regulator intends to sanction debt-ridden Thames Water 104 million pounds ($133 million), which is about the maximum penalty permitted (www.reuters.com/world/uk/ofwat-proposes-168-mln-stg-penalty-uk-three-water-firms-2024-08-06/). For failing to manage their wastewater treatment works and networks, including their operation of storm over flows, Thames, Yorkshire, and Northumbrian have been proposed to pay fines of £104 million, £47 million, and £17 million, respectively, by Ofwat, the water regulator for England and Wales. The UK government wants to clean up its rivers and oceans; therefore, it has introduced infinite punishments for polluters. The existing penalty cap of £250,000 will be removed, making way for the imposition of larger fines.

6.3.6 Less Garbage

Water quality is enhanced as a result of wastewater treatment since less waste is released into the environment. In doing so, the government mitigates water pollution's effects on human health and prevents further water scarcity. Reduced spending on environmental rehabilitation programmes to combat pollution is another benefit of wastewater treatment.

a. Fertilizer Manufacturing
   In "drying lagoons," any leftover biodegradable materials are dried before being processed into organic fertilizer. The resulting organic fertilizer is subsequently put to use in agriculture to boost harvests. This reduces the amount of harmful chemical fertilizers that are released into the environment.
b. Lowest Possible Odor Emissions
   Modern wastewater systems emit significantly fewer odors than older systems. Because of the unpleasant odors that can be released, many individuals are turned off from installing a septic tank or similar system. Once properly maintained, there is no need to worry about unpleasant odors when using current systems. Humans often contaminate water bodies such as rivers, lakes, oceans, and groundwater. Water treatment provides us with hygienic and safe water to consume (Aziz and Mustafa, 2019). The whole water treatment process, particularly for public water systems, comprises some 5 key phases, as outlined by the Centers for Disease Control and Prevention (CDC). The common steps of river water treatment and methods are given in Figures 6.3a and 6.3b, and the processes are demonstrated in the following Flow Chart.

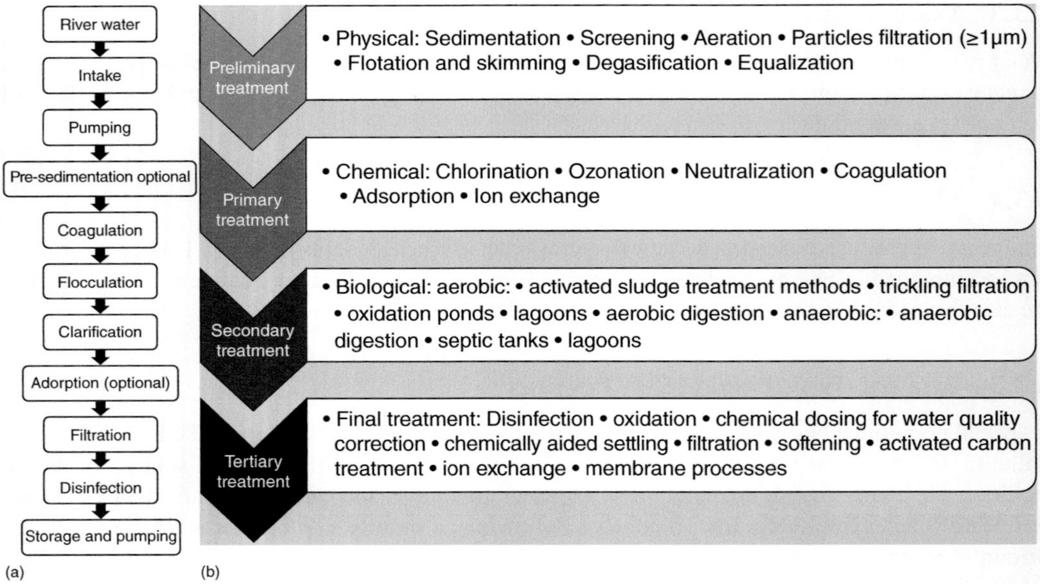

FIGURE 6.3a Water treatment methods. (Aziz and Mustafa, 2019. Under the terms of the Creative Commons Attribution License.)

FIGURE 6.3b Classification of wastewater treatment methods. (Poerio et al., 2019. http://creativecommons.org/licenses/by/4.0/.)

The wastewater processing can be grouped into four main treatments: preliminary treatment, primary treatment, secondary treatment, and tertiary treatment, also named final or advanced treatment. Figure 6.3b explains the classification of wastewater treatment methods.

6.4 FUNCTIONAL MATERIALS FOR WASTEWATER TREATMENT

Most wastewater comes from homes, factories, and municipal systems. Depending on where it comes from, wastewater may contain a wide range of inorganic and organic contaminants, as well as microorganisms. Heavy metal ions, textile dyes, and other harmful chemicals are routinely released by modern manufacturing and processing businesses, contaminating groundwater and surface water. Exploration, processing, and treatment of oil and gas, as well as unintentional oil spills, all contribute significantly to the volume of oily wastewater generated by the offshore oil and gas industry. Wastewater's increasing environmental repercussions in recent years have sparked rising alarm on a global scale. The accumulated wastewater can cause lasting damage to natural ecosystems and pose major risks to human health if it is not cleaned promptly and effectively. In the fight against environmental pollution, specially engineered functional materials are crucial. Several functional materials have been applied to wastewater treatment technologies in recent years, resulting in significant advancements in this field (Zhang et al., 2023). Modifications of traditional organic and inorganic materials, as well as sustainable biomass and natural polymers, are the starting points for a wide variety of innovative functional materials (Mustafa et al., 2021; Chen et al., 2022). The rapid progress of human society depends on effective wastewater management and drinkable water purification to limit environmental pollution and health risks. The contamination caused by the leaching of toxic substances has become an issue in society and is directly responsible for the health problems experienced by millions of individuals.

6.4.1 Carrier Biofilms

Biofilms are widely employed in the wastewater treatment industry. It promotes the development and adherence of microorganisms and serves as a biological filler in an aerated tank or wetland. While the biofilm itself isn't very good at filtering out contaminants, it does have a high resistance to microbial deterioration thanks to its rough and hard exterior. Transporters in biofilms are typically elastic fillers like polymers or ceramic sites like shale, volcanic rock, and clay. The most common biofilm carrier used for wastewater treatment is ceramsite (Khan et al., 2022).

Chen (2004) reviewed the development, design, and applications of electrochemical technologies in water and wastewater treatment. Particular focus was given to electrodeposition, electrocoagulation (EC), electroflotation (EF), and electrooxidation. Chen (2004) mentioned that the newly developed stable and active electrodes for oxygen evolution would definitely boost the adoption of this technology. Electrooxidation is finding its application in wastewater treatment in combination with other technologies.

Biofilm carriers, such as carbon fibers, polyurethane, or non-woven textiles, have been employed by Wang and colleagues (2021) and Makisha (2021) to create a fixed-bed baffled reactor for microbial nitrification and denitrification reactions to remove nitrogen from synthetic aquaculture wastewater. Carbon fibers were shown to have significant potential in treating wastewater from aquaculture, and their research revealed that the biofilm formation is affected differently depending on the biofilm carrier material.

6.4.2 Filtering Sand

Water purification procedures, including those involving the treatment of wastewater and the production of potable water, have long made use of sand as a filter. Sand is inexpensive and easy to work with, but it can't filter out dangerous chemicals. Improvements to the surface, mechanical, and adsorption properties of sand are the focus of the current study. Quartz sand's potential as a functional carrier was studied by Chen et al. (2012).

They used a technique involving repeated heating and steaming to generate a sort of aluminum salt-modified quartz sand, which outperformed an untreated quartz sand filter in terms of its ability to remove algae. Saini et al. (2021) developed a straightforward and inexpensive strategy for cleaning up the water left over from the kitchens and laboratories of universities and colleges.

6.4.3 Three Biomass Materials

Rice-husk (RH), coconut shells, and sawdust are excellent sources of biomass material (Chen et al., 2018). RH is an agricultural by-product that can be used for the removal of cationic dyes (e.g., methylene blue) and crystal violet (methyl violet) from wastewater. Compared to other adsorbents, RH shows low adsorption capacity for various dyes, but its low cost and abundant availability are advantageous for large-scale industrial applications; for example, the removal of dyes from textile wastewater. Another important application of this type of material can be found in the production of biochar using palm kernel shell as an efficient and cost-effective adsorbent to remove dyes from wastewater. An intriguing honeycomb-like carbon foam was recently produced by Tan et al. (2018) from larch sawdust, and this material can serve as a hydrophobic adsorbent for oil spill cleanup. A comprehensive review was published on this aspect by Vishnu et al. (2022).

6.4.4 Chitosan-Based Biopolymers

Chitosan is a mucopolysaccharide found in the shell of crabs and shrimps, usually obtained by treating the shell with sodium hydroxide. In this reaction, the acetyl groups of the biopolymer are hydrolyzed, leading to the formation of free primary amino groups in the molecular structure of the biopolymer. Chitosan has been recognized as an alternative adsorbent for the removal of copper, chromium, and dyes from wastewater (https://encyclopedia.pub/entry/22263). Due to the presence of free primary amino sites, the adsorption of dyes and the adsorptive removal of contaminants from wastewater is possible using chitosan due to its ability to form chelates with metal ions in acidic circumstances. Chitosan dissolves easily in aqueous solutions, including those with low acidity. In recent years, many studies on chitosan have aimed to find ways to lessen its solubility. Making low-solubility chitosan involves cross-linking the chitosan chains by reacting their amino sites with cross-linking chemicals. Many scientists have studied the strategies for modifying chitosan and designing chitosan-based adsorbents for use in fixed-bed column packing for industrial wastewater treatment (https://encyclopedia.pub/entry/22263).

6.5 WASTE-DERIVED FUNCTIONAL MATERIALS FOR WASTEWATER REMEDIATION

The earliest inorganic materials likely used by humans were stones and animal bones. Inorganic materials have been utilized for a wide range of applications since their discovery. These include, but are not limited to, high strength and toughness; dielectricity; semi-conductivity; magnetism; superconductivity, etc. In today's world, functional inorganic materials are absolutely crucial.

Superconductors are just one example of a class of functional materials whose discovery and "design" have remained mysterious despite the promise of transformative applications (efficient power grids, powerful magnets, computation, etc.). Diffusion barriers need high-temperature reactions and physical proximity to phases determined by thermodynamic equilibrium in much of synthetic solid-state chemistry. Many useful materials, however, are inherently metastable (i.e., not stable at ambient conditions or not found on equilibrium phase diagrams), as is the case with high-temperature superconductors (https://neilsonlab.colostate.edu/research.html).

Innovative technologies, crucial to the growth of many sectors of industry, rely heavily on functional inorganic materials (Zarkov et al., 2021). Electronics, optoelectronics, catalysis, biomedicine, and many other fields are concerned with many facets of human life, and technological advancement is facilitated by the employment of novel materials, nanostructures, or multicomponent composites with specific chemical or physical properties. The development of superior synthesis pathways, reliable characterization, and a deep understanding of the structure-property relationships in materials are rightly considered to be fundamentally important scientific issues due to the wide and varied range of potential applications of functional inorganic materials.

Science and technology related to catalysis are one of the most promising avenues for addressing our planet's pressing energy and ecological issues (Wang et al., 2023). Supported metal catalysts, metal oxides, and ordered porous materials are just some of the catalytic materials that have been designed (www.eng.kyushu-u.ac.jp/e/lab interdisciplinary01.html) by scientists. The materials' chemical and physical properties can be modified by adjusting their composition, crystal structure, surface chemistry, and morphology. Since the study of functional inorganic materials encompasses such a broad and rapidly developing area, it would be impossible to summarize all of the current research in a single Forum.

Maintaining a population that has access to clean water is crucial to the long-term health of our civilization. Water contamination is a major problem today due to the effects of growing industrialization and urbanization on the environment and human health. There has been a lot of research done on the importance of getting rid of dangerous pollutants (including heavy metals, microplastics, antibiotics, and viruses) in water. Adsorption, photocatalytic degradation, electrochemical treatment, and alternative operational procedures are just some of the efficient and novel technologies that have been developed for wastewater remediation (Chen et al., 2022). To achieve effective pollutant

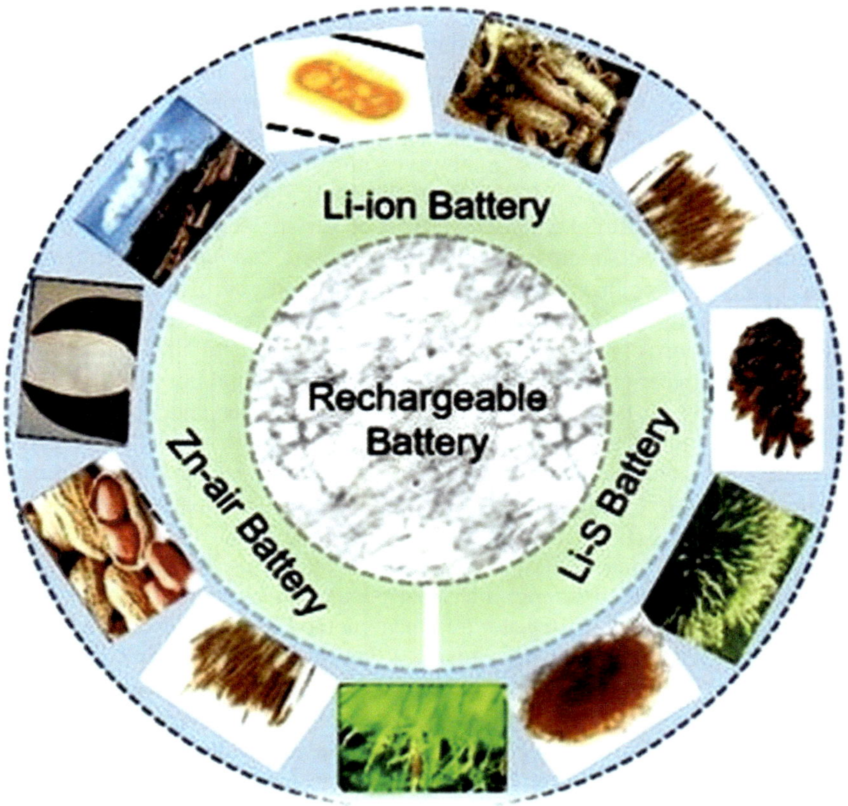

FIGURE 6.4 Sustainable batteries like lithium-ion batteries (Feng et al., 2024. https://doi.org/10.1002/celc.202400086 .http://creativecommons.org/licenses/by/4.0/.)

degradation/removal, all of these techniques rely on functional micro/nanomaterials, notwithstanding their vast differences in mechanisms and operating processes.

Pollution of water sources poses serious risks to human health and to the future of the planet. Reliable wastewater purification techniques are needed immediately to protect the quality of our drinking water. Cost-effective materials are very desirable because most wastewater cleanup methods rely primarily on functional materials. The wastewater cleanup industry has been expanding rapidly in recent years, which is both environmentally and economically significant. Chen et al. (2022) provide a comprehensive analysis of the uses of waste (e.g., bio-wastes, electronic wastes, and industrial wastes) derived materials for wastewater purification. Creating functional materials from waste via effective strategies is of utmost importance. The "waste-to-value" principle is of great economic and environmental prominence. Batteries that use biomass materials are a representative example of the use of renewable resources for energy storage. Currently, sustainable batteries are some of the most widely used energy storage devices on the market (Figure 6.4).

6.6 FUNCTIONAL MATERIALS IN A BIOLOGICAL AERATED FILTER

In municipal wastewater treatment, the biological aerated filter (BAF) has gradually replaced the older activated sludge technique during the past two decades. The BAF has numerous benefits, such as the simultaneous removal of sediments, organic materials, and nutrients. In addition to these benefits, BAF systems decrease sludge generation, shorten the residence time of biomass, increase resilience to environmental shock, and lessen biomass loading and pollution. When compared to an activated sludge system, BAF has the potential to be more space- and labor-efficient in terms of carbonaceous and ammonia removal. When it comes to the secondary treatment of municipal wastewater, the biological aerated filter (BAF) is a highly effective biological treatment method for removing contaminants. Little is known, however, about how the bacteria in the BAF interact with the process of pollutant removal. Researchers Ren et al. (2018) used an up-flow biologically activated filter (UBAF) reactor to examine the effects of varying aeration rates and filter medium heights on pollutant removal and microbial community structure. *Illumina* pyrosequencing was used to examine the biofilm's microbial composition. At an aeration rate of 65 L/h, Ren et al. (2018) showed that the UBAF was more effective at removing chemical oxygen demand (COD), NH4+-N, NO3-N, and total phosphorus (TP). In addition, most of the COD and NH4+-N removal occurred between 0 and 25 centimeters into the filter medium. The microbial community structure in the UBAF showed that at 10 cm of filter media height, Planctomycetes and Comamonadaceae were more abundant than in any other treatment type by a factor of 11% and 48.1%, respectively. As a conclusion, shifts in the abundance of Proteobacteria, Planctomycetes, and Nitrospirae presumably shed light on the mechanism by which nitrogen and phosphorus were removed.

Biological aerated filters (BAFs) are starting to gain traction as a viable option for treating wastewater because of their reduced footprint and increased efficacy (Xu et al., 2022). The BAF's low-cost biological fillers enabled a dense microbial population and high treatment efficacy with a short dwell period. As a potential wastewater treatment method, BAF involves pumping wastewater up or down through a filter medium that contains a bed of filter media providing a high surface area to grow biomass. BAF is structurally similar to a three-stage sewage treatment filter, but uses a single homogenous filter material instead (www.chinadicheng.com/news/structure-of-biological-aerated-filter.html). Filter body, filter material, supporting layer, water distribution system, gas distribution system, backwash system, effluent system, pipeline, and automatic control system are the nine basic components of the BAF (Chen et al., 2016). Two biological aerated filters named BAF1 and BAF2 were constructed by Ding et al. (2023) (Figure 6.5).

Different parts of the Biological Aerated Filter are given below:

6.6.1 The Filter Tank Itself

Its job is to sustain the weight of the filter material and aeration device while also holding the treated water and the filter material. Steel machinery and reinforced concrete structures are used to create the various-sized and shaped objects.

6.6.2 Layer 2: Biological Insulation

The biofilm itself is carried by the filler layer, which also serves to hold suspended particles. Honeycomb tubular filters, bundle filters, circular radiation filters, shield filters, mesh filters, cylinder filters, etc., are now in use in BAF. Materials, including porous ceramsite, quartz sand, expanded shale, lightweight plastic, expanded silicon aluminate, plastic modules, and FRP, are typically used as filters.

Particle fillers can vary in their physical and chemical characteristics in interesting and sometimes related ways. The key to BAF technology's success lies in the selection of bio-carrier filler. It indicates whether or not the BAF filter material is functional. The following criteria should be used to determine which filler to use. Surfaces should have certain porosity and roughness, which is conducive to the attachment and growth of microbial members a. Good mechanical strength; b. Porous inert carriers with large specific surface area and high porosity are generally selected, which is

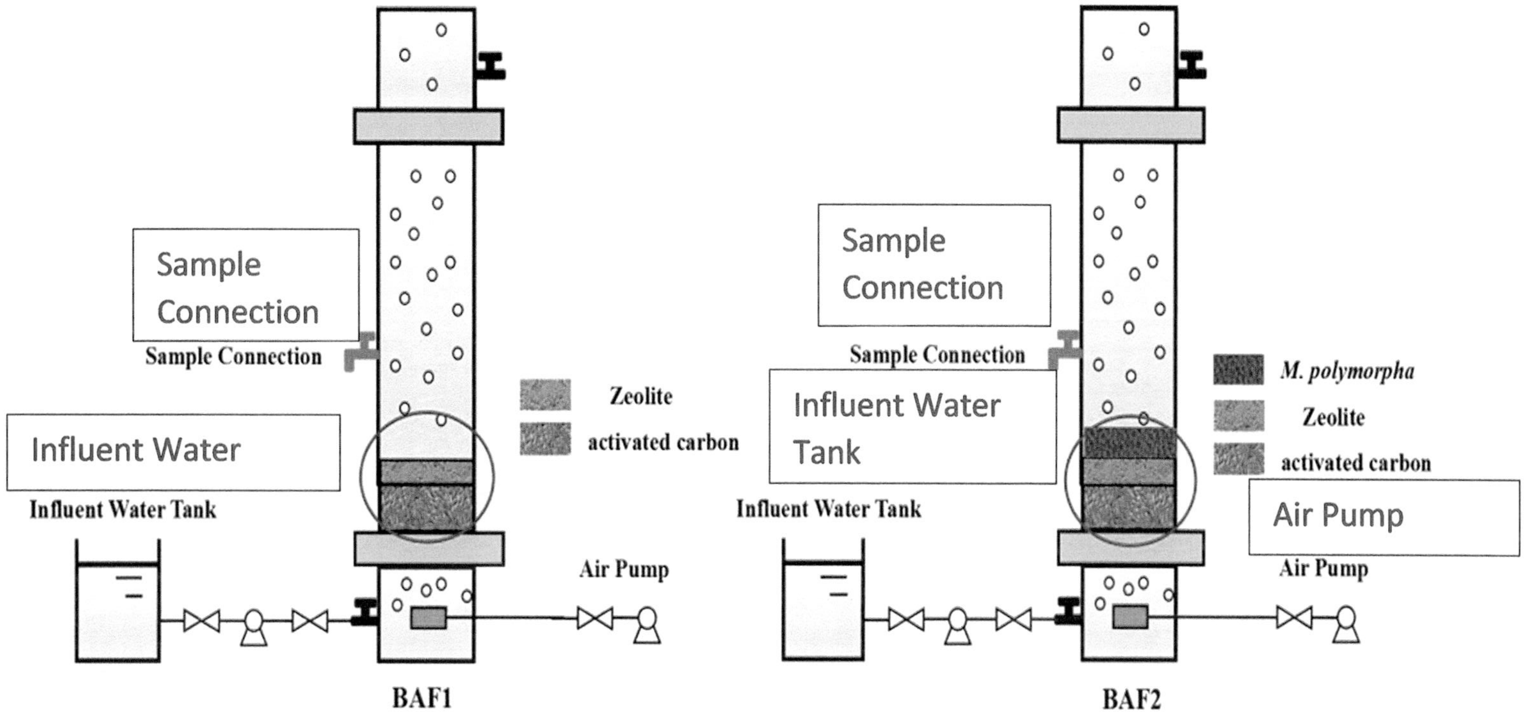

FIGURE 6.5 Biological aerated filters (BAFs): Sample Schematic diagram of BAF1 and BAF2. (Ding et al., 2023. Open Access Article.)

conducive to the adsorption, sustainable growth, and formation of biofilm; c. The regular spherical packing is selected to make the distribution of air and water even, and the flow resistance is small.

6.6.3 Layer 3 Bearing

The primary functions of the supporting layer are to prevent the biological filler from settling to the bottom of the filter and the filter head from becoming blocked, and to allow for consistent backwashing. The foundation is a layer of stones. Bearing layer heights range from 400 mm to 600 mm.

6.6.4 The Water Supply Network

The water distribution chamber at the bottom of the filter and the water distribution filter head on the filter plate make up the bulk of BAF's distribution system.

6.6.5 Aeration and Ventilation

The aerated biological filter's gas distribution system serves two primary functions: (i) ensuring aeration requirements during normal operation, and (ii) ensuring gas distribution during air and water backwashing. The required air supply is calculated during the process and then used in the aeration system's design. A perforated pipe can be used in the basic aeration system of BAF. In practice, oxygen aeration and backwash aeration share the same distribution air pipe. However, oxygen-filled aeration has a lower air demand than backwash, making uniform gas distribution more challenging.

6.6.6 Backwashing Method

To keep a BAF running smoothly, backwashing is essential. The goal of the backwash process is to thoroughly clean the filter material so that it can resume its wastewater-interception role with little downtime. Airwater combination backwashing often entails performing a separate backwash with airwater, then a backwash with gaswater, and finally a backwash with clean water. Backwashing requires expert control over the water pressure and duration of the flush.

6.6.7 Water Distribution System

The effluent system of BAF can be treated by peripheral water or a single weir (www.chinadicheng.com/news/structure-of-biological-aerated-filter.html). BAFs are known for producing high-quality effluent due to the efficient removal of organic matter and nutrients by the biofilm that develops on the media. Biological aerated filters (BAFs) are used to treat wastewater and drinking water. They can be used in water distribution systems to remove ammonia, iron, and manganese.

For large-scale engineering projects, various state-of-the-art BAF technologies have been developed and proven to date (Chen et al., 2016). Not only can using treated wastewater help with water scarcity, but it also improves soil fertility and decreases the need for artificial fertilizers. Xu et al. (2022) improved the effectiveness of wastewater recycling by lowering treatment costs and increasing the volume of recycled water. All tests were performed on an innovative integrated membrane biological aerated filter (IMBAF), which consists of a down-flow cylindrical biological aerated filter (BAF) filled with silica and a novel sand-coated polystyrene granules (SCP), followed by ultrafiltration (UF) and reverse osmosis (RO) membranes. The 73.6 L IMBAF reactor was run for 270 days (in three 90-day stages) under a variety of circumstances for recycling its backwash water. With removal rates of 94.2%, 68.4%, 54.4%, 91.24%, and 99.95% for turbidity, 5-day biochemical oxygen demand (BOD5), chemical oxygen demand (COD), oil and grease (O&G), and fecal coliform (FC), respectively, BAF produced excellent quality water for feeding the UF membrane. By the end of the third stage, UF had recycled 99.88% of the influent, and only 0.12% had been sent to the sludge pile. The BAF and UF module do an excellent job of improving the water quality that is fed into the RO system. After 75 days of continuous operation, the RO membrane exhibited modest clogging as evidenced by a 14% increase in trans-membrane pressure (TMP) and a 9.4% drop in membrane permeability. Increased wastewater recycling, less membrane clogging, and decreased concentrate disposal and chemical cleaning expenses are all results of using a BAF construction.

6.6.8 Using a Biological Aerated Filter in Conjunction With Ultrafiltration to Clean Rural Water Supplies

Many rural areas rely on lakes and reservoirs for their water supply, which are vulnerable to runoff and non-point source pollution and often of lower quality than their urban counterparts. Due to the dispersed nature of villages, which are often located in varied and complex topographies, drinking water supply in remote rural areas typically suffers from various challenges, such as the high cost of construction and maintenance of centralized drinking water treatment plants and pipe networks. According to research by Wu et al. (2023), dirty reservoir water can be treated using a hybrid technique consisting of a biological aerated filter (BAF) and ultrafiltration. Turbidity, chroma, and the organic matter index were utilized as performance measures. The authors investigated the combined process across a range of crucial operational parameters, such as filler kinds and empty bed contact time, over the course of 260 days in a long-term experiment (EBCT). Microorganisms with specified roles were found to be enriched in distinct BAF reactors after a comprehensive evaluation of the microbial communities at different stages. Using granule-activated carbon (GAC) as filler and an EBCT of 45 minutes, an 85.5% DOC removal rate was achieved in the combined process. In China, most BAF reactor effluents are safe for human consumption. Some rural locations may be able to use the combined system to treat dirty water.

6.7 BIOCHAR-BASED FUNCTIONAL MATERIALS IN THE PURIFICATION OF AGRICULTURAL WASTEWATER

Biochar, a carbon-rich solid created by pyrolyzing biomass with little or no oxygen, is widely acknowledged as a versatile material with numerous uses in the fields of energy generation and environmental remediation. Biochar's chemical and physical properties are highly sensitive to its starting materials, production processes, and pyrolysis temperatures and pressures. Several thermochemical methods, such as fast pyrolysis, slow pyrolysis, flash carbonization, hydrothermal carbonization, torrefaction, and gasification, are used to create it from raw materials such as wood biomass, animal manure, crop leftovers, and solid wastes. The fast expansion of the agricultural business in recent decades has resulted in an increase in the prevalence of agricultural contamination caused by the release of antibiotics, pesticides, or hazardous metals into farmlands. Many scientists have tried using biochar-based functional materials to clean up contaminated agricultural wastewater. However, due to its many useful characteristics, biochar has also shown great promise as a technique for purifying water (e.g., cost-effective, high specific surface area, and surface reactive groups). Wei et al. (2018) wrote a comprehensive review on this topic, focusing on the ways in which biochar-based functional materials are made and used to purge various contaminants from agricultural settings. With its potential uses in bio-energy generation, pollutant immobilization, soil fertility improvement, carbon sequestration, soil amendment, and environmental remediation, biochar application has emerged as a promising biotechnology in recent years. While biochar-based functional materials show promise, their use is still in its infancy due to a number of factors, including the absence of quality standards, the immaturity of designed biochar production systems, and the gap between laboratory scale and field use. Wei et al. (2018) honed in on this gap in the literature and showed its significance. The scientific community agrees that biochar has promising applications for cleaning up polluted water. The specific reactive area, functional groups, and stability of the raw biochar have all been shown in studies to be significantly improved after suitable alteration. Wei et al. (2018) mentioned that not all pollutants can be eliminated with biochar.

Biomass decomposition yields coke, also known as charcoal, as a byproduct during the thermochemical process. Catalysis, adsorption, fuel cells, soil improvement, etc., are just some of the numerous fields where its low cost and high specific surface area (which can reach over 1,000 m^2g^1) might be of use. As a result, biochar has emerged as a means to boost the efficacy and financial benefits of thermochemical processes through its application in water treatment. Biochar material has great potential for use in wastewater treatment, and Li et al. (2023) analyzed the manufacturing methods of several novel biomass materials, categorized the various modification tactics, and talked about their implications. According to their review, biochar is a byproduct of the thermochemical conversion of biomass that is gaining popularity for a number of uses. Biochar has several advantages as an environmentally friendly material, such as low cost, ease of obtaining, stability, etc., when compared to other materials like MOFs. Biochar is used for a variety of environmental purposes, including lowering greenhouse gas emissions, decreasing soil pollution, increasing soil fertility, treating wastewater, and more (Thompson et al., 2016).

According to Li et al. (2023), the following factors are particularly relevant to the study of new biochar's development in light of the needs of the sustainable development concept. To better control the preparation of high-performance catalysts, researchers are (1) developing new methods for material synthesis; (2) optimizing preparation processes; (3) using multiple composite modifications to increase biochar materials' functionality; and (4) exploring new catalysts that are recyclable, low-cost, and environmentally friendly for use in environmental pollution control projects.

6.8 BIOINSPIRED MATERIALS FOR WATER PURIFICATION

Inadequate access to safe drinking water and sanitation is a major contributor to water scarcity issues around the world. Water purification and environmental remediation technology advancements, along with appropriate conservation regulations, will be needed to meet future difficulties. More effective and sustainable methods will be developed with the help of new bioinspired materials. Reverse osmosis and other membrane-based separation processes rely heavily on amphiphilic self-assembly (Gonzalez-Perez and Persson, 2016). Water contaminants, such as heavy metals, can be removed and processed with the use of mesoporous support materials for semiconductor growth in photocatalytic degradation of pollutants and novel carriers for immobilization of bacteria in bioreactors. Improving and optimizing the fabrication process, as well as learning more about their effectiveness in pilot and small-scale purification systems, are challenges that need to be overcome. These novel biomimetic materials will be integrated into existing water purification technologies to enhance their purification/removal performance in a way that is both economical and gentle on the environment. Future water purification technologies will rely heavily on the creation of biomimetic or bioinspired materials (Yunus et al., 2012).

Many rapidly expanding subfields of water science, such as membrane separation technologies, have relied heavily on heuristics in their technical growth. The development of new materials is becoming more methodical, with many taking cues from existing materials of biological and non-biological origin. Many of these cutting-edge materials are based on the well-known self-assembly capabilities of amphiphilic molecules, which are found practically everywhere in living beings and include block copolymers and surfactants (Yunus et al., 2012). This will make the existing heuristic approach to material creation obsolete by allowing for more precise tuning based on a deeper understanding of the science driving self-assembly.

More and more uses have been found for biomimetic materials as they have been refined over time. Biomimetic approaches are essential for future water purification and environmental remediation, and recent advances include biomimetic block copolymer membranes with water permeation motifs, biocontinuous liquid crystalline templates for TiO_2 deposition for advanced oxidation processes, and new artificial biofilms under controlled environments for pollutant removal.

6.9 UTILIZABLE MATERIALS DERIVED FROM BIOMASS WASTE TREATMENT

The growing generation of biomass waste and the absence of practical and efficient techniques to recycle these biomass waste resources pose a growing challenge to cities, especially in developing nations. As a result, there is a lot of interest in finding effective ways to recycle them. Through photosynthesis, plants absorb solar energy and store it as biomass by incorporating carbon dioxide into their tissues. The term "biomass" refers to a broad category of organic materials that includes everything from plants to the byproducts of animals that eat plants. Biomass energy has just surpassed coal, oil, and natural gas as the world's fourth-largest energy source.

Forestry residues, agricultural wastes, animal wastes, industrial wastes, Municipal solid waste (MSW), food processing wastes, and other waste streams are all examples of biomass wastes. Biomass feedstock from agricultural and forestry wastes, animal wastes, industrial wastes, and municipal solid waste (MSW) has all been the subject of extensive research over the past five years to extract natural polymers or convert them into functional materials for value-added applications (Zhou and Wang, 2020). Most biomass wastes are either left in the field to decompose, sent to a landfill, or incinerated outdoors for purposes such as cooking, drying, and charcoal production. These methods are inefficient at best and cause serious environmental pollution, such as greenhouse gas emissions and air quality deterioration. As a result, there has been renewed interest in finding ways to turn biomass wastes into useful products in fields as diverse as medicine, materials, and food packaging.

The exponential growth of biomass wastes has spawned a slew of disposal and management challenges. As a result, there has been a rise in interest in finding ways to turn biomass wastes into useful products. Research on the use of biomass waste and functional materials made from such waste was summarized by Zhou and Wang (2020). Recent research efforts center on these three primary areas: (1) Extracting natural polymers from biomass wastes; (2) Reusing biomass wastes; (3) Preparing carbon-based materials as novel adsorbents, catalyst carriers, electrode materials, and functional composites. Agricultural and forestry wastes, animal wastes, industrial wastes, and municipal solid wastes have all been collected as low-cost raw materials for industrial-scale production in the form of biomass wastes; however, further research is needed to evaluate the quality and safety of biomass wastes derived products and to develop highly feasible and cost-effective methods for the conversion of biomass wastes. Products recovered from Industrial Wastes are mentioned by Taneja et al. (2023). From food Waste (FW), various types of value-added products are recovered, viz, single cell protein, enzymes, bio-fertilizers, biochar, and bioplastics (Figure 6.6). Food waste-related matters in developing countries are currently a prominent threatening factor for sustainable development. Preventing food waste in the food processing and agricultural sectors requires improved technological solutions and infrastructure.

6.10 NANOTECHNOLOGY FOR WATER PURIFICATION

The problems of ensuring that people have access to clean water are highly interdisciplinary, necessitating solutions that also span traditional academic boundaries. Water's origin is connected to the weather, the food supply, human health, and many other facets of existence. Approximately 83% of freshwater supplies have dropped in the last 50 years, demonstrating the urgent and coordinated action that is needed. Sustainable nanotechnology has made significant advances toward the goal of providing safe drinking water to all people. One of the most crucial areas of study across

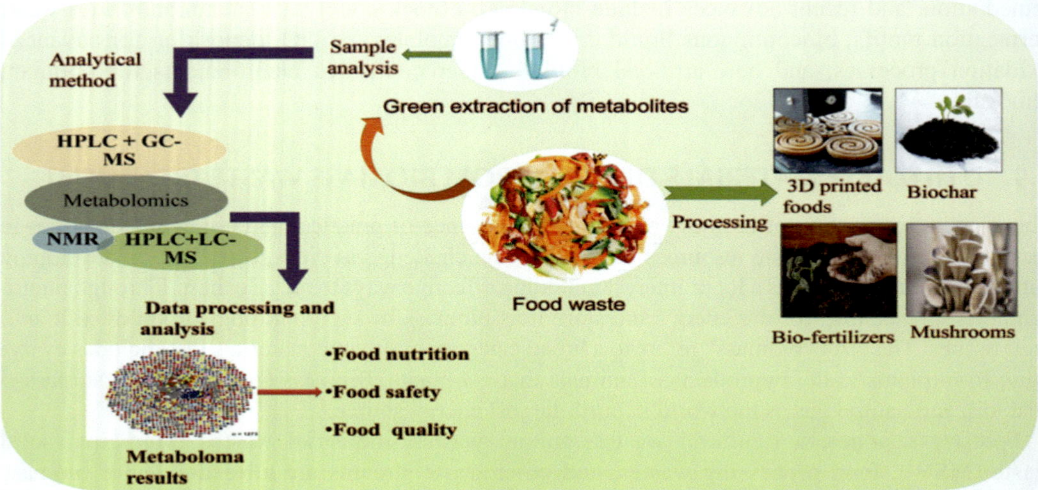

FIGURE 6.6 Some materials, like bio-fertilizers and biochar recovered from Food waste (Taneja et al., 2023. Open access.)

disciplines is water, and this will only increase in the future. The urgent need to provide access to clean water afforded by nanotechnology solutions and the wide gaps in ensuring their implementation were highlighted by Nagar and Pradeep (2020). They talked on the cutting edge of nanotechnology in several subfields of the clean water space, with an eye toward practical applications and recommendations for future study. Nagar and Pradeep (2020) remarked that cities may live and breathe easily by embracing such technology, expanding the idea of sustainable and economical clean water to the environment in general. Sustainability in clean water can be achieved by an awareness of global environmental concerns and an investigation of potential solutions provided by new nanotechnologies. Nagar and Pradeep (2020) provided concrete recommendations and attempted to measure the influence of such technologies.

The availability of water is crucial for the survival of all forms of life. However, there is a critical shortage of potable water globally. Water pollution is caused by the introduction of various gases, microbes, and other pollutants (chemicals, heavy metals, etc.) into water during rain, flowing water, etc. The unique qualities of nanomaterials, such as increased surface area, low concentration efficiency, etc., make them promising candidates for the efficient treatment of polluted water (including metal toxins, organic and inorganic pollutants, and more). Water purification using nanostructured catalytic membranes, nano sorbents, and nano photocatalysts is energy and cost-intensive, but it also has a low environmental impact (www.mdpi.com/2073-4441/12/2/495).

To examine the nanoparticles or nanomaterials utilized in nanotechnology, a microscope is required due to their minute size. Nanomaterials have replaced more traditional options in the water treatment industry (Saleem and Zaidi, 2020). Traditional water purification techniques have the following drawbacks: Screening, filtration, sedimentation, gravity separation, reverse osmosis, ion exchange, neutralization, remineralization, micro- and ultrafiltration are only a few of the most prevalent water purification techniques. Despite their value, many of these procedures are prohibitively expensive due to their excessive use of resources like water and electricity, as well as wasteful use of reagents (www.azonano.com/article.aspx?ArticleID=4918).

Soil, water, and air remediation with nanotechnology is a growing field. Sensing and detecting, treating and remediating, and preventing pollution are the three main elements of nanoparticle-related water remediation. The combination of conventional remediation techniques with nanobiotechnology is also being studied (Kunduru et al., 2017). Because of this, we can increase productivity. Recent research has shown that nanotechnology offers bright prospects, both on its own and when combined with other fields.

The use and impact of nanoparticles in wastewater treatment (www.sciencedirect.com/science/article/abs/pii/S0960852421013420) is a cutting-edge technique made possible by nanotechnology to be more economical than other techniques (Figure 6.7).

One can examine how nanotechnology is used in wastewater treatment (www.earthreminder.com/nanotechnology-in-wastewater-treatment/). Nanotechnology can be used to purify wastewater by removing a wide range of impurities. Because nanoparticles are purpose-built, there is no way for any pollutants to survive nano-treatment. TiO_2, ZnO, ceramic membranes, nanowire membranes, polymer membranes, carbon nanotubes, submicron nano powder, metal (oxides), magnetic nanoparticles, nanostructure boron-doped diamond, and so on are just some of the nanoparticles used in wastewater nano processing. Wastewater treatment using nanotechnology is cost-effective because nanoparticles have a large surface area, dissolve quickly, respond quickly, and absorb strongly, all of which contribute to the water's efficient purification. Nanotechnology is typically implemented in the five main stages of wastewater treatment, which are given below.

6.10.1 Assimilation

The first stage, known as absorption, involves cleaning the wastewater by filtering out any organic or inorganic debris. There are several elements that influence an adsorbent's efficiency, including the kinetics of absorption, the degree of selectivity, and the active site. High specific surface area, easily accessible adsorption sites, a wide variety of contaminant-CNT interactions, controllable surface chemistry, and the ability to be reused are only some of the appealing features of carbon-based nano-adsorbents. Adsorption of persistent pollutants and preconcentration of less obvious ones are two applications of carbon nanotubes in the field of environmental science. Similar benefits can be found in nanoscale metal oxides, including a high specific surface area, short intraparticle diffusion distance, an increase in available adsorption sites, the ability to be compressed without sacrificing much surface area, and a low tendency toward agglomeration during reuse. These particles are used in slurry reactors and absorption media filters. Nanofibers with a core-shell structure are another example of a representative nanomaterial because of the unique qualities they exhibit, such as a reactive core for degradation, a short internal diffusion distance, and a tailored shell surface chemistry for selective adsorption. A common component of reactive nano-adsorbents is nanofibers. Rapid mixing of nano-adsorbents facilitates mass transfer in slurry reactors and absorbers. Commercial application and proven effectiveness of nano-adsorbents for arsenic removal. Nanotechnology and Nanoparticles in Wastewater Treatment process (www.earthreminder.com/nanotechnology-in-wastewater-treatment/) is highlighted in Figure 6.8a and 6.8b.

6.10.2 Membranes and Membrane Processes

Membranes are the physical barriers that allow for size-based separation of solutes and solvents. Wastewater contains a wide range of solid materials of varying sizes; membrane techniques help to cleanse it with minimal land and chemical use, advanced automation, and adaptable architecture. Here, however, the structural adaptation between membrane selectivity and permeability presents the greatest challenge. Nano-zeolites, Nano-Ag, Carbon nanotubes, Aquaporin, nano-TiO_2, and nano-magnetite are all examples of nanomaterials used in membranes and membrane processes. Due to

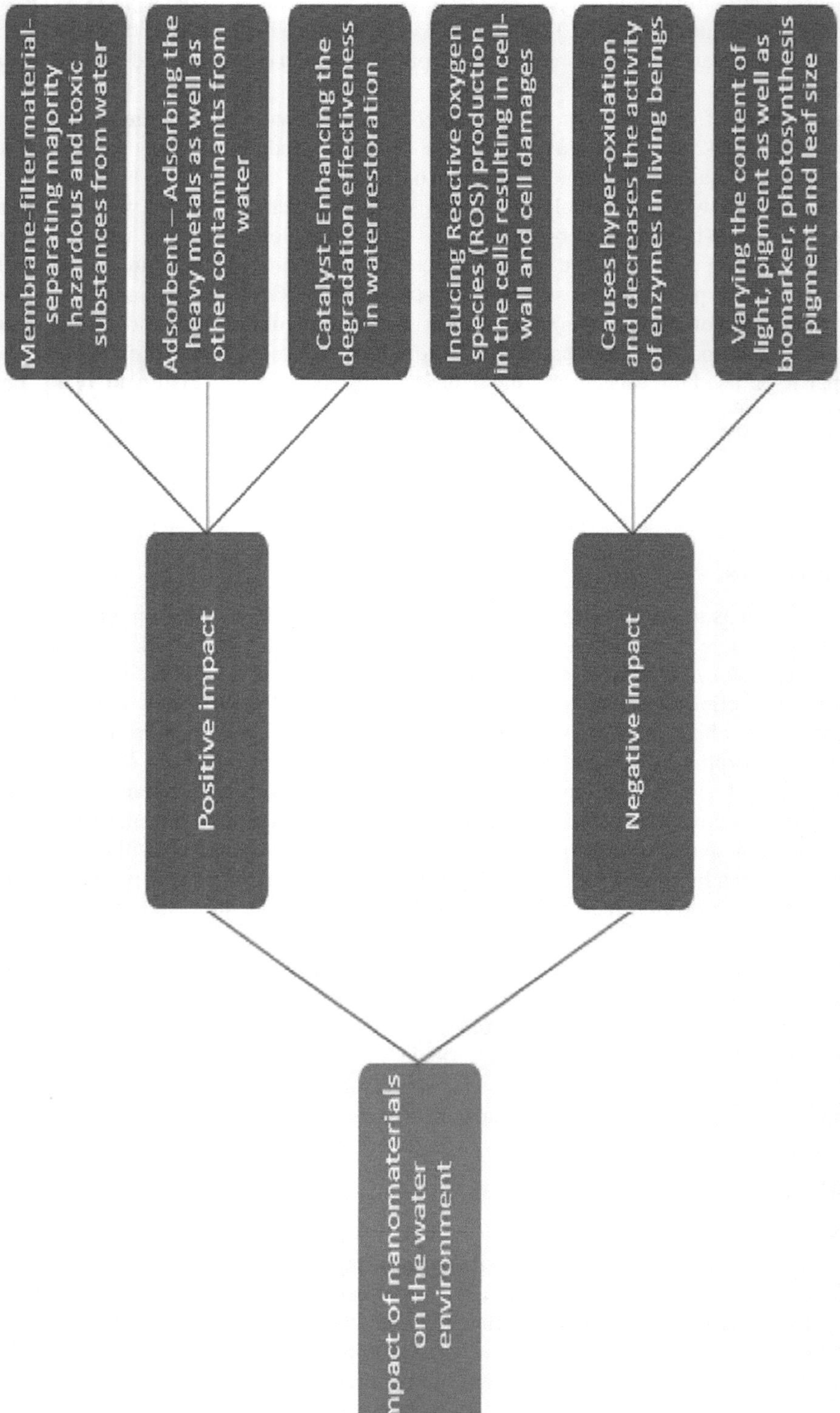

FIGURE 6.7 Nanoparticle utilization and impact in the context of wastewater treatment (Saleem and Zaidi, 2020. https://doi.org/10.3390/nano10091764. Creative Commons Attribution License.)

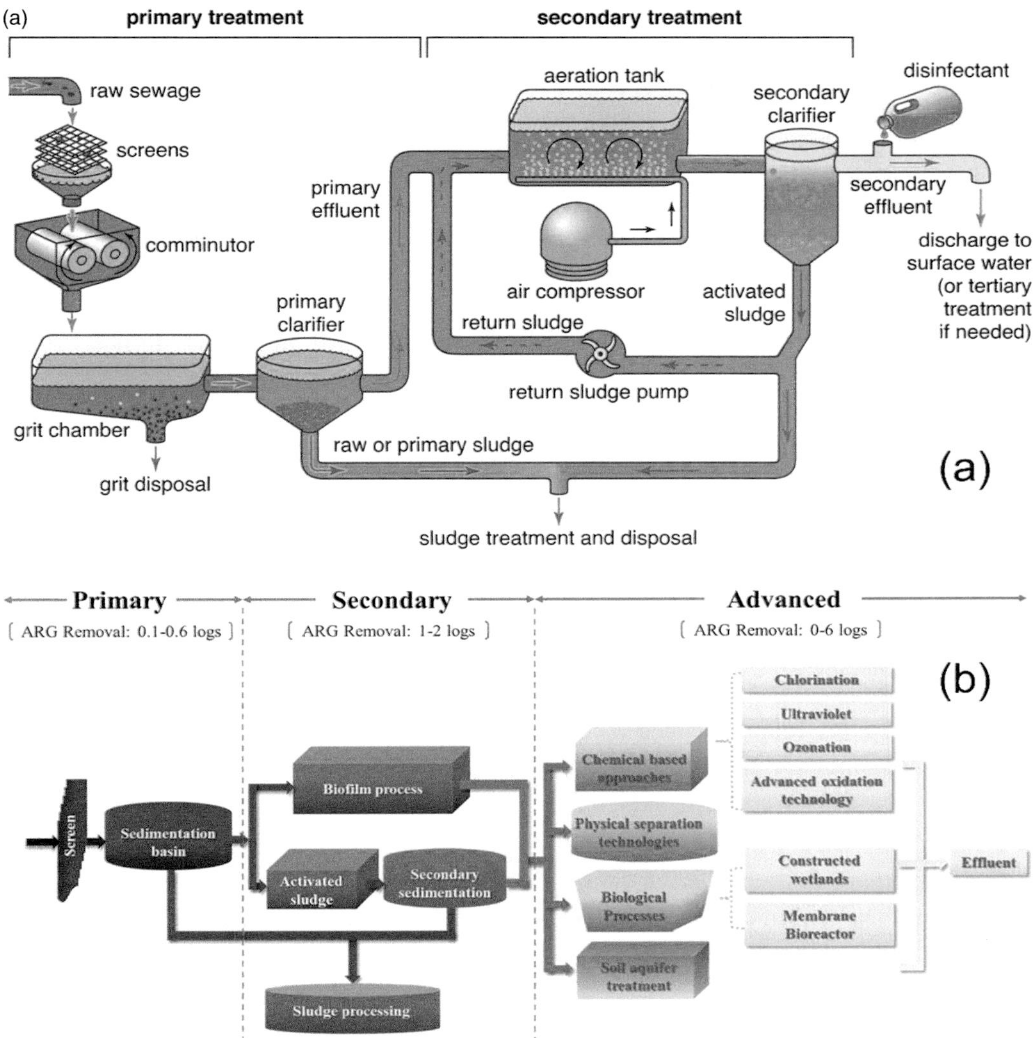

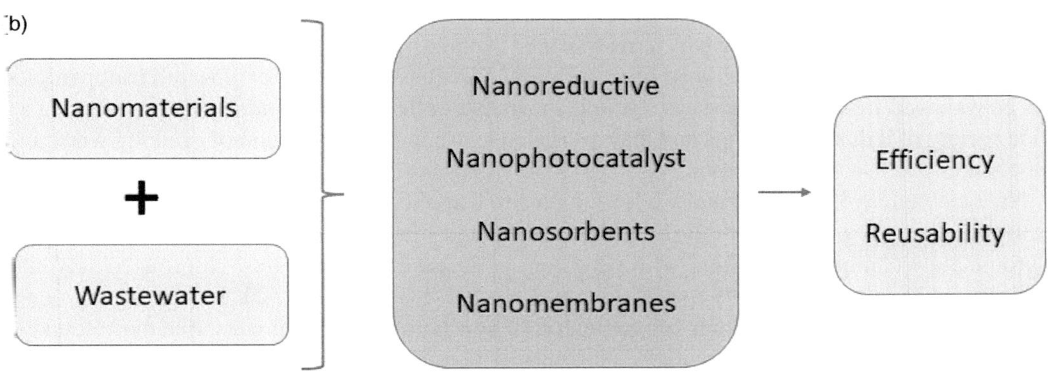

FIGURE 6.8a Nanotechnology in the Wastewater Treatment Process (Epelle et al., 2022. Creative Commons Attribution License.)

FIGURE 6.8b Nanomaterials and processes in the wastewater treatment (Mbarek et al., 2022. Open access.)

their molecular sieve and hydrophilic properties, nano zeolites are used in high-permeability thin film nanocomposite membranes. Nano-Ag, used in anti-biofouling membranes, is highly effective against a wide variety of microorganisms while posing no risk to human health. Antimicrobial activity, a tiny diameter, atomically smooth inner surface, variable opening chemistry, and great mechanical and chemical stability are just a few of the features shared by carbon nanotubes. These are employed in aligned carbon nanotube membranes and those that prevent biofouling. Aquaporin membranes are very selective and have high permeability. Reactive membranes, high-performance thin-film nanocomposite membranes, and forward osmosis technologies utilize nano-TiO_2 and nano-magnetite, respectively. The former has been designed for its photocatalytic activity, hydrophilicity, and excellent chemical stability, while the latter's customizable surface chemistry and superparamagnetic properties make it an attractive alternative.

6.10.3 Distinctive Filtration by Means of Nanotechnology

There has been a recent uptick in the prominence of membrane technologies like Nano filtration (NF) in the field of water purification. Nano-filtration membranes (NF membranes) are commonly used in water purification systems. Separating substances with sizes between 0.001 and 0.1 micrometers, Nano-filtration is a low-pressure membrane method. Nano-filtration membranes typically have pore diameters of 0.2-4 nm. These membranes are pressure-driven and share properties with both ultrafiltration and reverse osmosis membranes. Inorganic ions (Na and Ca), bacteria, and turbidity can all be removed from water using nano-filtration membranes (https://nanografi.com/blog/water-purification-and-nanotechnology/). Carbon nanotube walls line the interior of the new filtered membranes' hollow cylinders, which are arranged radially. Filters successfully removed pathogens, including *S. aureus* and *E. coli,* from the tainted water supply. Carbon nanotube filters are simple to clean using either an autoclave or ultrasonic cleaning.

6.10.4 Nano Ceramic Filters

Nano ceramic filters, made from a combination of positively charged micro glass and negatively charged nano alumina fiber, can capture and hold negatively charged particles. Nano ceramic filters provide excellent bacterial and viral removal performance. In terms of particle capacity and resistance to clogging, Nano ceramic filters excel. The Nano ceramic fillers can chemisorb dissolved heavy metals.

Nanoparticles in water should disappear after treatment. When used for environmental purposes, nanoparticles always leak into the surrounding medium. To evaluate the environmental dangers posed by these substances, it is important to comprehend their permanence, toxicity, bioavailability, and mobility. What will happen to terrestrial and aquatic species when these nanoparticles are released into the land and water is an open question. The question of how engineered nanoparticles can be removed from the urban water cycle is inevitable in light of their increasing prevalence in a wide variety of industrial settings and their prospective use in the purification of drinking water and wastewater.

Nanotechnology is a new area of study with potential applications across disciplines. Particularly, nano-remediation, which refers to the use of engineered materials for environmental cleanup, is a promising technique for dealing with persistent pollutants like pesticides, chlorinated solvents, halogenated chemicals, or heavy metals (Ganie et al., 2021; Guerra et al., 2018). In addition, nano-remediation is a long-term strategy for getting rid of new types of contaminants, like those found in pharmaceuticals and personal care items. Nanomaterials can be used in a wide range of media, from water and soil to air and even space.

6.11 QUALITY OF LIFE WHEN LIVING CLOSE TO A MUNICIPAL WASTEWATER TREATMENT PLANT

There is concern that the working conditions at a wastewater treatment plant (WTP) pose a significant health risk to both the WTP's employees and the surrounding community. Both aerobic and anaerobic WTPs have been found to harbor distinct bacterial and fungal populations. According to Vantarakis et al. (2016), the number of bacteria in WTP droplets is 10-1,000 times higher than in a water source. Temperature, wind speed, smog, and specific humidity all play a role in the dispersion of aerosols and the viability of airborne microorganisms (Grisoli et al., 2009). Assessment of airborne microorganism contamination in an industrial region characterized by an open composting facility and a wastewater treatment plant (Kruczek and Olańczuk-Neyman, 2004).

Vantarakis et al. (2016) looked into how close proximity to a sewage treatment facility affected residents' quality of life. 235 people who lived within a 500-metre radius of a municipal wastewater treatment facility (cases) and 97 people who lived in a separate region (controls) participated in a case-control study. Participants self-reported their perceptions of their own health and overall happiness using a standardized questionnaire. Aerosol samples were also collected near a wastewater treatment facility to examine the concentration of airborne harmful bacteria. Increased risk for respiratory and skin disorders was also noted, along with a significant risk for symptoms like headache, unusual weariness, and attention issues. Irritability and low mood seemed to be common symptoms. The number of cases reporting gastrointestinal problems was likewise significantly higher than that of the control group. Near the wastewater treatment plant, a high concentration of harmful airborne bacteria was detected during air monitoring. More in-depth epidemiological studies are required to identify the root cause and assess the disease burden for those who live near the wastewater treatment facility.

6.12 SMELLS COMING FROM SEWAGE TREATMENT FACILITIES (WWTPS) AND EFFECTS

A wide variety of human-made activities—from refining and petrochemical production to livestock farming, food processing, chemical manufacturing, and even sewage treatment plants—generate odors. Due to the development of companies and the absence of environmental protection legislation, this issue is particularly common in developing countries. At the same time, the introduction of EU environment and climate change policies show a growing concern for the quality of the environment by concentrating on improving air quality and decreasing GHG emissions. Odor emissions are viewed as a serious problem in the realm of air quality management. Wastewater treatment plants (WWTPs), sanitary landfills, composting operations, and similar facilities are major contributors to environmental odors. Historically, inorganic and organic pollutants in the influent were the primary focus of WWTP engineering and design (Zarra et al., 2019; Wysocka et al., 2019). Odors have the power to spoil your day and make you feel ill. Although there are many various origins of unpleasant odors, personal sewage treatment is one of the more frequent ones. The accumulation of organic materials can foster an atmosphere that is ideal for the growth of bacteria, which can give off some really unpleasant odors! Odors have major negative effects on human health, too. Headaches, dizziness, nausea, and vomiting can all result from breathing in offensive odors. In certain instances, it may even result in breathing difficulties and respiratory issues.

Large-scale water pollution is caused by human activities. People are irresponsibly wasting this natural resource, putting the very survival of humanity at risk. Wastewater odor is a global issue that needs immediate attention. Humans have a limited threshold for unpleasantness, and odor quickly crosses that threshold. The scent is recognized by the individual through the nervous system's physiological and psychological reaction. It hurts the natural environment. Humans are negatively impacted by wastewater's pungent odor.

Odorous chemical compounds are produced during wastewater collection and treatment due to the anaerobic breakdown of organic materials (Conti et al., 2020). Odors are produced by a synergistic interaction between many volatile chemical species that might elicit the olfactory response. That's because of the complex chemical reactions going on between many species of volatile chemicals, such as sulphides, mercaptans, ammonia, amines, and volatile organic compounds (VOCs) like esters, acids, aldehydes, ketones, and alcohols. The term "volatile organic compound" (VOC) refers to a broad class of chemicals that share a volatility trait, including a wide variety of fatty acids, alcohols, aldehydes, amines, carbonates, esters, sulphides, disulfides, mercaptans, and heterocyclic nitrogen compounds.

Wastewater treatment plants smell; in most cases, these odors are the result of the anaerobic decomposition of organic materials. The strong smell that most people associate with these facilities comes from hydrogen sulfide, a by-product of decomposition. Other compounds that cause odors include amines and mercaptans, which are both rich in sulfur.

The odors created at wastewater treatment plants are often likened to ammonia, rotten eggs, and various organic substances. In most cases, these smells are the result of the anaerobic decomposition of organic materials. The strong smell that most people associate with these facilities comes from hydrogen sulfide, a by-product of decomposition. Other compounds that cause odors include amines and mercaptans, which are both rich in sulfur. Unfortunately, these smells can be made worse unless the appropriate preventive steps are taken. Weather conditions, including heat, humidity, and high winds, can exacerbate the problem. Wastewater plants tend to receive most complaints during the summer months, but maintenance and control measures need to work well throughout the year.

6.13 REMEDIES

6.13.1 Lift Oxygen Limits

Ensuring the aerobic bacteria have enough oxygen to aerobically break down BOD requires a higher oxygen supply. The slowed rate of anaerobic digestion will aid with odor management.

6.13.2 Recycled Water from Waste

The anaerobic digestion that causes odor needs to be prevented, and this can only be done if dissolved oxygen is delivered all the way to the bottom of an aeration basin. Dissolved oxygen levels in the wastewater system can be made almost uniform through proper circulation. The result will be improved aerobic digestion without unpleasant odors.

Current aeration solutions are not always reliable for maximizing oxygen supply and ensuring uniform circulation. Dissolved oxygen levels in deeper water bodies are not improved by surface aerators because of their limited depth of circulation. Although they are effective at boosting dissolved oxygen, coarse and fine bubble diffusers have trouble eliminating hypoxic zones. With an oxygen transfer efficiency of up to 99 percent, the OxyMem MABR generates up to 50 percent less sludge than conventional methods, thus cutting down on odor.

Effluent flow from municipal and industrial sources is collected by treatment plants. Each plant requires a different amount of care, but they all have one thing in common: they all stink! Disease-causing microbes like those listed below may be acquired by contact with wastewater or its byproducts.

1. Gastroenteritis (diarrhea or vomiting)
2. Giardiasis and cryptosporidiosis (severe stomach cramps, diarrhea, or vomiting)
3. Hepatitis and other viral illnesses
4. Liver infections
5. Skin and eye infections

Hydrogen sulphide (H_2S), also known as “rotten egg” gas, is the most talked-about odiferous chemical related to wastewater. H_2S has an odor detection threshold in air of about 0.5 parts per billion (this is the lowest quantity at which humans can detect it). Inhaling or being exposed to concentrations as low as five times the scent detection limit can induce discomfort. H_2S at high quantities is deadly and can render a person unconscious after just a few breaths. Anaerobic (oxygen-free) reactions that involve Sulphur can also produce mercaptans (or thiols). There’s no doubt that wastewater treatment plants are a major source of noxious odorous substances, which can have serious consequences for human health at even trace concentrations, including throat and eye irritation, nausea, coughing, and even death at extremely high levels (Czarnota et al., 2023).

Airborne contaminants below the defined toxicity criteria are a common cause of odor emissions. Depending on the type of pollutant being discussed, there may or may not be a correlation between odor and toxicity. According to research (Piccardo et al., 2022), some hazardous substances do not cause olfactory nuisance while others do. Accordingly, it is always important to conduct a thorough analysis of the toxicity/odor nuisance connection, taking into account the unique properties of the airborne combination and the nature of the source.

Dosing wastewater with bacteria is an easy way to improve water quality without breaking the bank. Microorganisms are crucial to the treatment and purification of wastewater before it can be safely released into the environment. Wastewater treatment facilities may significantly minimize odor generation when treated with the right microbes. Bacteria break down organic materials at the source, resulting in lower BOD (biological oxygen demand), COD (chemical oxygen demand), and TSS (total suspended solids), and fewer or no odors being formed. Adding bacteria with the necessary metabolic characteristics is crucial for effective sludge and organic waste degradation.

Treatment of wastewater is a time-consuming and intricate procedure that must also meet a number of regulations (https://organicabiotech.com/odour-control-in-wastewater-treatment/). To eradicate the foul odors and the need to add toxic chemicals to wastewater that make it unsafe for the environment, it is necessary to address the problem at its source, which is not possible with chemical treatments alone. The moment has come to look to nature for answers, and to study its processes so that we can restore order without endangering any of the living beings inside it.

Aerosol inhalation has been linked to an increase in antibodies against Gram-negative bacteria and intestinal viruses in a small number of investigations. The virus that causes depression, fatigue, coughing, and fever that characterizes “the sewage worker’s syndrome” is responsible for these symptoms. Thorn et al. (2002) and Kruczalak and Olaczuk-Neyman (2004) describe the condition as having overall malaise, weakness, acute rhinitis, and fever, along with gastrointestinal symptoms.

6.14 SEWAGE WASTE ANALYSIS

Wastewater analysis is an important business operation, and all manufacturing facilities should keep close tabs on their wastewater levels. During the production or processing of goods, industrial companies frequently produce wastewater. Before being released to industrial estate, public water resources, or community resources, all industrial factories must conduct a Wastewater Analysis to ensure compliance with the water quality laws on legal standards, monitor the wastewater treatment system, and control the pollutants in the water within the scope of the Department of Industrial Works’ criteria (www.thaiparker.co.th/en/articles/laboratory-services/the-importance-of-wastewater-analysis) for more on the merits and significance of wastewater analysis.

6.14.1 Be Familiar with Water’s Characteristics

With the help of water quality monitoring, you may find out whether or not the wastewater from your plant satisfies the requirements set by the regulation standards, as well as learn more about the water’s components, impurities, and contaminants than you ever imagined possible.

6.14.2 Recognize the Origin of Pollution or the Rationale Behind Wastewater Production

Once the water properties are determined by the water quality monitor, several parameters will be displayed that will immediately point out the source of pollution or the cause of your industrial wastewater, such as polluted wastewater that contained chemicals that can be determined by COD (Chemical Oxygen Demand), wastewater in the form of suspended solids with high TDS (Total Dissolved Solids), or wastewater that is typically found in industrial plants, such as wastewater that contains chemicals that are toxic to aquatic life.

6.14.3 Formulate a Strategy for Enhancing Water Quality and Treating Wastewater

Once we know what's causing water pollution, thanks to the water quality test, we can start searching for ways to fix the problem and decide on the best wastewater treatment system, whether it's a physical treatment that removes contaminants like grease and oil or a biological one that uses bacteria to break down harmful substances. A biological wastewater treatment is one that uses microorganisms to break down contaminants in the wastewater and then gets rid of the resulting sludge by drying it off.

6.15 LOWERING THE LEVELS OF CONTAMINANTS IN SYSTEMS THAT USE RECYCLED WATER

Most industrial wastewater come from the chemical and petrochemical industry, the paper and pulp industry, the food processing industry, the tanning industry, and other manufacturing industries. Typical characteristics of these effluents include a high organic strength (1-200 g/L), a pH outside of neutrality, varying temperatures, salinity, turbidity, and a high concentration of heavy metals. High salt concentrations have been found in wastewater from sources including leather production, food processing and preservation, textile processing, and petroleum refining. It is difficult to categorize industrial wastewater since its composition varies depending on the chemicals employed in the upstream processes and the type of treatment it has undergone (Ahmed et al., 2022).

High removal efficiency for several hazardous contaminants has emerged from recent advancements in wastewater treatment technologies (Mutami et al., 2013). However, some technologies are better suited than others for zeroing in on certain pollutants. Recent developments in the use of membrane technologies, adsorption, Fenton-based processes, advanced oxidation processes (AOP), and hybrid systems like electrically-enhanced membrane bioreactors (eMBRs) and integrated eMBR-adsorption systems for the removal of hazardous pollutants from industrial wastewater have been reviewed in detail by Ahmed et al. (2022). These Scientists emphasized the benefits and drawbacks of each technology, compared their essential features, and discussed the most recent shifts in traditional treatment approaches and developments in advanced treatment systems. Ahmed et al. (2022) also cover the process of removing pharmaceuticals and other new toxins from wastewater.

Chemical precipitation, flotation, and ion exchange are the traditional procedures for extracting heavy metals from wastewater. However, these methods generate hazardous sludge, use excessive amounts of energy, and have low removal efficiencies, all of which restrict their use. To improve the quality of the treated effluent, many other alternative treatment procedures have been studied recently. Some examples are membrane separation, electro-technologies, photocatalytic processes, and adsorption employing inexpensive materials. Wastewater with significant concentrations of heavy metals is often treated using adsorption or membrane separation.

6.16 CONCLUSION

The study of water's physics, chemistry, and biology is fascinating, as it's essential for human survival, transportation, industry, and recreation. Contamination from point and diffuse sources poses a threat to human health. Filtration techniques improve water quality by removing undesirable materials, improving flavor, fragrance, and clarity. Wastewater treatment facilities are crucial for maintaining clean, safe water for humans and ecosystems. They must cleanse wastewater from cities, businesses, and industries, which contains high concentrations of pollutants, human waste, environmental contaminants, and ecosystem-damaging substances. Maintaining treatment systems and using top-notch equipment can increase a plant's capacity to filter chemicals and pollutants. On-site treatment is typically used for wastewater from refineries, manufacturing plants, and industrial effluent, making wastewater treatment a crucial environmental protection measure that should be supported internationally.

Treatment plants are crucial for releasing clean, healthy water into the environment and surrounding communities. The introduction of industrial waste into freshwater bodies reduces the supply of unpolluted water. High-quality equipment and efficient treatment systems can improve plant efficiency. The development of technologies for industrial wastewater treatment, genetically engineered cultures, and green regeneration techniques is essential for meeting growing water demands and providing water security. Wastewater treatment plants are essential for maintaining safe, healthy water for communities and environmental habitats. Highly toxic industrial waste contaminants, which can be odorless and difficult to detect, require chemical and biological treatment to prevent undetected leakage into groundwater.

REFERENCES

Ahmed, M., M.O. Mavukkandy, A. Giwa, E. Maria, K. Evina, K. Olaf, Vincenzo, and W.H. Shadi. 2022. Recent developments in hazardous pollutants removal from wastewater and water reuse within a circular economy. *npj Clean Water*. 5: 12. https://doi.org/10.1038/s41545-022-00154-5

Amin, M. M., H. Hashemi, B. Bina, A. Ebrahimi, H. R. Pourzamani, and A. Ebrahimi. 2014. Environmental pollutants removal from composting leachate using anaerobic biological treatment process. *Int. J. Health Sys. Dis. Manage*. 2: 136–141. www.ijhsdm.org.

Aziz, S.Q., and J.S. Mustafa. 2019. Step-by-step design and calculations for water treatment plant units. *Adv. Environ. Biol.* 13(8): 1–16. https://doi.org/10.22587/aeb.2019.13.8.1

Chen, C.-J., X.-X. Huang, C.-X. Lei, W.-J. Zhu, Y.-X. Chen and W.-X. Wu. 2012. Improving anammox start-up with bamboo charcoal. *Chemosphere*. 89(10): 1224–1229. https://doi.org/10.1016/j.chemosphere.2012.07.045

Chen, G. 2004. Electrochemical technologies in wastewater treatment. *Separation Purification Technol.* 38: 11–41.

Chen, J., J. Luo, Q. Luo, and Z. Pang. 2018. *Wastewater Treatment: Application of New Functional Materials*. Walter de Gruyter GmbH & Co KG.

Chen, T., T. Peng, C. Feng, N. Chen, Q. Hu, and C. Hao. 2016. The feasibility of an up-flow partially aerated biological filter (U-PABF) for nitrogen and COD removal from domestic wastewater. *Bioresource. Technol.* 218: 307–317.

Chen, Z., W. Wei, H. Chen, and B.J. Ni. 2022. Recent advances in waste-derived functional materials for wastewater remediation. *Eco-Environ. Health*. 1(2): 86–104. https://doi.org/10.1016/j.eehl.2022.05.00

Conti, C., M. Guarino, and J. Bacenetti. 2020. Measurements techniques and models to assess odor annoyance: A review. *Environ. Int.* 134: 105261.

Czarnota, J., A. Masło´n, and R. Pajura. 2023. Wastewater treatment plants as a source of malodorous substances hazardous to health, including a case study from Poland. *Int. J. Environ. Res. Public Health*. 20: 5379. https://doi.org/10.3390/ijerph20075379

Desai, C., K.R. Jain, R. Boopathy, E.D. van Hellebusch, and D. Madamwar. 2021. Editorial: Eco-sustainable bioremediation of textile dye wastewaters: Innovative microbial treatment technologies and mechanistic insights of textile dye biodegradation. *Front. Microbiol.* 12: 707083. https://doi.org/10.3389/fmicb.2021.707083

Ding, J., Y. Meng, S. Lu, Y. Peng, W. Yan, W. Li, J. Hu, T. Ye, Y. Zhong, and H. Zhang. 2023. The treatment of aquaculture wastewater with biological aerated filters: From the treatment process to the microbial mechanism. *Toxics.* 11(6): 478. https://doi.org/10.3390/toxics11060478

Epelle, E.I., P.U. Okoye, S. Roddy, B. Gunes, and J.A. Okolie. 2022. Advances in the applications of nanomaterials for wastewater treatment. *Environments.* 9: 141. https://doi.org/10.3390/environments9110141

Feng, A., X. Zhu, Y. Chen, P. Liu, F. Han, Y. Zu, X. Li, and P. Bi. 2024. Functional biomass-derived materials for the development of sustainable batteries. *ChemElectrochem.* 13: e202400086.

Ganie, A.S., S. Bano, N. Khan, S. Sultana, Z. Rehman, M.M. Rahman, F. Sabir, Foulon, and M.Z. Khan. 2021. Nanoremediation technologies for sustainable remediation of contaminated environments: Recent advances and challenges. *Chemosphere.* 275: 130065. https://doi.org/10.1016/j.chemosphere.2021.130065

Gonzalez-Perez, A., and K.M. Persson. 2016. Bioinspired materials for water purification. *Materials* (Basel). 9(6): 447. https://doi.org/10.3390/ma9060447

Grisoli, P., M. Rodolfi, S. Villani, E. Grignani, D. Cottica, A. Berri, and C. Dacarro. 2009. Assessment of airborne microorganism contamination in an industrial area characterized by an open composting facility and a wastewater treatment plant. *Environ Res.* 109(2): 135–142.

Guerra, F.D., M.F. Attia, D.C. Whitehead, and F. Alexis. 2018. Nanotechnology for environmental remediation: Materials and applications. *Molecules.* 23: 1–23. https://doi.org/10.3390/molecules23071760

Jain, K., A.S. Patel, V.P. Pardhi, and S.J.S. Flora. 2021. Nanotechnology in wastewater management: A new paradigm towards wastewater treatment. *Molecules* 26: 1797. https://doi.org/10.3390/molecules26061797

Khan, N., Z.A. Tabasi, J. Liu, B.H. Zhang, and Y. Zhao. 2022. Recent advances in functional materials for wastewater treatment: From materials to technological innovations. *J. Mar. Sci. Eng.* 10(4): 534. https://doi.org/10.3390/jmse10040534

Kruczek, K., and K. Olańczuk-Neyman. 2004. Microorganisms in the air over wastewater treatment plants. *Polish J. Environ. Stud.* 13(5): 537–542.

Kumar, V., M. Sharma, S. Sondhi, K. Kaur, D. Sharma, S. Sharma, and D. Utreja. 2023. Removal of inorganic pollutants from wastewater: Innovative technologies and toxicity assessment. *Sustainability.* 15: 16376. https://doi.org/10.3390/su152316376

Kunduru, K.R., M. Nazarkovsky, S. Farah, R.P. Pawar, A. Basu and A.J. Domb. 2017. Nanotechnology for water purification: Applications of nanotechnology methods in wastewater treatment. *Water Purification.* https://doi.org/10.1016/B978-0-12-804300-4.00002-2

Li, R., Y. Wu, X. Lou, H. Li, J. Cheng, B. Shen, and L. Qin. 2023. Porous biochar materials for sustainable water treatment: Synthesis, modification, and application. *Water.* 15: 395. https://doi.org/10.3390/w15030395

Li, Z-H., L. Yuan, S-X. Gao, L. Wang, and G-P. Sheng. 2019. Mitigated membrane fouling and enhanced removal of extracellular antibiotic resistance genes from wastewater effluent via an integrated pre-coagulation and microfiltration process, *Water Res.*, 159: 145–152. https://doi.org/10.1016/j.watres.2019.05.005

Makisha, N. 2021. Application of biofilm carrier in aerobic reactors as a method to improve quality of wastewater treatment. *Hydrology.* 8: 77. https://doi.org/10.3390/hydrology.8020077

Mathew, M.M., K. Khatana, V. Vats, R. Danker, R. Kumar, H.U. Dahms, and J.S. Wang. 2022. Biological approaches integrating algae and bacteria for the degradation of wastewater contaminants—A review. *Front. Microbiol.* 12: 801051. https://doi.org/10.3389/fmicb.2021.801051

Mbarek, W.B., L. Escoda, J. Saurina, E. Pineda, F.M. Alminderej, M. Khitouni and J-J. Suñol. 2022. Nanomaterials as a sustainable choice for treating wastewater: A review. *Materials.* 15: 8576. https://doi.org/10.3390/ma15238576

Molazadeh, M., H. Ahmadzadeh, H.R. Pourianfar, S. Lyon, and P.H. Rampelotto. 2019. The use of microalgae for coupling wastewater treatment with CO_2 bio fixation. *Front. Bioeng. Biotechnol.* 7: 42. https://doi.org/10.3389/fbioe.2019.00042

Mustafa, S., H.N. Bhatti, M. Maqbool, and M. Iqbal. 2021. Microalgae biosorption, bioaccumulation and biodegradation efficiency for the remediation of wastewater and carbon dioxide mitigation: Prospects, challenges and opportunities. *J. Water Process Eng.* 41: 102009. https://doi.org/10.1016/j.jwpe.2021.102009

Mutamim, N.S.A., Z.Z. Noor, M.M.A. Hassan, A. Yuniarto and G. Olsson. 2013. Membrane bioreactor: Applications and limitations in treating high strength industrial wastewater. *Chem. Eng. J.* 225: 109–119.

Nagar, A., and T. Pradeep. 2020. Clean water through nanotechnology: Needs, gaps, and fulfillment. *ACS Nano.* 14(6):6420–6435. doi: 10.1021/acsnano.9b01730.

Piccardo, M.T., M. Geretto, A. Pulpier, and A. Izzotti. 2022. Odor emissions: A public health concern for health risk perception. *Environ. Res.* 204(Part B): 112121. https://doi.org/10.1016/j.envres.2021.112121

Poerio, T., Piacentini, E and R. Mazzei. 2019. Membrane processes for microplastic removal. *Molecules.* 24: 4148. https://doi.org/10.3390/molecules24224148

Ren, J., W. Cheng, T. Wan, M. Wang and C. Zhang. 2018. Effect of aeration rates and filter media heights on the performance of pollutant removal in an up-flow biological aerated filter. *Water.* 10(9): 1244. https://doi.org/10.3390/w10091244

Saini, G., S. Kalra, and U. Kaur. 2021. The purification of wastewater on a small scale by using plants and sand filter. *Appl. Water Sci.* 11: 1–6.

Saleem, H., and S.J. Zaidi. 2020. Developments in the Application of nanomaterials for water treatment and their impact on the environment. *Nanomaterials.* 10: 1764. https://doi.org/10.3390/nano10091764

Tan, J., W. Li, C. Ma, Q. Wu, Z. Xu, and S. Liu. 2018. Synthesis of honeycomb-like carbon foam from larch sawdust as efficient absorbents for oil spills cleanup and recovery. *Materials.* 11: 1106.

Taneja, A., R. Sharma, S. Khetrapal, A. Sharma, R. Nagraik, B. Venkidasamy, M.N. Ghate, S. Azizov, S. Sharma and D. Kumar. 2023. Value addition employing waste bio-materials in environmental remedies and food sector. *Metabolites* 13: 624. https://doi.org/10.3390/metabo13050624

Thompson, K.A., K.K. Shimabuku, J.P. Kearns, D.R.U. Knappe, R.S. Summers, and S.M. Cook. 2016. Environmental comparison of biochar and activated carbon for tertiary wastewater treatment. *Environ. Sci. Technol.* 50: 11253–11262.

Thorn, J., L. Beijer, and R. Rylander. 2002. Work related symptoms among sewage workers: A nationwide survey in Sweden. *Occupational Environ. Med.* 59(8): 562–566.

Vantarakis, A., S. Paparrodopoulos, P. Kokkinos, G. Vantarakis, K. Fragou, and I. Detorakis. 2016. Impact on the quality of life when living close to a municipal wastewater treatment plant. *J. Environ. Public Health.* 2016(2): Article ID 8467023: 8. https://doi.org/10.1155/2016/8467023

Vishnu, D., B. Dhandapani, S. Authilingam, and S.V. Sivakumar. 2022 A comprehensive review of effective adsorbents used for the removal of dyes from wastewater. *Curr. Anal. Chem.* 18: 255–268.

Wang, R., Q. Xu, C. Chen, X. Li, C. Zhang, and D. Zhang. 2021. Microbial nitrogen removal in synthetic aquaculture wastewater by fixed-bed baffled reactors packed with different biofilm carrier materials. *Bio Resource. Technol.* 331: 125045.

Wang, S., C. Hu, F. Cheng, and X. Lu. 2023. Performance of a combined low-consumption bio treatment system with cost-effective ecological treatment technology for sewage treatment. *J. Water Process Eng.* 51:103380.

Wei, D., B. Li, H. Huang, L. Luo, J. Zhang, Y. Yang, J. Guo, L. Tang, L. Tang, G. Zeng, and Y Zhou. 2018. Biochar-based functional materials in the purification of agricultural wastewater: Fabrication, application and future research needs. *Chemosphere.* 197: 165–180.

Wu, Q., C. Chen, Y. Zhang, and P. Tang. 2023. Safe purification of rural drinking water by biological aerated filter coupled with ultrafiltration. *Sci. Total Environ.* 868: 161632. https://doi.org/10.1016/j.scitotenv.161632

Wysocka, I., J. Gębicki, and J. Namieśnik, J. 2019. Technologies for deodorization of malodorous gases. *Environ. Sci. Pollut. Res.* 26: 9409–9434.

Xu, H., B. Yang, L. Yanbiao, X. Song, and W. Sand. 2022. Recent advances on the treatment of domestic wastewater by biological aerated filter. In: Shah, M., S. Rodriguez-Couto, and J Biswas (eds.) *An Innovative Role of Biofiltration in Wastewater Treatment Plants (*WWTPs) (pp. 155–170). Elsevier. https://doi.org/10.1016/B978-0-12-823946-9.00023

Yunus I.S., H.A. Kurniawan, D. Adityawarman, and A. Indarto. 2012. Nanotechnologies in water and air pollution treatment. *Environ. Technol. Rev.* 1: 136–148. https://doi.org/10.1080/21622515.2012.733966

Zarra, T., M.G. Galang, F. Ballesteros, V. Belgiorno, and V. Naddeo. 2019. Environmental odour management by artificial neural network—A review. *Environ. Int.* 133: 105189.

Zarkov, A., A. Kareiva, and L. Tamasauskaite-Tamasiunaite. 2021. Advances in functional inorganic materials prepared by wet chemical methods. *Crystals*. 11: 943.

Zhang, Y., Y. Hong, and X. Wang. 2023. Recent advances on using functional materials to increase the pollutant removal capabilities of microalgae and bacteria: Especially for their symbiotic systems. *Curr. Pollution Rep*. https://doi.org/10.1007/s40726-023-00259-6

Zhou, C., and Y. Wang 2020 .Recent progress in the conversion of biomass wastes into functional materials for value-added applications. *Sci. Technol. Adv. Mater.* 21, 787–804. DOI: 10.1080/14686996.2020.1848213

7 Consequences of Heavy Metal Pollution on Ecosystems and Populations

7.1 INTRODUCTION

The planet is seriously threatened by air, land, and water pollution as a result of various human activities. Particularly, water contamination has had detrimental effects on the ecosystem. The creation of massive amounts of wastewater has placed a great deal of strain on humanity, in addition to the scarcity of water resources brought on by drought and abuse. Water pollutants come in a variety of forms. The inorganic pollutants are the most dangerous because of their high toxicity and lack of biodegradability. Generally speaking, heavy metals are non-essential elements that are also harmful in small doses. Due to their potentially disastrous effects on human health and the environment, these metals have been the subject of extensive study. Because of their toxicity and persistence in the environment, they are well recognized as a leading cause of pollution. Heavy metal concentrations in many ecosystems have increased in comparison to their respective natural background levels as a result of industrialization, urbanization, and agricultural activities (Hadia-e-Fatima and Ahmed, 2018). The bio magnification and bioaccumulation of heavy metals in foods like fresh vegetables and fruits pose serious dangers to human health, including cancer, kidney dysfunction, immune system imbalance, and even mortality. The poisoning of the environment with heavy metals is a major problem because of the dangers it poses to human, animal, and plant life. Heavy metals, unlike other metals and metalloids, are not degraded or destroyed in the Earth's crust because of their persistent and stable nature. Due to their bioaccumulative properties, these heavy metals and metalloids have the potential to infiltrate plants, animals, and humans over time via atmospheric deposition, water ingestion, and the chain reaction of the food web.

Heavy metal poisoning of soils, water, and crops is a severe global environmental concern caused by improper waste disposal. Metals with a specific gravity greater than 5 g/cm^3 are considered heavy metals. Copper (Cu), nickel (Ni), chromium (Cr), lead (Pb), cadmium (Cd), mercury (Hg), iron (Fe), and arsenic (As) are the most prevalent ambient heavy metals. Iron and nickel, among other heavy metals, are required in minute quantities for the maintenance of all known forms of life. Heavy metals are harmful to living creatures not only when present in large quantities, but also when present in minute quantities, as is the case with lead, cadmium, and mercury. People who eat plants and other foods cultivated in contaminated soil may be at increased risk of developing metabolic disorders because of these substances.

Heavy metals, which constitute a serious threat to the environment, tend to accumulate in the Earth's crust's mantle. Heavy metals include mercury (Hg), cadmium (Cd), chromium (Cr), nickel (Ni), lead (Pb), and many more because they are chemical elements that are metallic with a relatively larger density and are hazardous over a safe range (Nkwunonwo et al., 2020). Zinc, lead, iron, and copper are all heavy metals that can be found naturally and play important roles in maintaining

DOI: 10.1201/9781003618164-7

ecological stability. However, these metals pollute the food chain and constitute a source of toxicity for humans and the overall ecological function as a result of their bioaccumulation, degradability, and excessive levels. In the fields of environmental science and geochemistry, this is one of the most pressing problems of the day. When it comes to maintaining cellular equilibrium, heavy metals are crucial players. Heavy metal contamination is a worldwide problem that poses significant threats to human health and the environment (Sarker et al., 2022). However, these metals cause numerous harmful impacts on the environment and have devastating consequences for all forms of life. It has come to light in recent years that exposure to some heavy metal pollution can have negative impacts on crop quality, with knock-on effects on food security and human health (Jessica et al., 2020). Natural foods were found to contain traces of metals such as chromium, cadmium, copper, lead, and mercury. Singh and Kumar (2006) found evidence that these substances were environmental pollutants in naturally occurring foods. However, some heavy metals are toxic to both ecosystems and individual species. Heavy metals can enter the human body through the lungs, the digestive system, or the skin. In addition, heavy metal concentrations in live cells determine the relative importance of their positive and negative effects (Munir et al., 2022). Because low quantities of metal ions can lead to nutritional deficiencies, while high concentrations can be harmful, it is important to keep them within a safe range.

7.2 HEAVY METALS

A heavy metal is any chemical element with a high density and low-concentration toxicity or poisonousness. Heavy metals include elements such as mercury, cadmium, arsenic, chromium, thallium, and lead (Pb). Metals with a high density compared to water are considered heavy metals. According to the density criteria, heavy metals are those with a value of more than 5 g/cm^3. Cobalt, copper, zinc, and manganese are all examples of heavy metals that the human body needs in very small amounts, but too much of any of them can be dangerous. Because of the assumed relationship between weight and toxicity, metalloids like arsenic, which can cause toxicity even at low doses, are also included in the category of heavy metals. In recent years, environmental poisoning by these metals has become an increasingly urgent ecological and worldwide public health issue. The exponential expansion in their use has led to an increase in human exposure in a variety of settings, including manufacturing, agriculture, households, and technology. The crust of our planet contains heavy metals in their natural forms. They are indestructible (www.lenntech.com/processes/heavy/heavy-metals/heavy-metals.htm) and cannot be broken down. They are ingested, in trace amounts, via food, water, and air. Some heavy metals, like copper, selenium, and zinc, play crucial roles in human metabolism as trace elements. However, they can be poisonous if consumed in large quantities. Some potential origins of heavy metal poisoning include contaminated drinking water (from, for example, lead pipes), high ambient air concentrations close to emission sources, and food ingestion. Bioaccumulation is a major problem with heavy metals. When a chemical's concentration in a living thing rises over time, relative to its environmental concentration, this process is called bioaccumulation. When substances are taken in and stored by an organism faster than they are digested or expelled, an accumulation occurs. Industrial and consumer waste, as well as acid rain's breakdown of soils and subsequent release of heavy metals into waterways, are all potential entry points for these contaminants. Contamination can be defined as the presence of any metal (or metalloid) species in an undesirable location, or in a sufficient quantity to harm human health or the natural environment (Singh et al., 2011). Due to their carcinogenic and poisonous properties, the detection of heavy metals in air, water, and soil is of utmost relevance. Most scientists throughout the world are reporting on the determination and health implications of heavy metals in the environment because of their toxicity and carcinogenic tendency.

7.3 ORIGINS OF POLLUTANT METALS

Natural heavy metals are not susceptible to biological degradation or destruction. Because of their inorganic nature, metal ions are essential to the growth and maintenance of all forms of life. Concern over the deteriorating quality of the environment led to a trace element designation for the elements that occur in low amounts in natural biological systems (Asati et al., 2016; Das and Poater, 2021; Pande et al., 2022). The fundamental ingredients of plant, animal, and human life can be broken down into two categories: major elements and trace elements (including toxic elements). Figures 7.1a, 7.1b, and 7.2 show the sources of heavy metal contamination in water, agriculture, and aquaculture, as well as the toxicity of the metals to plants.

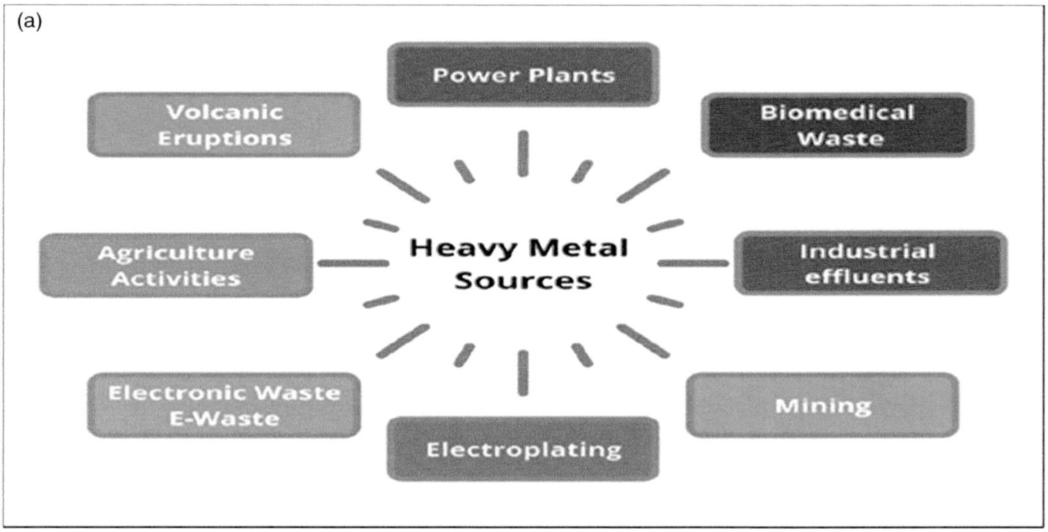

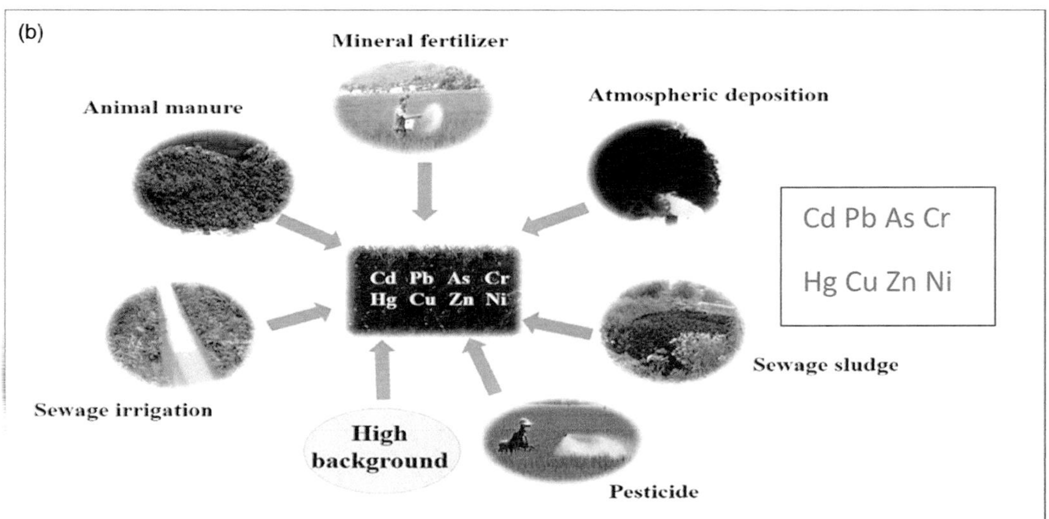

FIGURE. 7.1a Heavy metal contamination in water and aquaculture and their sources (Sonone et al., 2021, https://creativecommons.org/licenses/by/4.0/.)

FIGURE 7.1b Sources of Heavy Metals in Agricultural Soils (Wan et al., 2024, Open Access.)

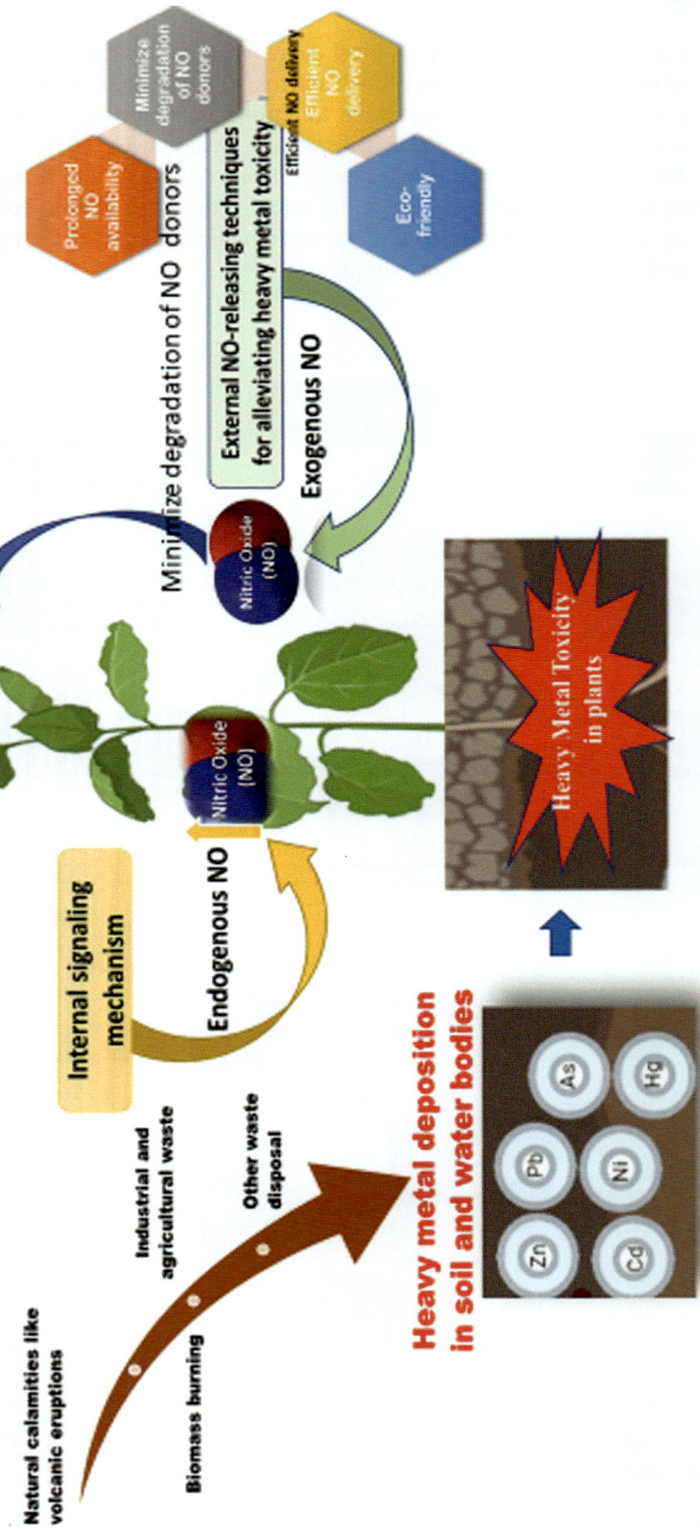

FIGURE 7.2 Heavy metal toxicity in plants (Pande et al., 2022. http://creativecommons.org/licenses/by/4.0.)

7.3.1 Metals of Essential Importance

Even while some heavy metals (Fe, Cu, and Zn) are necessary for plant and animal life, their availability in the medium fluctuates, and metals like Cu, Zn, Fe, Mn, Mo, Ni, and Co are critical micronutrients, whose uptake in excess to the plant's requirements result in toxic effects (Asati et al., 2016). Heavy metal concentrations in terrestrial plants, include As (0.02–7.0), Cd (0.1–2.4), Hg (0.005–0.02), Pb (1.1–13.3), Sb (0.02–0.06), Co (0.5–1.0), Cr (0.2–1.1), Fe (4.1–14.15), Mn (15–100), Mo (1.0–10.0), Ni (1.1–8.0), Sr (0.3–100), and Zn (8–100) in g g-1 dry weight.

Heavy metals have been split into two categories—essential and non-essential—based on research conducted by Kim et al. (2019). Essential Heavy metals are coenzymes in biological processes and are less harmful at low quantities. Iron is present in hemoglobin and myoglobin; cobalt is present in vitamin B12. In addition to being non-biodegradable and causing severe toxicity to living creatures, non-essential heavy metals are also very hazardous, even at very low quantities.

7.3.2 Heavy Metals Used as Examples

Lead, a metal and one of the densest elements known to man. It has an atomic mass of 207.2 and an atomic number of 82. The storage of extremely corrosive compounds typically necessitates the use of this metal (https://study.com/academy/lesson/heavy-metals-definition-examples.html). Copper, a ductile and easily worked metal with high electrical conductivity. It has an atomic number of 29 and an atomic weight of 63 u. The atomic number and weight of mercury, a metal, are 80 and 200 u, respectively. Its name, "quicksilver," comes from the way it looks and how quickly it moves. Group 15 and Period 4 on the periodic table contain the metalloid element arsenic. Both metallic and nonmetallic properties can be found in this metalloid. The number of its atoms is 33, and its mass is 74.92 u.

7.4 HEAVY METALS IN DRINKING WATER CAN CAUSE HEALTH PROBLEMS

Humans are at risk from heavy metal poisoning in drinking water, which can lead to serious health problems like cancer and organ damage. Heavy metals' toxic effect on human health is shown in Figure 7.3.

Heavy metals are invisible to the human eye but can be discovered with a simple water test. While exposure to heavy metals can have long-term negative health impacts, you may not notice any symptoms right away (www.kent.co.in/blog/harmful-effects-of-heavy-metal-contamination-in-drinking-water/). Heavier metals are those with a high density, specific gravity, or atomic weight, all of which make them poisonous. They occur naturally in the Earth's crust, but their geochemical and biochemical balance has been severely altered by human activity, leading to their introduction into our drinking water. Cobalt, copper, zinc, and manganese are all examples of heavy metals that the human body needs in very small amounts, but too much of any of them can be dangerous. The danger posed by a metal may vary depending on factors such as the amount ingested and the route of exposure. However, certain heavy metals are of extreme concern because of the damage they can do to various organs at even minute concentrations. Mercury, lead, cadmium, arsenic, nickel, chromium, and zinc contamination in drinking water pose serious health problems for the general public and medical professionals.

Heavy metals' deleterious effects on the human body are detailed. The body is unable to excrete these metals, so they build up over time. It has no immediate effect on the body but can cause serious long-term problems, many of which affect the brain (www.kent.co.in/blog/harmful-effects-of-heavy-metal-contamination-in-drinking-water/). It has the potential to impair mental and CNS performance. The lungs, liver, kidneys, and other important organs are all negatively impacted. Heavy metals have been linked to cancer and are therefore classified as carcinogens. Heavy metals

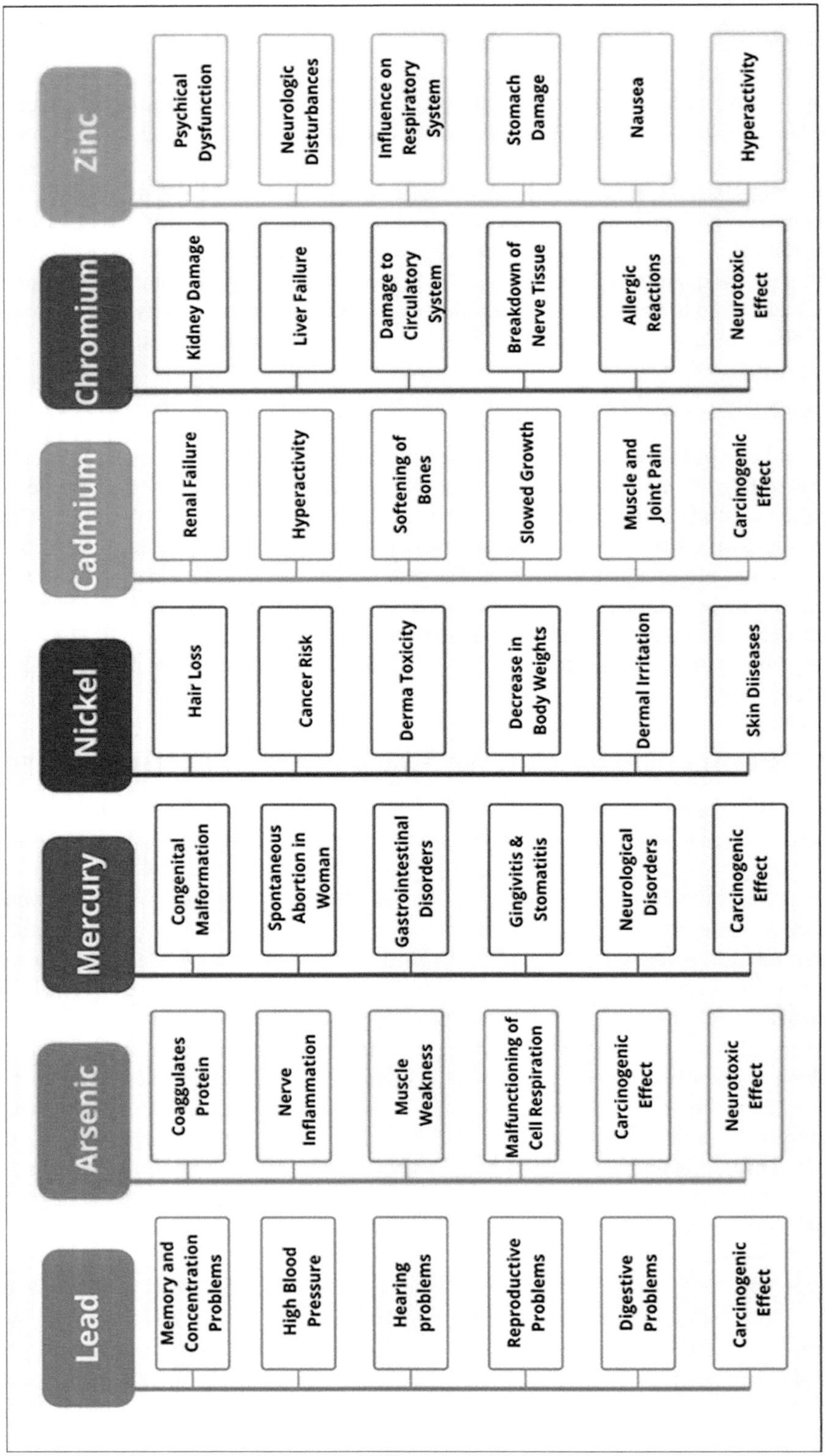

FIGURE 7.3 Heavy metal toxic effects on human health (Sonone et al., 2021, https://creativecommons.org/licenses/by/4.0/)

have a more negative impact on the health of young children because their bodies are still in the process of maturation. Children's neural systems are particularly vulnerable to the long-term effects of heavy metal exposure, which can include learning and memory impairments, as well as aggression and hyperactivity. The list below provides the specific ways in which certain metals are hazardous to human health.

7.4.1 Mercury

Acid rain, burning fossil fuels, and the disposal of municipal, industrial, and mining waste all contribute to the leaching of soil, where it can then enter waterways. Mercury in drinking water is harmful to the kidneys and the neurological system.

7.4.2 Lead and Cadmium

Corrosion of ageing pipes is a common source of lead in drinking water. Kidney and neurological system damage, as well as cognitive impairment, have all been linked to lead in drinking water. Cadmium (Cd) is a metal that can be found in low concentrations in acidic water and low total dissolved solids (TDS) environments, such as those found in surface and groundwater. Consistent exposure to cadmium at concentrations above the MCL (Maximum Contaminant Level) can be harmful to the kidneys, bones, liver, and blood, according to the Water Quality Association.

7.4.3 Toxic Metals: Arsenic and Nickel

When it comes to adult heavy metal poisoning, arsenic is by far the most common culprit. The production of chemicals and glassware, as well as the refining of copper, zinc, and lead, all release arsenic into the environment. Arsenic poisoning in drinking water has been linked to cancer, as well as damage to the skin, liver, and eyes.

Nickel is an alloying metal that is hard and silvery white in color. It is highly conductive to both heat and electricity and is highly resistant to corrosion. Nickel is also an important component of several alloying agents, including stainless steel. Nickel is an essential nutrient for many organisms, but it may also be poisonous and even carcinogenic in large quantities. Nickel allergies affect more women than men. Contact with skin may result in dermatitis.

7.4.4 Chromite

Many foods contain chromium because it is a trace mineral. Use of carbohydrates and fats by the body for energy production requires the mineral chromium. Diabetes and cardiovascular disease are two of the complications that might result from not getting enough chromium. It is unclear how trivalent chromium works to regulate sugar and fat metabolism. Only in hospitalized patients who are unable to consume solid food and who receive long-term intravenous nourishment have cases of trivalent chromium insufficiency been reported.

7.4.5 Zinc

All living things require the trace metal zinc to thrive. Collagen, insulin, and testosterone can't be made without zinc, and neither can vitamin A, iron, or copper. The immune system also relies on zinc to work effectively. Due to the body's ability to regulate zinc absorption, zinc insufficiency is extremely uncommon in healthy people. Zinc insufficiency, however, can develop when either the overall food intake is insufficient or intestinal absorption is diminished due to disease (Crohn's, celiac).

7.5 BIOACCUMULATION AND TOXIC EFFECTS OF HEAVY METALS

One of the most pressing problems facing contemporary humanity is contamination of the natural environment. Heavy metals pose a severe hazard to the environment through pollution and contamination. Heavy metals have entered the environment as a result of rapid industrialization and urbanization, and their mobilization and transport rates in the environment have dramatically increased since the 1940s. Weathering of metal-containing rocks and volcanic eruptions are among their natural environmental origins, whereas industrial emissions, mining, smelting, and agricultural activities, including the application of pesticides and phosphate fertilizers, are among their primary human origins. Heavy metals like cadmium (Cd) are released into the environment as a result of the combustion of fossil fuels (Ali and Khan, 2017; Khan et al., 2004). Toxic heavy metals remain in the environment, pollute food supplies, and contribute to a variety of health issues. Heavy metals in the environment pose a serious risk to living things if they are exposed to them over long periods.

Heavy Metals (HMs) and metalloids are agricultural soil pollutants since they have the potential to harm crop health and production if they are present in the soil at high concentrations. HMs are resistant to degradation, and if plants do not absorb them or leach them out, they can accumulate in the soil and last for a very long time. Anthropogenic sources, including but not limited to the nonferrous metallurgical industry, mining, mineral processing, electroplating, leather tanning, and chemical industries, are commonly attributed to the discharge of hazardous metals such as HMs. High quantities of elements that commonly contaminate agricultural soils and have detrimental effects on plants include Ni, Cd, Hg, Cu, Cr, As, Pb, and Zn. Anthropogenic, industrial, mining, and mineral processing are commonly attributed to the discharge of hazardous metals such as HMs, and enter humans through the food chain, as depicted in Figure 7.4 (Abd Elnabi et al., 2023).

The soil has already absorbed the majority of these heavy metals. On the other hand, long-term exposure to heavy metals can cause lung cancer and bone fractures in the human body. Regarding the use of ordinary foodstuffs like fruits and vegetables that have been tainted with heavy metals, human health concerns have grown significantly over the past few decades. Cd and Pb are rapidly absorbed by plants and introduced into the food chain and causing toxicity to all organisms, including plants, animals, and even humans. The toxicity, environmental endurance, and bioaccumulate characteristics of heavy metals make them well-known environmental contaminants. The weathering of igneous rocks containing metals and volcanic eruptions are two examples of their natural origins, while mining and other anthropogenic activities provide additional examples of their anthropogenic origins. An increase in the mobilization of these elements in the environment and disturbance of their biogeochemical cycles has resulted from mining and industrial processing for the extraction of mineral resources and their subsequent applications for industrial, agricultural, and economic development. It is a public health issue because poisonous heavy metals have contaminated both aquatic and terrestrial environments (Ali et al., 2017, 2019). Heavy metals are persistent contaminants that can be found in the environment and have a domino effect, contaminating all the food sources in their path. Heavy metals can accumulate in biota, posing a risk to the health of anyone who consumes them. Focusing on environmental permanence, toxicity to living creatures, and bioaccumulative potential, this chapter provides a thorough discussion of the various features of heavy metals as hazardous elements. Bioaccumulation of these elements and their effects on human health are explored, with fish, rice, and cigarettes receiving particular attention. Ali et al.'s (2019) work will provide a significant pedagogical resource for environmental scientists. Some of the most harmful heavy metals and metalloids to the environment are chromium, nickel, copper, zinc, cadmium, lead, mercury, and arsenic. These components are extremely vital to animal and human health, and their trophic transmission in aquatic and terrestrial food chains/webs is crucial. The amounts of potentially harmful heavy metals and metalloids in various environmental compartments and the native biota must be evaluated and monitored. Steps should be taken to reduce the negative effects of dangerous heavy metals and metalloids on human health and the environment, according to a review of the relevant environmental chemistry and ecotoxicology literature.

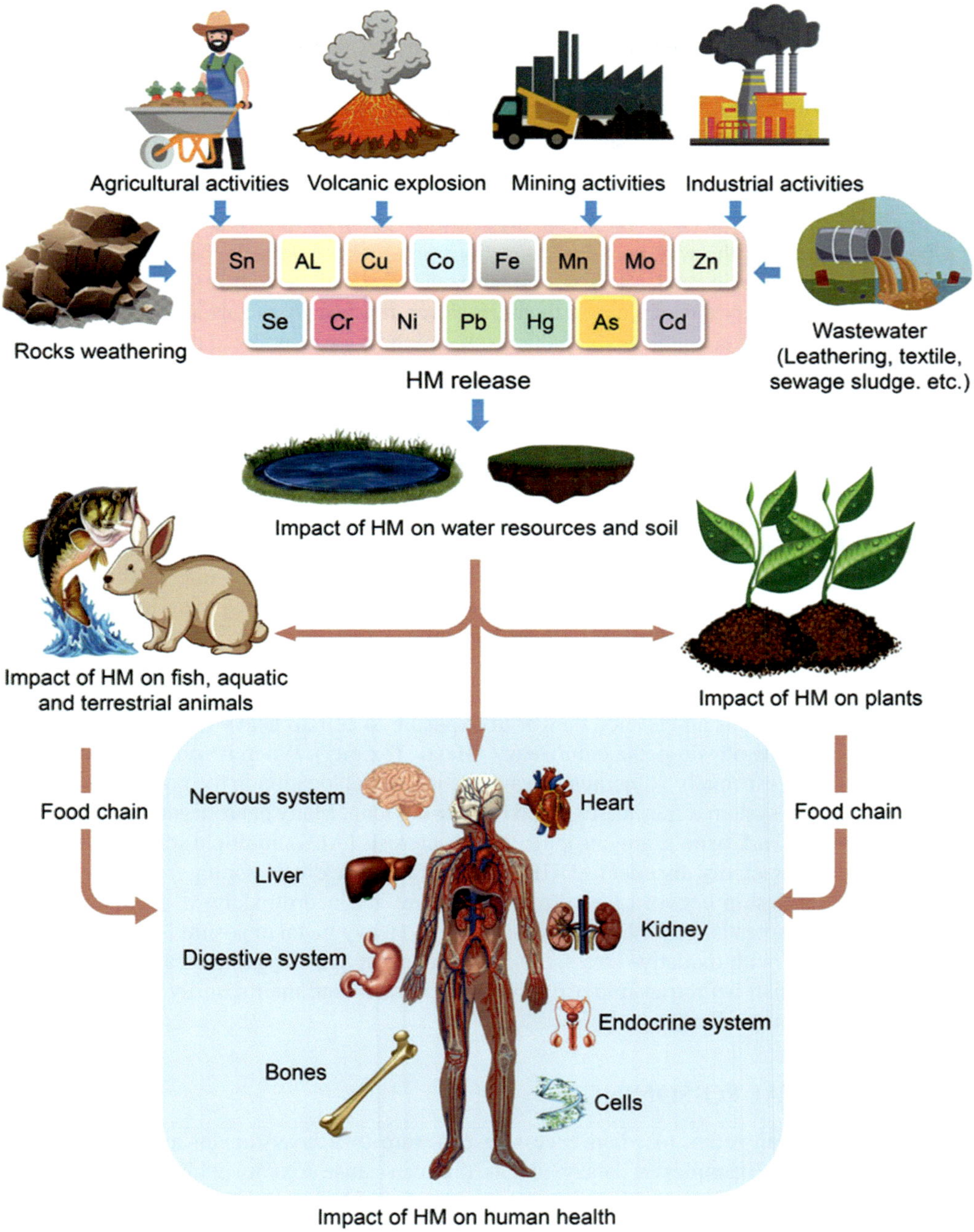

FIGURE. 7.4 Impacts of heavy metals on humans through the food chain (Abd Elnabi et al., 2023. Open access.)

7.6 INGESTION OF TOXIC METALS

The periodic table includes a group of elements known as "heavy metals." Naturally occurring metallic elements with a high density (or "heavy for their size") relative to their size are referred to by this term. Of the 35 naturally occurring metals, 23 are classified as heavy metals due to their high specific density (more than 5 g/cm^3) and atomic weight (higher than 40.04). Antimony, tellurium, bismuth, tin, thallium, gold, arsenic, cerium, gallium, cadmium, chromium, cobalt, copper, iron, lead, mercury, manganese, nickel, platinum, silver, uranium, vanadium, and zinc are all examples of what are commonly referred to as heavy metals. Exposure to high concentrations of some metals can result in heavy metal poisoning. It alters the way your body functions and makes you sick. We are surrounded by heavy metals such as arsenic, lead, mercury, and others. The ground we walk on, the water we drink, and the goods we use every day all contain them. However, most heavy metals are harmful in excessive amounts. Ingestion of contaminated food or drink, or inhalation of polluted dust or fumes, can all lead to poisoning. Heavy metal poisoning of any kind is extremely uncommon in the United States. And professionals warn against using heavy metal tests or "detox" therapies discovered online without first consulting a physician. Some of these could even be harmful to your health (www.webmd.com/a-to-z-guides/). There are several potential routes of exposure, including the diet, medication, environment, and even leisure activities. Exposure to heavy metals can occur in three ways: through the skin, the lungs, and the digestive tract. Both short-term, repeated exposure and long-term, steady exposure can cause toxicity. Children's developing brains and higher rates of absorption make them especially vulnerable to the side effects of lead exposure. Acute metal poisoning typically manifests with nausea, vomiting, diarrhea, and abdominal pain. Long-term exposure has been linked to an increased risk of cancer and numerous symptoms caused by organ damage.

Polluting the environment with heavy metals can contaminate the air, water, sewage, ocean, waterways, and can build up in plants, crops, seafood, and meat, all of which can have negative effects on humans. There is an elevated danger of exposure to certain heavy metals and toxicity in some professions (www.physio-pedia.com/Heavy_Metal_Toxicity). When it comes to heavy metal pollution, the brain is especially vulnerable because of its central location in the body. Heavy metals can harm the nervous system if they are exposed to it for too long. Many neurodegenerative diseases, such as Alzheimer's, Parkinson's, amyotrophic lateral sclerosis (ALS), multiple sclerosis (MS), and attention-deficit hyperactivity disorder (ADHD), can be traced back to poisoning from heavy metals. Evidence of a relationship between heavy metal exposure and neurotoxicity is available (https://link.springer.com/chapter/10.1007/978-3-030-66376-64). Heavy metal exposure is thought to be to blame for this, along with oxidative stress, the involvement of specific proteins and enzymes, and the disruption of normal neurotransmitter production. Diseases and their severity can be mitigated if proper precautions are taken.

7.7 HEAVY METAL POISONING

Heavy metal poisoning refers to when excessive exposure to a heavy metal affects the normal function of the body. Examples of heavy metals that can cause toxicity include lead, mercury, arsenic, cadmium, and chromium. Exposure may occur through the diet, from medications, from the environment, or in the course of work or play. Heavy metals can enter the body through the skin, by inhalation, or by ingestion. Toxicity can result from sudden, severe exposure or from chronic exposure over time. Symptoms can vary depending on the metal involved, the amount absorbed, and the age of the person exposed. Nausea, vomiting, diarrhea, and abdominal pain are common symptoms of acute metal ingestion. Chronic exposure may cause various symptoms resulting from damage to body organs, and may increase the risk of cancer.

The metal involved, the amount absorbed, and the age of the individual exposed can all affect the symptoms. Acute metal consumption is frequently accompanied by nausea, vomiting, diarrhea, and abdominal discomfort. Numerous symptoms brought on by organ damage may arise from prolonged

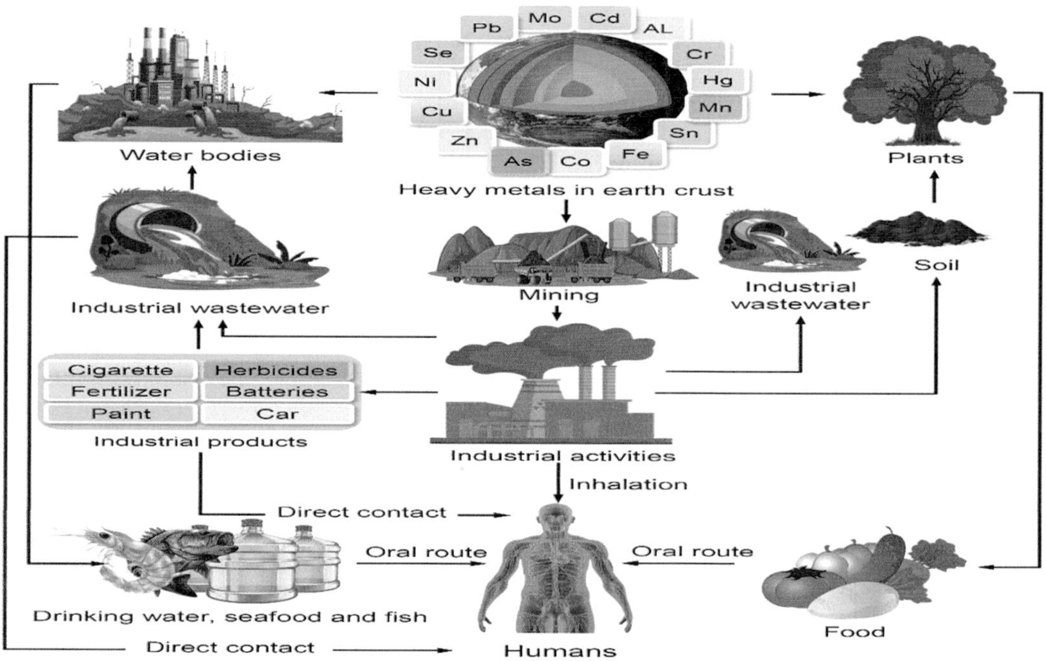

FIGURE 7.5 The diverse mechanisms involved in HM intoxication (Abd Elnabi et al., 2023.)

exposure. Figure 7.5 depicts the origins and impacts of HMs on humans through the food chain. Prolonged exposure to HMs through dermal contact, ingestion of contaminated food, and inhalation can result in the development of diverse disorders in both humans and other organisms (Abd Elnabi et al., 2023). Heavy metals are elements found in nature that accumulate in the environment mostly due to anthropogenic activities. Humans are exposed to them by occupational exposure or consuming foods that contain these elements. These substances can cause toxic effects that affect health and well-being.

It is noted that Selenium in the supplement is the probable cause of hair loss. Selenium is a trace mineral essential for good health. Selenium (Se) is obtained from food such as meat, vegetables, nuts, seafood, bread, eggs, and cereals. The chemical elements that might be most dangerous to human health are heavy metals, including lead, cadmium, mercury, and metalloid arsenic. The symptoms of selenium toxicity are referred to as selenosis. When comparing acute and chronic poisoning, selenosis from chronic intoxication is frequently caused by occupational exposure to inorganic selenium in an industrial setting, organic compounds found in splendid plants or cattle, or supplementation. Inhaling selenium dust and fumes or hydrogen selenide during an industrial accident can result in acute selenium poisoning, which irritates the respiratory system. Humans are exposed to HMs through inhalation, ingestion, or contact with the skin. Some studies indicate lead exposure through food reached the mother's breastmilk and their infant's body.

It is hypothesized that selenium coordinates into disulfide bridges, causing a structural defect that results in hair loss. Alopecia has been attributed to the breakdown of structural proteins in keratin. In contrast to alopecia areata or hair loss brought on by androgen, selenium-induced alopecia is widespread. The follicles stay intact, but the hair gets dry and brittle and breaks off readily at the scalp. The most common method of removing hair is via scratching, which causes an itchy rash on the scalp. Keshan disease, a juvenile multifocal myocarditis, and Kaschin-Beck disease, a chondrodystrophy, have been linked to extremely low selenium levels in humans. Hair and nail loss are a result of chronic selenium toxicity (Agarwal et al., 2016). Slight damage, such as scratching, causes hair to break at the scalp level (Figure 7.6), but the follicles stay intact. Hair loss can occur anywhere.

Heavy Metals have the potential to undergo transfer from aquatic invertebrates to higher trophic levels, ultimately leading to bio magnification. This process can result in harmful consequences for all organisms, including humans. The ingestion of various HMs by aquatic invertebrates has the potential to impact aquatic ecosystems, as depicted in Figure 7.7.

The following are some symptoms of metal poisoning: Experience disorientation, numbness, nausea, and vomiting. Drop Dead Symptoms include stomach pain, diarrhea, fainting, dehydration, anemia, damage to the kidneys or liver, and respiratory distress. Sputum changes in color, thickness, or quantity can indicate a health problem, such as a respiratory infection, lung disease, or lung cancer. Lung disease can increase the risk of cognitive decline, and memory loss can be a symptom of lung disease. The general rules can help prevent heavy metal poisoning (www.webmd.com/a-to-z-guides/what-is-heavy-metal-poisoning).

Before handling heavy metals, always use a mask and other protective gear. Find out if fish advisories are in effect in your region, have a professional check your house for lead paint, and get any trouble spots fixed to make sure the fish you eat is safe. Check product labels for heavy metal warnings (www.everydayhealth.com/heavy-metal-poisoning/guide/). Exposure to metals in the environment, whether through food, water, or industrial chemicals, can lead to heavy metal poisoning. Some heavy metals, like zinc, copper, chromium, iron, and manganese, are necessary in trace levels for proper bodily function, but are poisonous in large doses. The accumulation of these chemicals in the body's soft tissues can lead to poisoning and major health concerns.

Minamata disease is a methylmercury (MeHg) poisoning caused by ingesting fish and shellfish contaminated by MeHg discharged in wastewater from a chemical plant. The disease was first discovered in Minamata City, Japan, in 1956 (www.theatlantic.com/science/archive/2017/09/mercury-poisoning-minamata/541191/). Heavy metals, a unique class of naturally occurring elements, pose a severe health threat to living organisms and marine life. Heavy metal concentration in oceanic environments is an emerging issue due to anthropogenic activities and industrialization worldwide (Vallius, 2014). Some parts of freshwater, such as the rivers Ravi, Chenab, Kabul, and Indus, are highly contaminated, affecting fish and human health (Afzaal et al., 2022). Heavy metal contamination is considered a poisoning agent for fish health, affecting their vital operations, reproduction, and immune system, and causing pathological changes.

Lead poisoning causes high blood pressure and harm to the reproductive organs if exposed to it for too long. Long-term exposure to lead poisoning damages the reproductive organs and raises blood pressure. Mercury poisoning can cause damage to the intestines, skin, eyes, brain, and lungs. Low blood pressure, cancer, gastrointestinal issues, and nervous system abnormalities can all result from arsenic exposure. Cadmium toxicity can lead to decreased lung and renal function. Damage to the lungs, the brain, the eyes, the skin, and the intestines is the possible outcomes of mercury poisoning. Arsenic poisoning can lead to nervous system disorders, stomach difficulties, cancer, and low blood pressure. Reduced lung and renal function may result from cadmium poisoning.

Pesticides, disinfectants, antiseptics, pigments, dry batteries, explosives, and some therapeutic products all contain inorganic mercury salts. One of the most common ways people are exposed to mercury is through their diet, specifically fish that are high in methyl mercury. Production of chlorine and caustic soda, mining and metallurgy, electroplating, chemical and textile manufacture, paper and pharmaceutical production, and leather tanning are all possible sources of mercury-containing effluents. Microorganisms in the soil and water convert environmental mercury molecules into methyl mercury. Toxic levels of methyl mercury in the water are ingested by people through the consumption of fish (swordfish, king mackerel, fresh tuna, tile fish, shark). Mercury is a common element that can be found in both the ground and water. This means that over time, some food products may get contaminated with trace quantities of mercury. However, some foods contain significantly more mercury than others, even though the average quantity is still safe.

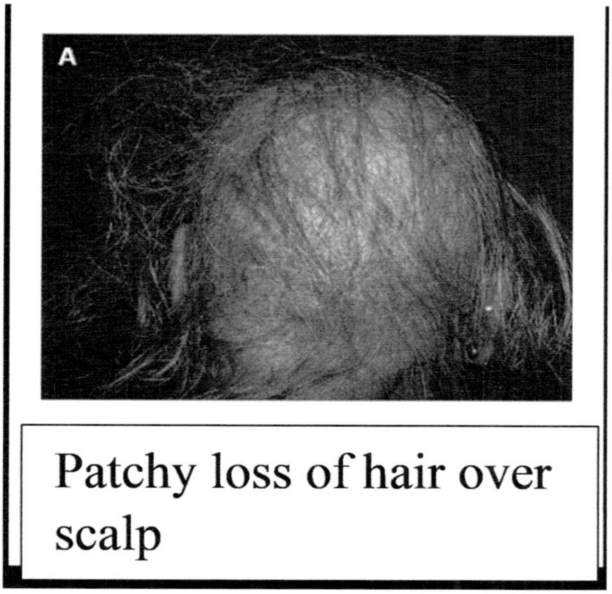

FIGURE 7.6 Patchy loss of hair over the scalp (Machado et al., 2023. https://creativecommons.org/licenses/by/4.0/.)

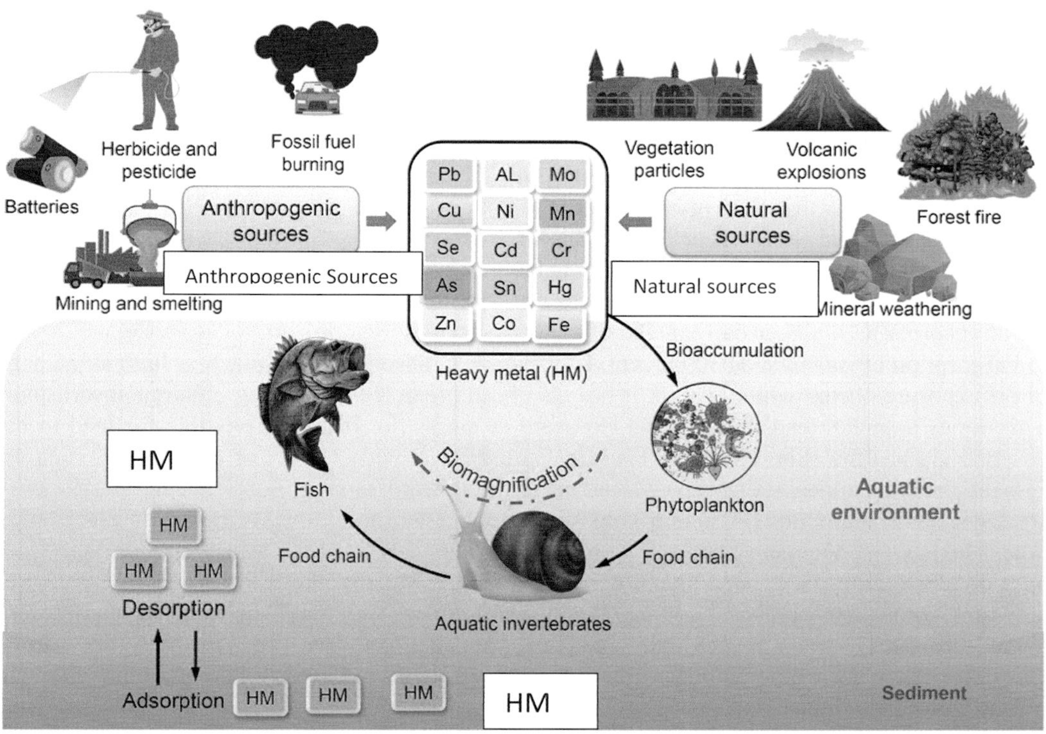

FIGURE 7.7 Distribution of Heavy Metals in Aquatic Environments (Abd Elnabi et al., 2023. https://creative commons.org/license.s/by/4.0/.)

Methylmercury, an organic form of mercury that can be found in seafood, is generally thought to be the culprit in cases of mercury poisoning. Symptoms of mercury poisoning may be exacerbated by a diet heavy in seafood consumed over a long period. Cadmium is a heavy metal that accumulates in human tissues since our bodies have no way of eliminating it. It is the leading cause of poisoning among people who don't smoke. Inorganic cadmium deposits can be seen in the kidneys, liver, pancreas, and lungs.

Cadmium is a major environmental pollutant and is known for its wide toxic manifestations. Naringenin is a naturally occurring citrus flavanone that has been reported to have a wide range of pharmacological properties. According to a study by Renugadevi and Prabu (2010), exposure to cadmium (Cd) resulted in oxidative stress and hepatotoxicity in liver tissues. In their investigation, it was noticed that cadmium (5mg/kg) was administered orally for 4 weeks to induce hepatotoxicity. Elevated hepatic enzyme and lipid peroxidation marker values demonstrated this. Renugadevi and Prabu (2010) used the following technique to harm rats' livers by exposing them to cadmium: Widespread hepatocyte deterioration, including necrosis, inflammation, centrilobular vacuolization, inflammatory cell infiltration, fatty degenerative alterations, and sinusoidal dilatation, was caused by oral treatment at a dose of 5 mg/kg for 28 days. Cadmium inhibits the nonenzymatic antioxidants glutathione (GSH), thyroid-stimulating hormone (TSH), vitamins C and E, and the antioxidant enzymes superoxide dismutase (SOD), catalase (CAT), glutathione (GSH), and glutaredoxin (GR). The effect of naringenin (NAR) on the acute nephrotoxicity brought on by cisplatin was investigated in rats. NAR reduced the extent of cisplatin-induced nephrotoxicity, as evidenced by significant reduction in serum urea and creatinine concentrations, decreased polyuria, reduction in body weight loss, marked reduction in urinary fractional sodium excretion and glutathione S-transferase (GST) activity, and increased creatinine clearance (Badarya et al., 2005).

Numerous batteries, vapor lamps, electroplating, and soldering items contain cadmium. Two to four hours after exposure, symptoms of cadmium poisoning can appear. Pain in the muscles and respiratory problems are two symptoms of cadmium buildup. Lead, like other heavy metals, is toxic to the liver because it accumulates over time in the environment (in the form of water, soil, dust, and man-made items) and damages cellular components through increased oxidative stress. Subcortical and thalamic areas of the brain are particularly vulnerable to changes in signal due to lead poisoning. The presence of lead in the blood at concentrations greater than 0.2 ppm indicates exposure, and concentrations greater than 0.35 ppm indicate lead toxicity.

30–50% of inhaled lead is deposited in the lungs, and lead deposited in the lower respiratory tract is virtually entirely absorbed, according to studies. Lead is absorbed largely in the duodenum of the digestive tract; however, factors such as age and nutritional state can significantly alter the rate of absorption. A higher daily intake of calcium, phosphorus, or iron, for instance, can lower lead absorption by as much as 30 percent. However, lead absorption can reach as high as 96 percent if exposure occurs while fasting. It has also been predicted that young children absorb four to six times as much lead through oral exposure as adults do. It is also possible for lead to be absorbed via healthy skin. However, lead is also used in a wide variety of other items, such as pigments, paints, solder, stained glass, lead crystal glassware, ammunition, ceramic glazes, jewelry, toys, some cosmetics (including kohl and Sindoor), and traditional remedies in places like India, Mexico, and Vietnam. Lead can be present in water that has been transported through lead pipes or where pipes have been linked using lead solder. Most of the lead used in international trade now comes from recycled sources (www.who.int/news-room/fact-sheets/detail/lead-poisoning-and-health). The brain, liver, kidneys, and bones are all affected by lead in the body. Over time, it accumulates in the teeth and bones where it is stored. Lead in blood is a common indicator of lead exposure in humans.

WHO estimates that of the 2 million lives lost in 2019 due to exposure to known chemicals, 1 million were caused by lead poisoning (www.who.int/news-room/fact-sheets/detail/lead-poisoning-and-health, www.cdc.gov/nceh/lead/news/cdc-updates-blood-lead-reference-value.html). Due to its long-term effects on health, the World Health Organization estimates that lead exposure results in 21.7 million disability-adjusted life years, or DALYs, of disability and mortality worldwide. This includes 4% of the global burden of cardiovascular disease and 3% of the global burden of chronic renal disease.

Chemical pesticides and other industrial uses account for the vast majority of arsenic's current usage. Arsenic compound overdose can cause lethargy, headaches, confusion, and convulsions. It can also contribute to neurological symptoms like brain damage, nerve disease in the limbs, and the breakdown of the myelin sheath, a fatty layer surrounding nerve fibers. Arsenic poisoning can also manifest externally, manifesting as edema (swelling) in the subcutaneous fatty tissue and subcutaneous fat. Fingernail discoloration and white bands caused by arsenic buildup have also been reported. Garlic breath is a common problem that might affect some people. A severe weakening of muscles, muscle pains, chills, and a fever can occur with prolonged arsenic exposure, typically as a result of working in locations where the heavy metal toxin is ubiquitous. They may also have an abnormal darkening or reddening of their skin, as well as a scaling-like irritation on their palms and feet (a condition known as hyperkeratosis). Two weeks to eight weeks after exposure, these symptoms may appear. Soil, sediment, and groundwater can all contain inorganic arsenic, which can build up in the body and cause problems for the liver, kidneys, lungs, and digestive system. Inorganic arsenic causes comparable signs, but it acts slightly differently and deposits a residue in human organs and common tissues like skin and hair. Symptoms of acute inorganic arsenic poisoning include nausea, vomiting, diarrhea, muscle spasms, a painfully burning sensation in the mouth and throat, and abdominal pain.

Locations around or inside of hazardous waste dumps are typical places where people are exposed to higher-than-normal arsenic levels. Glass infiltrates, pesticides, wood preservatives, paints, and the smelting of copper, zinc, and lead can all expose people to arsenic. Arsenic can be fatal if ingested in large quantities (www.ncbi.nlm.nih.gov/books/NBK560920/). Gross pigmentation with hyperkeratinization, warts, dermatitis, vaso spasticity, Raynaud's phenomenon, decreased nerve conduction velocity, lung cancer, conjunctivitis, peripheral neuropathies, encephalopathy, laryngitis, bronchitis, rhinitis, and death are all potential outcomes of prolonged arsenic exposure.

7.8 ENVIRONMENTAL CONTAMINATION BY HEAVY METALS

Industrial effluents, organic wastes, trash burning, transportation, and electricity generation all contribute to the widespread release of toxic metals into the environment. Whether in gaseous or particle form, they can be blown by the wind to locations several kilometers from their origination. Metal contaminants in the air eventually rain down onto land or the tops of water bodies (Mahurpawar, 2015). Thus, air is another medium via which the environment can become polluted. The term "environment" can be used to describe the physical and social conditions in which humans live. Physical, chemical, aesthetic, and cultural properties and conditions of the foregoing have an impact on human health and well-being. These include the land, water, and atmosphere of the Earth; microorganisms, plants, and animal life; any part or combination of the first two items on this list, and the interrelationships among and between them. It consists of the land, the atmosphere, and the water on Earth. Vegetable crops produced in a contaminated environment produce heavy metal buildup, which then enters and multiplies in the food chain. Heavy metals can be found in biota and groundwater, posing a threat to human health if ingested. Kidney illness, respiratory issues, neurological abnormalities, and cancer are only some of the major health effects of being exposed to heavy metals. Heavy metals have effects on a wide range of creatures, including plants, humans, water supplies, soil nutrients, and aquatic life. Many of these are dangerous even at low concentrations, but they are introduced irreversibly since they cannot be destroyed and are usually present in trace amounts. Thus, the current heightened worry and demand for enhanced remedial actions are in response to the rising amounts of heavy metals in the environment. Heavy metal mixtures can come from a variety of sources.

Below a particular threshold value (about 0.01 mg/L[1]), nickel, copper, lead, chromium, zinc, mercury, and cadmium become extremely damaging to humans. The US Environmental Protection Agency has set a limit of 2 ppb for mercury in drinking water. Water treatment plant effluents should have no more than 0.0019 parts per million of Hg, and agricultural soils should have no more than 0.05 parts per million, according to the World Health Organization. Severe implications for human and animal health can result from prolonged or excessive exposure to "metal trace elements" (MTEs), which are typically acquired through inhalation of airborne pollutants, ingestion of polluted water and food, and contact with contaminated soils and industrial sites. As more and more of a wide variety of metals are used in manufacturing and daily life, so too have the environmental problems caused by their contamination. There are several factors that contribute to an individual's susceptibility to the harmful effects of these substances. These include the chemical species, the route of exposure, the dose, and the gender, age, genetics, and nutritional state of the exposed individual.

Metal corrosion, air deposition, soil erosion of metal ions, and leaching of heavy metals to soil and groundwater are all potential sources of environmental pollution. Heavy metal pollution is largely attributable to natural processes like weathering and volcanic eruptions. Metal refining, coal-fired power plants, petroleum combustion, nuclear power plants, high-tension lines, plastics, textiles, microelectronics, wood-preservation, and paper-processing factories are all examples of industrial sources.

The biosphere includes all of Earth's living organisms, the atmosphere includes all of Earth's air and water, the lithosphere includes all of Earth's land and water, and the hydrosphere includes all of Earth's liquid water. Pollutants and environmental contaminants are substances found in the environment at concentrations above those found in their natural habitats (Briffa, 2020). Pollution and contamination of an ecosystem are possible. There is a distinction between pollution and contamination, yet contaminants can also be pollutants. According to the academic literature, pollution is "the introduction by man, directly or indirectly, of substances or energy into the environment that results in such deleterious effects as harm to living resources, hazards to human health, hindrance to environmental activities, impairment of quality for use of the environment, and reduction of amenities." However, contamination occurs when compounds in the environment reach quantities that are outside the range that would normally be found there and in the organism. The term "environmental pollution" is used to describe the negative alteration of the physical, chemical, and biological aspects of air, water, and soil. Chemicals and energy sources, including sound, heat, and light, are all examples of pollution.

The harmful impacts of heavy metal pollution on ecosystems around the world have made this issue of growing importance. The rapidly expanding agriculture and metal sectors, along with poor waste management, fertilizers, and pesticides, are all contributing to the release of these inorganic pollutants into our waterways, soils, and environment. Briffa (2020) details the pathways and ultimate destinations of many environmental contaminants. Some metals interfere with biological processes and development, while others collect in various body parts, leading to disorders as severe as cancer. Each metal's toxicity and pharmacokinetics in humans are given in this review. The severity and cause for concern of heavy metal bioaccumulation in humans are described in detail by Briffa (2020) in their review.

Human activities are the primary cause of heavy metal pollution in the environment. Only trace amounts of each are needed in the diet to maintain health. Depending on the metal and the dose, heavy metal poisoning can be either temporary or permanent, even fatal. Heavy metals can be dangerous to human health; therefore, workers in companies that produce them should take special precautions and wear protective gear to limit their daily exposure. Many of the symptoms are shared with other neurological illnesses and hence require careful diagnosis.

The primary strategy for dealing with heavy metals should be to keep them out of people's bodies in the first place. One area for improvement is reducing pollution from vehicles and factories. Heavy metals can also enter the body through the food we eat. Heavy metals can be removed from soils, sediments, and water by phytoremediation and intercropping. Heavy metals in the soil can be removed by planting hype accumulator plants. Selective uptake by the root systems of these plants allows the pollutant to be transported, bioaccumulated, and ultimately removed from the environment through decomposing plant matter (Sao et al., 2014).

Inorganic chemicals can be stabilized or extracted using plants. Heavy metals are extracted, or phyto-extracted, from the roots and taken up by the leaves. When it comes to Phyto stabilization, only plant parts that actively remove heavy metals are used. Phyto degradation, rhizofiltration, and rhizo degradation are all methods for eliminating organic pollutants. When the soil is wet, a process called rhizofiltration takes place in which contaminants in solution around the root are absorbed by, or precipitated into, the roots. Degradation of pollutants in the rhizosphere and enhancement of root quality through microbial activity constitute rhizo degradation (Sao et al., 2014).

It is important to consider the plant's species because various plants accumulate various metals. Temperature affects vegetative uptake through root length, while soil pH, organic matter, and phosphorus levels all have a role in how bioavailable metals like ethylenediaminetetraacetic acid (EDTA) are to plants (Chen et al., 2004a, 2004b). When hyperaccumulator plants take in harmful metals, they don't show any symptoms.

The complete plant of *Pteris vittata* L. absorbs arsenic at a rate of about 0.7 mg/g dry weight. Arsenic at a concentration of 0.2 mg/g dry weight is taken up by the roots of *Populus nigra*. Lead is absorbed at a rate of more than 50 mg/g in dry weight by the entire plant of *Brassica carinata* A. Braun. The tomato plant, *Lycopersicon esculentum,* takes up cadmium through its leaves and lead through its roots. The plant's fruit is effective at chelating cadmium. An et al. (2011) found that *Thlaspi caerulescens* shoots absorb 10 mg/g of zinc and 0.1 mg/g of cadmium in dry weight.

Intercropping, on the other hand, involves simultaneously cultivating two plant species. Heavy metal accumulation is assisted, and plant biomass is increased. Using intercropping in tandem with phytoremediation is preferable to the usage of chelators from an environmental perspective. Choosing the right species for use as an intercropping plant is essential. It has been observed that intercropping improves field diversity and stability while decreasing fertilizer use. This technique helps lessen the prevalence of weeds, diseases, and insects (Wan and Lei, 2018). There are three different types of intercropping: The practice of harvesting many crops at once is known as "mixed intercropping." Row intercropping, in which crops of different types are grown in consecutive rows. Relay cropping, in which a second crop is planted after the first has matured.

Common intercropping combinations include maize and chickpeas. When cultivated alongside chickpeas, maize increases its soil cadmium absorption. Iron absorption is enhanced when tomato roots are intercropped with the Japan clover plant. This intercropping also boosted cadmium absorption (Sao et al., 2014). Organic farming can be viewed as an additional strategy for decreasing the number of toxic metals we take into our systems. The water is filtered and contains no or very few impurities, and no pesticides are employed. Instead of utilizing pesticides, biodiversity is employed, which has the added benefit of increasing the number of species of wild animals and plants.

Heavy metals can be removed from water, with temperature, pH, ionic strength, and the presence of natural organic matter serving as the most critical components in the removal process. Removal of heavy metal ions from wastewater is of prime importance for a clean environment and human health. Different reported methods were devoted to heavy metal ions removal from various wastewater sources (Tonk et al., 2022). Some heavy metal removal techniques (Figure 7.8a and 7.8b) from wastewaters are suggested by Razzak et al. (2022).

The greatest byproducts and agricultural waste to get rid of heavy metals appear to be dairy manure, the leftover waste material from rice, and peanuts. Mineral deposits and natural soil, on the other hand, appear to be the least successful at removing heavy metals (Sall et al., 2020). Some foods are known to chelate with heavy metals once they have entered the body. Some foods and supplements can help the body gradually eliminate metals from the system, including cilantro, wild blueberries, garlic, lemon water, chlorella, spirulina, curry, green tea, Atlantic dulse, barley grass, tomatoes, and probiotics (Joseph et al., 2019).

Many human activities, including the use of fossil fuels, the release of vehicle exhaust fumes, mining, agriculture, and the incineration of solid and liquid wastes, contribute to the presence of heavy metals in the water supply. Volcanic eruptions, thermal springs, erosion, infiltration, etc., all contribute to the natural occurrence of heavy metals. Most heavy metals are hazardous to people and aquatic life, making this water poisoning a threat to all forms of life. Therefore, to lessen the degree of contamination of the natural waters, it is crucial to discover efficient methods for eliminating these toxins. Mercury, lead, cadmium, chromium, and nickel are just a few examples of the dangerous heavy metals that have been examined by Sall et al. (2020), along with their effects on the environment and human health, and the synthesis and characterization techniques used to remove these metals. Their primary goal in conducting this review was to highlight the great interest of COPs (conducting organic polymers) for the removal of toxic heavy metals from waters and to present recent works and methods (2000–2020) on the environmental impact and toxicity of heavy metals and the removal of toxic heavy metals, using chemically and/or electrochemically synthesized COPs.

Heavy metal poisoning of soils, water, and crops is a severe global environmental concern caused by improper waste disposal. Metals with a specific gravity greater than 5 g/cm^{-3} are considered heavy metals. Copper (Cu), nickel (Ni), chromium (Cr), lead (Pb), cadmium (Cd), mercury (Hg), iron (Fe), and arsenic (As) are the most prevalent ambient heavy metals. Iron and nickel, among other heavy metals, are required in extremely trace amounts by all known forms of life (Bakshi et al., 2018). Heavy metals are harmful to living creatures not only when present in large quantities, but also when present in minute quantities, as is the case with lead, cadmium, and mercury. Consumption of plants and other foods cultivated in contaminated soil is linked to metabolic disorders in animals and humans.

A healthy environment is linked to healthy people. People who work with or eat food grown in soil contaminated with heavy metals from improper waste management should be concerned about the disposal of certain products containing heavy metals in open dumpsites. Heavy metal contamination in the soil has been linked to waste generation and disposal practices. Despite recent reports linking landfills to soil contamination, no viable solution has been provided for waste disposal management in several developing countries (Nyiramigisha et al., 2021). This poses the greatest danger to the surrounding agricultural land and human areas near landfills. Heavy metal contamination of soil poses threats to human health, pollutes groundwater, causes phytotoxicity in plants, and reduces crop yields. Heavy metals pose a serious hazard to animal and human health because they are absorbed by plants through their roots and then concentrated at higher and higher levels as they move up the food chain. Soil, vegetables, and other food crops near dumpsites were found to be contaminated with heavy metals, according to certain researchers (Nyiramigisha et al., 2021). Heavy metal-contaminated soil is a major contributor to the contamination of vegetables and other crops, which has been linked to negative health effects in urban areas in developing countries. Heavy metals in various open dumpsites must, therefore, be monitored constantly. Heavy metals in soil and crops are too high; hence, steps must be taken to lower their concentration. Nyiramigisha et al. (2021) explored the negative effects of heavy metal contamination on soil and crops planted near dumpsites, as well as several methods for addressing heavy metal pollution in soil.

7.9 PESTICIDES AND OTHER HEAVY METALS IN PLANT AND SOIL TOXICOLOGY IN THE AGRICULTURAL SECTOR

Massive economic growth and rapid expansion in many sectors, including agriculture and manufacturing, are contributing to a more polluted environment. Pollutants in the environment can be either naturally occurring or the result of human activity. Dangerous issues arise for abiotic (water, air, soil) and biotic (animals, plants, and humans) ecosystems as a result of certain environmental activities, such as synthetic industries, coal conversion, and garbage burning. Heavy metals and

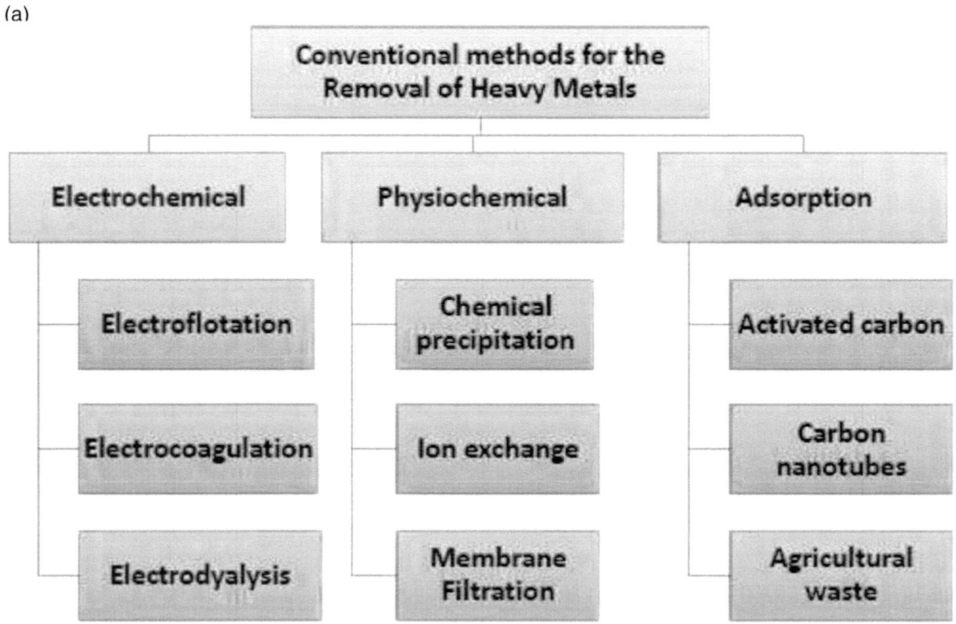

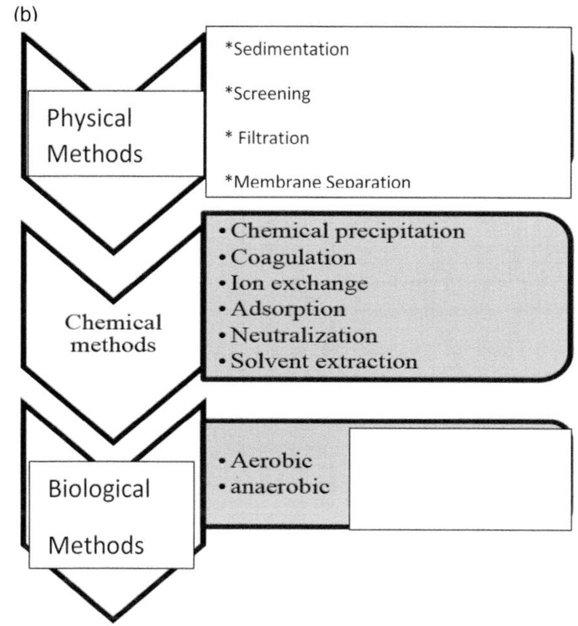

FIGURE 7.8a a Conventional methods used in the removal of heavy metals from wastewater (Razzak et al., 2022. https://creativecommons.org/licenses/by/4.0/.)

FIGURE 7.8b Conventional technologies for heavy metal removal (Abdel-Raouf and Abdul-Raheim, 2017. Creative Commons Attribution License.)

pesticides are just two examples of the environmental toxicants that pose a threat to the health of the entire ecosystem (Alengebawy et al., 2021).

Scientists' focus has always been disproportionately large on environmental issues. Pollution caused by toxic substances is a major environmental problem that has a negative impact on human health and agricultural output. Pesticides and heavy metals top the list of environmental toxicants that are harmful to wildlife. Insecticides, herbicides, and fungicides, together with other heavy metals and pesticides, have been shown to have negative effects on plant and soil life, as well as on human health. Heavy metal buildup and pesticide residues were studied by Alengebawy et al. (2021), who reported their findings. Plant physiological factors and features of polluted soil have also been discussed. Heavy metal and pesticide exposure-related illnesses in humans have also been documented (Chin, 2010). Heavy metal bioaccumulation, pesticide mechanisms of action, and pesticide transmission are highlighted. The bioavailability of these pollutants in soil and their uptake by plants have also been taken into account. Meanwhile, the harmful effects of heavy metals and insecticides together have been discussed, along with their synergistic and antagonistic interactions. This review incorporates all previously relevant studies. This review contains extensive data that significantly advances our knowledge of environmental toxicants and their potentially devastating impacts.

Honey bees, predators, birds, plants, and small mammals are all negatively impacted by the excessive and unchecked use of pesticides on various crop types. Furthermore, these repercussions reduce biodiversity throughout the entire biological system. High rates of death in honey bees and parasitoids that feed on nectar result from systemic pesticides' contamination of floral and extrafloral nectar components, which are transported throughout the plant via the vascular system. Without fully understanding the risks, many pesticides are used extensively in agriculture to reduce pest damage and boost yields. Pesticide residues accumulate in the soil as a result of this careless use, adding to the level of pollution and posing a threat to both plant and animal life. According to Xiao et al. (2017), heavy metal pollution in agricultural soil and plants is largely influenced by both agriculture and industry, particularly soils near cement and electroplating companies. In other words, the soil's surface is an ideal location for accumulating heavy metals, which are then taken up by plants via their roots and vascular systems when water is absorbed.

7.10 HEAVY METALS ANALYSIS USING ATOMIC ABSORPTION SPECTROMETRY (AAS), INDUCTIVELY COUPLED PLASMA OPTICAL EMISSION SPECTROMETRY (ICP-OES), AND INDUCTIVELY COUPLED PLASMA MASS SPECTROMETRY (ICP-MS)

Cadmium, chromium, and lead are just a few of the heavy metals that can be found in the earth's crust in variable amounts and thus in our environment. They can be taken orally or inhaled and absorbed into the body. While the human body only needs trace amounts of some of these metals, they play an essential role in metabolism (www.shimadzu.eu/heavy-metals-analysis). However, when consumed in high quantities, they become deadly and harmful to human beings. Heavy metal poisoning is common due to a variety of sources, including lead transfer pipes in water systems, factory air pollution, and contaminated meat, poultry, and shellfish. Drinking water is constantly monitored in Europe because it is considered the most important "food" in the region. Water, tap water, tobacco plants, food, and food packaging can all contain harmful levels of heavy metals; this chapter will explain how to test for their presence using atomic absorption spectrometry (AAS), inductively coupled plasma optical emission spectrometry (ICP-OES), and inductively coupled plasma mass spectrometry (ICP-MS). The heavy metal pollutants in cannabis flower were analyzed using the Shimadzu ICPMS-2030 in application SSI-ICP-006 (www.shimadzu.eu/sites/shimadzu.seg/files/006 cannabisicpms.pdf).

Research (www.shimadzu.com/an/literature/icp/apb420008.html) provides information on the transfer of heavy metals from plastic food packaging materials into aqueous simulants using

ICP-MS. The Shimadzu group has employed a refined ICP-MS technique to investigate the leaching of heavy metals from three distinct types of plastics often used in food packaging. When acetic acid (at 3% concentration) was utilized as a food surrogate, the leaching quantities of 14 metallic elements from various polymers varied greatly. The leachate contained trace amounts of toxic metals such as As, Cd, and Pb. Heavy metal leaching was found to be much higher in food simulant B compared to other aqueous solutions. Contact times longer than 1 hour resulted in a statistically significant increase in metal migration. The determinations were accurate and precise, as evidenced by the high spike recovery rates and low RSD percent across many tests.

Sewage sludge (SS) was analyzed by Tytła et al. (2023) for its concentrations of seven heavy metals (HMs): Cd, Cr, Cu, Ni, Pb, Zn, and Hg. Using ultrasonic disintegration (UD) of thickened excess sludge to enhance the benefits of anaerobic digestion, the researchers sought to undertake the first complete examination of the potential ecological risk of heavy metals in SS (sewage sludge) collected from two municipal wastewater treatment plants (WWTPs) (Malwina et al., 2023). Potential ecological danger was evaluated using two sets of indices, one for the total amount of heavy metals and one for their chemical forms. Tyta et al. (2023) used a variant of the BCR sequential extraction technique for this purpose. Their research found that Cd, Cu, and Zn (and to a lesser extent Ni and Cr) posed the greatest ecological risk based on total index values, whereas Ni, Zn, Cd, and Cu posed the greatest risk based on speciation index values (and in some cases Cr). Wastewater treatment plants (WWTPs) generate sewage sludge (SS), which can be a solid, semi-solid, or liquid waste. Production of SS has risen dramatically during the past few decades, particularly in emerging nations. Population expansion, improved wastewater treatment methods, and expanded industrial activity are major contributors to the rise in SS production (Turek et al., 2019).

Organic substances and minerals essential for plant growth are abundant in sewage sludge (Kowalik et al., 2022). Because of this, sludge can be an effective additive to organic fertilizer. However, it also contains contaminants such as harmful viruses, toxic organic compounds, and possibly bioaccumulative, toxic, and carcinogenic heavy metals (HMs), all of which can hurt soil quality and plant growth. One of the most important problems restricting the use of SS in agriculture is the presence of heavy metals. Limits on the use of HMs in SS (sewage sludge) applications have been established by law in Poland and other EU countries. Using SS in farming remains the best sustainable and long-term option for dealing with SS. However, HM-containing sludge may pose risks to ecosystems. For this reason, sludge should be stabilized before application to lessen its negative impact on aquatic ecosystems and plant growth. This is crucial because HMs, which can have harmful effects on human health, can enter the food chain through the consumption of plants grown on fields fed with sewage sludge.

7.11 CONTAMINATION OF FOOD WITH HEAVY METALS AND METALLOIDS

Although many metals are present in the foods we eat and the cells in our bodies, only a select few are essential to our well-being and survival. Biochemical damage in our cells and obvious clinical symptoms are possible outcomes of insufficient inorganic nutrient intake. When the deficiency is remedied by administering enough of the nutrient in question, symptoms typically subside. Beyond ensuring their products don't contain any dangerous metals or essential trace elements in concentrations high enough to pose health problems to consumers, manufacturers and processors should care about more than just meeting the many regulatory requirements and standards of practice related to metals in food (Mukherjee et al., 2023). Metals that could hasten deterioration or reduce the quality of food should also be looked for. Even foods with trace amounts of metals might change in unfavorable ways during the cooking and preserving procedures. It just takes a few milligrams per kilogram for metal ions to form complexes with organic molecules. Iron pollution can lessen the flavor and scent of many liquid meals, such as fruit juices, milk, beer, and wine.

Since it poses serious risks to ecosystems and human health, heavy metal and metalloid toxicity in the environment and food have captured the public's attention. Several metals and metalloids, such as arsenic, antimony, and selenium, pose a serious hazard to human health, as noted by Mukherjee et al. (2023). These metals and metalloids can be found in food. Determining how dangerous these potentially dangerous elements are, requires detecting even trace amounts of them. Atomic absorption spectrometry (AAS), spectrofluorimetric, inductively coupled plasma spectrometry, e-tongues, electrochemical apt sensors, Raman spectroscopy, and fluorescence sensors are just some of the conventional and cutting-edge technologies that have proven useful in determining even trace amounts of heavy metals and metalloids in food and nutritional supplements. Food may include metals and metalloids that are extremely toxic to humans. Among these are the metals cadmium, lead, mercury, tin, manganese, aluminum, and the metalloids arsenic, antimony, and selenium (Se). To fully assess the threat caused by these dangerous substances, even minute concentrations must be detected (Rai et al., 2019).

Atomic absorption spectrometry and other conventional techniques for detecting heavy metals have a low detection limit but come with a number of drawbacks. Thus, it is crucial to provide a speedy technique for online and real-time detection of heavy metals. With its high sensitivity, specificity, and accuracy, the electrochemical apt sensor-based technique holds great promise in the detection of heavy metals. It's progressing quickly, but more study is needed before it can be utilized for on-site detection. This chapter provides a brief history, overview, and evolution of electrochemical apt sensors. Wang et al. (2020) provide a concise summary of how nanomaterials and electrochemically suitable sensors have been used to detect heavy metals (mostly mercury, lead, cadmium, and arsenic).

Environmental pollution with harmful heavy metals, as noted by Huang et al. (2013), can potentially contaminate the rice. The possible health risk to local adults and children was assessed by studying the concentrations of four metals (arsenic, cadmium, mercury, and lead) in rice samples obtained in Zhejiang, China. Arsenic (As) was discovered at a mean concentration of 0.080 mg/kg, Cd at 0.037 mg/kg, Hg at 0.005 mg/kg, and Pb at 0.060 mg/kg. With the help of the rice consumption statistics, EDIs were determined. Adults were expected to consume 0.49 g/kg bw/day of As through rice, whereas children consumed 0.34 g/kg bw/day of As through rice. Adults consumed 1.02, 0.64, 0.37, and 1.26 g/kg bw/day of As, Cd, Hg, and Pb, respectively, whereas children consumed 0.63, 0.83, 0.47, and 1.63 g/kg bw/day, respectively. Based on the results of the risk assessment, eating rice poses no danger to your health due to the presence of these elements. However, it is estimated that the levels of Cd and Pb in children and of Hg in adults are higher than what is considered safe.

7.12 POPULATION EXPOSURE TO HEAVY METALS IN A SOUTH CHINESE MINING REGION

However, mining-related enterprises have often been carried out in an unrestrained manner, leading to serious soil erosion and environmental difficulties, especially heavy metal contamination, which has hampered local and national economies. Acidic mine drainage (AMD) poses a threat to the health of those living in the vicinity of mining operations since its discharge can contaminate downstream water, agricultural soils, food crops, and biota. The mining industry is responsible for a significant amount of environmental damage, especially in underdeveloped countries. China has long been a major player in the global mineral resources market. The country produces and consumes a wide variety of metalliferous and nonmetallic mineral commodities. Each year, China generates hundreds of millions of tonnes of mining waste from its roughly 8,000 state-controlled mining businesses and 230,000 collectively held mines. Since 2005, heavy metal pollution has caused a dramatic increase in health-related occurrences in China, with the most serious accidents receiving widespread media coverage. High levels of Cd in rice sold in the Chinese market have been linked to Acidic mine drainage (AMD) affected fields, according to a recent study (Zhuang et al., 2014). Millions of individuals throughout the world are projected to be exposed to metals in mine sites (Wciso et al., 2002), and the dangers to their health are not limited to mine workers.

7.13 EFFECTS OF CEMENT DUST ON SOIL AND CROP HEAVY METAL POLLUTION

The production of cement requires a lot of energy and is a major cause of global warming. Cement manufacturing poses significant risks to public health and the environment due to air pollutants and energy consumption. Raw materials and fossil fuels are two examples of nonrenewable resources used extensively in cement production. It is believed that cement manufacture accounts for between 5 and 6 percent of all human-caused carbon dioxide greenhouse emissions (Potgieter, 2012). Emissions of dust and gases into the atmosphere occur from the use of raw materials and energy. Nitrogen oxides (NOx), carbon dioxide, water, oxygen, dust, chlorides, fluorides, sulphur dioxide, carbon monoxide, and even lesser amounts of organic compounds and heavy metals are all found in the exhaust fumes from a cement kiln (Marlowe and Mansfield, 2002). Cement kiln incineration of industrial waste releases toxic metals and organic chemicals. In addition to the clinker cooler, other sources of dust emissions include crushers, grinders, and materials-handling machinery. Heavy metals and other air pollutants are dispersed by the wind and settle on nearby soil, vegetation, and water sources as a byproduct of the cement manufacturing process. Wind speed, particle size, and stack emissions all play a role in how much cement production dust gets deposited nearby. Environmental studies on cement plant emissions have identified several metals, including arsenic, cadmium, lead, mercury, thallium, aluminum, beryllium, chromium, copper, manganese, nickel, and zinc, as having a particularly hazardous effect on soil (Addo et al., 2012). Top soils near a cement factory are enriched in Pb, Zn, Cr, Cd, V, Pb, and Hg due to emissions from nearby cement kilns (Arpita and Mitko, 2011).

Multiple studies have found that cement production is a major contributor to soil pollution from heavy metals. Soil and barley crop heavy metal pollution near an Abyssinian cement factory was studied (Yitagesu and Bekele, 2019). Holetta Chemistry Laboratory received and examined a total of 36 soil samples and 22 barley samples. Soil samples are tested for the following factors: Major physicochemical features, total available heavy metals, and total heavy metals. The concentration of metals in a barley sample was measured using the dry aching technique. Atomic absorption spectrometer analysis was used to determine metal content. Mean values were compared between treatments using SPSS's one-way ANOVA function (Young-Chull, and Jae-Min., 2004). For each metal, they calculated and analyzed its Geo-accumulation Index and transfer factor.

If the Igeo readings are below 0, then it means the soil is relatively uncontaminated. Soil physicochemical parameters showed that the pH was neutral to slightly alkaline, the texture was clay to clay loam, and the levels of phosphorus, nitrogen, and potassium were all in the moderate to high range (Ahn et al., 2004). If the transfer factors for heavy metals are less than one, then the crop has not been affected by cement dust pollution. The concentration of heavy metals in the studied barley crop may not currently constitute a health danger and can also serve as supplies of necessary trace metals to the population; however, the levels discovered were below the legal limit specified by FAO/WHO.

7.14 CONCLUSION

Heavy metals, which are natural elements with high atomic weight and density, are prevalent in various sectors, including industry, households, farms, hospitals, and technical endeavors. These metals pose serious health risks, particularly arsenic, cadmium, chromium, lead, and mercury. Exposure to heavy metals through the food chain can lead to poisonous combinations with other environmental factors. Chronic exposure to these elements is a major concern in regions affected by metal pollution. Understanding the molecular basis of heavy metal interactions is crucial for evaluating and managing health hazards associated with chemical combinations. Heavy metals pose a significant health risk due to their accumulation in the human body and difficulty in natural elimination. They are emitted into the atmosphere through discharges from activities like industries, transport, and agriculture. Cleanup of polluted sites involves removing these particles, which can lead to water pollution and food transfer. Mechanisms are being studied to clean water and soil directly at the source, with solutions focusing on transforming heavy metals into less toxic forms. Soil remediation through phytoremediation and chelation is a potential option for addressing this issue.

Arsenic contamination is often treated using technologies like coagulation, flocculation, adsorption, and membrane. For mild cases of heavy metal poisoning, eliminating exposure can be sufficient. Advanced water treatment technologies are crucial for high-quality water and for reducing impurities. Detoxifying treatments and supplements can be incorporated into a natural heavy metal detox diet plan to support vital organs. Anti-inflammatory herbs and spices like basil, parsley, oregano, rosemary, thyme, ginger, turmeric, cinnamon, and cilantro can help remove heavy metals, including lead, mercury, arsenic, and cadmium. Heavy metals, such as arsenic, cadmium, chromium, lead, and mercury, pose significant health risks to living things due to their high toxicity. Advanced water treatment technologies are needed to ensure high-quality water and reduce contaminants. Microbial biosorbents are a cost-effective and environmentally acceptable solution for cleaning up heavy metal-affected environments. These biosorbents use microorganisms to absorb, precipitate, oxidize, and reduce heavy metals in soil. A sustainable strategy is needed to choose the best biosorbent, operational parameters, and an effective heavy metal removal mechanism in industrial effluent. Current knowledge and techniques are needed to reduce heavy metal toxicity and detoxify the environment. Although heavy metals cannot be completely eliminated, various remediation approaches can be used to reduce pollution and transform them into harmless forms. Utilizing living organisms like plants, fungi, algae, and bacteria is a sustainable method.

REFERENCES

Abd Elnabi, M.K., N.E. Elkaliny, M.M. Elyazied, S.H. Azab, S.A. Elkhalifa, S. Elmasry, M.S. Mouhamed, E.M. Shalamesh, N.A. Alhorieny, A.E. Abd Elaty, I.M. Elgendy, A.E. Etman, K.E. Saad, K. Tsigkou, S.S. Ali, M. Kornaros, and Y.A.-G. Mahmoud. 2023. Toxicity of heavy metals and recent advances in their removal: A review. *Toxics*. 11: 580. https://doi.org/10.3390/toxics11070580

Abdel-Raouf, M.S., and A.R.M. Abdul-Raheim. 2017. Removal of heavy metals from industrial waste water by biomass-based materials: A review. *J. Pollut. Eff. Cont*. 5: 180. https://doi.org/10.4172/2375-4397.1000180

Addo, M.A., E.O. Darko, C. Gordon, B.J.B. Nyarko, J.K. Gbadago, E. Nyarko, H.A. Affum, and B.O. Botwe. 2012. Evaluation of heavy metals contamination of soil and vegetation in the vicinity of a cement factory in the Volta region, Ghana. *Int. J. Sci. Technol.* 2(1): 40–50.

Afzaal, M., S. Hameed, I. Liaqat, A.A. Ali Khan, H. Abdul Manan, R. Shahid, and M. Altaf. 2022. Heavy metals contamination in water, sediments and fish of freshwater ecosystems in Pakistan. *Water Pract. Technol.* 17(5): 1253–1272. https://doi.org/10.2166/wpt.2022.039

Agarwal, P., Sharma, and U.S. Agarwal. 2016. Selenium toxicity: A rare diagnosis. *Indian J. Dermatol. Venereol. Leprol.* 82: 690–693.

Ahn, Y.-C., J.-M. Cho, G.-T. Kim, S.-R. Cha, J.-K. Lee, Y.-O. Park, S.-D. Kim, and S.-H. Lee. 2004. Physical, chemical, and electrical analysis of dust generated from cement plants for dust removal with an electrostatic precipitator. *Korean J. Chem. Eng.* 21: 182–186. https://doi.org/10.1007/BF02705396

Alengebawy, A., S.T. Abdelkhalek, S.R. Qureshi, and M.Q. Wang. 2021. Heavy metals and pesticides toxicity in agricultural soil and plants: Ecological risks and human health implications. *Toxics*. 9(3): 42. https://doi.org/10.3390/toxics9030042

Ali, H., and E. Khan. 2017. Environmental chemistry in the twenty-first century. *Environ. Chem. Lett.* 15(2): 329–346.

Ali, H., E. Khan, and I. Ilahi. 2019. Environmental chemistry and ecotoxicology of hazardous heavy metals: Environmental persistence, toxicity, and bioaccumulation. *J. Chem.* 2019: Article ID 6730305: 14. https://doi.org/10.1155/2019/6730305

An, L., Y. Pan, Z. Wang, and C. Zhu. 2011. Heavy metal absorption status of five plant species in monoculture and intercropping. *Plant Soil*. 345: 237–245.

Arpita, M., and V. Mitko. 2011. Heavy metals in soils around the cement factory in Rockfort, Kingston, Jamaica. *Int. J. Geosciences*. 2: 48–54.

Asati, A., M. Pichhode, and K. Nikhil. 2016. Effect of heavy metals on plants: An overview. *Intern. J. Appl. Innovation Eng. Manag.* 5: 56.

Badarya, O.A., S. Abdel-Maksoudb, W.A. Ahmedd, and G H. Owieda. 2005. Naringenin attenuates cisplatin nephrotoxicity in rats. *Life Sci.* 76: 2125–2135.

Bakshi, S., C. Banik, and Z. He. 2018. The impact of heavy metal contamination on soil health. In: Reicosky (ed.) *Managing Soil Health for Sustainable Agriculture* (Vol. 2, no. 8, pp. 1–36), Burleigh Dodds Science Publishing Ltd., Sawston.

Briffa, J. 2020. *Heavy metals in Maltese agricultural soil* (Master's dissertation). www.um.edu.mt/library/oar/handle/123456789/73391

Chen, Y., X. Li, and Z. Shen. 2004a. Leaching and uptake of heavy metals by ten different species of plants during an EDTA-assisted phytoextraction process. *Chemosphere.* 57: 187–196.

Chen, Y., Z. Shen, and X. Li. 2004b. The use of vetiver grass (*Vetiveria zizanioides*) in the phytoremediation of soils contaminated with heavy metals. *Appl. Geochem.* 19(10): 1553–1565.

Chin, N.P. 2010. Environmental toxins: Physical, social, and emotional. *Breastfeed. Med.* 5: 223–224. https://doi.org/10.1089/bfm.2010.0050

Das, T.K., and A. Poater. 2021. Review on the use of heavy metal deposits from water treatment waste towards catalytic chemical syntheses. *Int. J. Mol. Sci.* 22: 13383. https://doi.org/10.3390/ijms222413383

Hadia-e-Fatima, and A. Ahmed. 2018. Heavy metal pollution–A mini review. *J. Bacteriol. Mycol. Open Access.* 6(3):179–181.

Huang, Z., X.D. Pan., P. G Wu, J.L. Han, and Q. Chen. 2013. Health risk assessment of heavy metals in rice to the population in Zhejiang, China. *PLoS ONE.* 8(9): e75007. https://doi.org/10.1371/journal.pone.0075007

Jessica, B., E. Sinagra, and R. Blundell. 2020. Heavy metal pollution in the environment and their toxicological effects on humans. *Heliyon.* 6(9): e04691. https://doi.org/10.1016/j.heliyon.2020.e04691

Joseph, L., B.M. Jun, J.R.V. Flora, C.M. Park, and Y. Yoon. 2019. Removal of heavy metals from water sources in the developing world using low-cost materials: A review. *Chemosphere.* 229: 142–159.

Khan, F.U., A.U. Rahman, A. Jan, and M. Riaz. 2004. Toxic and trace metals (Pb, Cd, Zn, Cu, Mn, Ni, Co and Cr) in dust, dust fall/soil. *J. Chem. Soc. Pakistan.* 26(4): 453–456.

Kim, J.J., Y.S. Kim, and V. Kumar. 2019. Heavy metal toxicity: An update of chelating therapeutic strategies. *J. Trace Elements in Med. Biol.* 54: 226–231.

Kowalik, R., J. Gawdzik, P. Bak-Patyna, P. Ramiączek, and N. Jurišević. 2022. Risk Analysis of heavy metals migration from sewage sludge of wastewater treatment plants. *Int. J. Environ. Res. Public Health.* 19: 11829.

Machado, de C. and R.J. Dinis-Oliveira. 2023. Clinical and forensic signs resulting from exposure to heavy metals and other chemical elements of the periodic table. *J. Clin. Med.* 12: 2591. https://doi.org/10.3390/jcm12072591

Mahurpawar, M. 2015. Effects of heavy metals on Human health. *Intern. J. Res. Granthaalayah.* 3(9): 1–7.

Malwina, T., K. Widziewicz-Rzońca, J. Kernert, Z. Bernaś, and K. Słaby. 2023. First comprehensive analysis of potential ecological risk and factors influencing heavy metals binding in sewage sludge from WWTPs using the ultrasonic disintegration process. *Water.* 15(4): 666. https://doi.org/10.3390/w15040666

Marlowe, I., and D. Mansfield. 2002. *Toward a Sustainable Cement Industry Sub Study 10: Environment, Health & Safety Performance Improvement.* An Independent Study Commissioned by WBCSD.

Mukherjee, A.G., K. Renu, A.V. Gopalakrishnan, V.P. Veeraraghavan, S. Vinayagam, S. Paz-Montelongo, A. Dey, B. Vellingiri, A. George, H. Madhyastha, and R. Ganesan. 2023. Heavy metal and metalloid contamination in food and emerging technologies for its detection. *Sustainability.* 15(2): 1195. https://doi.org/10.3390/su15021195

Munir, N., M. Jahangeer, A. Bouyahya, El. Omari, N. Ghchime, R. Balahbib, A. Aboulaghras, S. Mahmood, Z. Akram, S.M. Ali Shah, I.N. Mikolaychik, M. Derkho, M. Rebezov, B. Venkidasamy, M. Thiruvengadam, and M. Ali Shariati. 2022. Heavy metal contamination of natural foods is a serious health issue: A review. *Sustainability.* 14: 161. https://doi.org/10.3390/su14010161

Nkwunonwo, U.C., P.O. Odika, and N. Onyia. 2020. A review of the health implications of heavy metals in food chain in Nigeria. *Sci World J.* 2020: 11. https://doi.org/10.1155/2020/6594109

Nyiramigisha, P., A. Komariah, and Sajidan. 2021. Harmful impacts of heavy metal contamination in the soil and crops grown around dumpsites. *Rev. Agric. Sci.* 9: 271–282. https://dx.doi.org/10.7831/ras.9.0_271

Pande, A., B. Mun, N.J. Methela, and W. Rahim. 2022. Heavy metal toxicity in plants and the potential NO-releasing novel techniques as the impending mitigation alternatives. *Front. Plant Sci., Sec. Plant Physiol.* 13: 2022. https://doi.org/10.3389/fpls.1019647

Potgieter, J.H. 2012. An overview of cement production: How "green" and sustainable is the industry? *Environ. Manage. Sustain. Develop.* 1(2). https://doi.org/10.5296/emsd. v1i2.1872

Rai, P.K., Lee, S.S., Zhang, M., Tsang, Y.F., Kim, K.-H. 2019. Heavy metals in food crops: Health risks, fate, mechanisms, and management. *Environ. Int.* 2019, 125: 365–385.

Razzak, S.A., M.O. Faruque, Z. Alsheikh, L. Alsheikhmohamad, D. Alkuroud, A. Alfayez, S.M. Zakir Hossain, and M.M. Hossain. 2022. Conventional methods used in removal of heavy metals from wastewater. *Environ. Adv.* 7: 100168. https://doi.org/10.1016/j.envadv.2022.100168

Renugadevi, J. and S.M. Prabu. 2010. Cadmium-induced hepatotoxicity in rats and the protective effect of naringenin. *Exp. Toxicol. Pathol.* 62(2): 171–181. https://doi.org/10.1016/j.etp.2009.03.010

Sall, M.L., A.K.D. Diaw, and D. Gningue-Sall. 2020. Toxic heavy metals: Impact on the environment and human health, and treatment with conducting organic polymers, a review. *Environ. Sci. Pollut. Res.* 27: 29927–29942. https://doi.org/10.1007/s11356-020-09354-3

Sao, K., F. Khan, P.K. Pandey, and M. Pandey. 2014. A review on heavy metals uptake by plants through biosorption. *Intern. Proc. Econ. Dev. Res.* 75: 78.

Sarker, A., J.E. Kim, A.R.M. Towfiqul, I. Bilal, M.A. Rakib, Md. Refat Jahan, N. Rakhi, R.M. Mohammed, and I. Tofazzal. 2022. Heavy metals contamination and associated health risks in food webs—A review focuses on food safety and environmental sustainability in Bangladesh. *Environ. Sci. Pollut. Res.* 29: 3230–3245 https://doi.org/10.1007/s11356-021-17153-7

Singh, R., N. Gautam, A. Mishra, and R. Gupta. 2011. Heavy metals and living systems: An overview. *Indian J. Pharmacol.* 43(3): 246–253. https://doi.org/10.4103/0253-7613.81505

Singh, S., and M. Kumar. 2006. Heavy metal load of soil, water and vegetables in peri-urban Delhi. *Environ. Monit. Assess.* 120(1–3): 79–91. https://doi.org/10.1007/s10661-005-9050-3

Sonone, S.S., S. Jadhav, M.S. Sankhla, and R. Kumar. 2021. Water contamination by heavy metals and their toxic effect on aquaculture and human health through food chain. *Lett. Appl. NanoBioScience.* 10(2): 2148–2166.

Tonk, S., L.E. Aradi, G. Kovács, A. Turza, and E. Rápó. 2022. Effectiveness and characterization of novel mineral clay in Cd2+ adsorption process: Linear and non-linear isotherm regression analysis. *Water.* 14: 279. https://doi.org/10.3390/w14030279

Turek, A., K. Wieczorek, and W.M. Wolf. 2019. Digestion procedure and determination of heavy metals in sewage sludge—An analytical problem. *Sustainability.* 11: 1753.

Tytła, M., K. Widziewicz-Rzońca, J. Kernert, Z. Bernaś, and K. Słaby. 2023. First comprehensive analysis of potential ecological risk and factors influencing heavy metals binding in sewage sludge from WWTPs using the ultrasonic disintegration process. *Water.* 15: 666. https://doi.org/10.3390/w15040666

Vallius, H. 2014. Heavy metal concentrations in sediment cores from the northern Baltic Sea: Declines over the last two decades. *Mar. Pollut. Bull.* 79: 359–364. https://doi.org/10.1016/j.marpolbul.2013.11.017

Wan, X, and M. Lei. 2018. Intercropping efficiency of four arsenic hyperaccumulator *Pteris vittata* populations as intercrops with *Morus alba*. *Environ. Sci. Pollut. Res.* 25: 12600–12611.

Wan, Y., J. Liu, Z. Zhuang, Q. Wang, and H. Li. 2024. Heavy metals in agricultural soils: Sources, influencing factors, and remediation strategies. *Toxics.* 12: 63. https://doi.org/10.3390/toxics12010063

Wang, L., X. Peng, H. Fu, C. Huang, Y. Li, and Z. Liu. 2020. Recent advances in the development of electrochemical apt sensors for detection of heavy metals in food. *Biosens. Bioelectron.* 147: 111777.

Wcisło, E., D. Ioven, E. Cisco, D. Ioven, R. Kucharski, and J. Szdzuj. 2002. Human health risk assessment case study: An abandoned metal smelter site in Poland. *Chemosphere.* 47: 507–515.

Xiao, R., S. Wang, R. Li, J.J. Wang, and Z. Zhang. 2017. Soil heavy metal contamination and health risks associated with artisanal gold mining in Tongguan, Shaanxi, China. *Ecotoxicol. Environ. Saf.* 141: 17–24.

Yitagesu, Y.H., and E. Bekele. 2019. Impacts of cement dust deposition on heavy metal pollution in soil and barley crop grown around Abyssinia cement factory. *Chem. Mater. Res.* 11(2): 1–11. https://doi.org/10.7176/CMR

Young-Chull, A and Jae-Min , C. 2004. Physical, chemical & electrical analysis of dust generated from cement plants for dust removal with an electrostatic precipitator. *Korean Journal on Chemical Engineering.* 21(1): 182.

Zhuang, P., H. Lu, Z. Li, Zou, and M.B. McBride. 2014. Multiple exposure and effects assessment of heavy metals in the population near mining area in South China. *PLoS One.* 9(4): e94484. https://doi.org/10.1371/journal.pone. 0094484

8 Production of Rare Earth Elements and its Effects on the Environment and Human Health

8.1 INTRODUCTION

Rare earth elements (REEs) are essential building blocks for numerous cutting-edge energy and computer systems. You can't have a strong magnet without them, and they're used in everything from electric power generation to rechargeable battery components to lasers. High demand, geopolitical conflicts, a lack of sustainable sources, and insufficient recovery efforts have all put significant strain on the worldwide supply, mining, and distribution of REEs in recent years. The mining, refining, and use of REEs (rare earth elements) raise serious concerns for human and environmental health. The term "rare earth elements" (henceforth "REEs") is used to refer to a group of 17 elements in the periodic table: yttrium (Y), scandium (Sc), and the 15 elements in the lanthanide series. Rare Earth Elements (REEs) are abundant in the Earth's continental top crust, much more so than gold (Au) and platinum (Pt).

To highlight the major knowledge gaps and point to the roles of three relevant phenomena involved in REE-related effects—oxidative stress, hormesis, and medium acidification—this review looks at the limited research into the biological effects of REEs (Feng and Xu, 2000) in animals and plants, as well as the lack of case reports or geographic studies from human exposures. There is limited evidence on the health consequences of the other REE, notably heavy REE, and the toxicological database is primarily limited to Ce and La, with some data also available for Gd and Nd. Iatrogenic exposure, occupational exposure, and possible or suspect environmental exposure are all ways in which humans are exposed to REE (Pagano et al., 2015a). The use of Gd in MRI contrast agents is an example of an iatrogenic exposure. Pneumoconiosis has been observed in workers who were exposed to REE dust at work. REE bioaccumulation was found to be inversely proportional to the distance from mining sites in studies of environmental exposures in communities living near REE mining areas. Rare earth pneumoconiosis is an uncommonly reported disease caused by the inhalation of dust containing lanthanides, also known as rare earth metals, which are common industrial materials. The pathologic manifestations and natural history of this disorder are incompletely understood (McDonald et al., 1995).

The world is teeming with rare earth elements. They are an essential aspect of the framework upon which our contemporary civilization rests. They are the lifeblood of our cellphones, automobiles, and countless other devices. Environmental contamination, geopolitical monopolies, and future sustainability are just some of the problems brought about by the mining and distribution of rare earth elements, and they need to be addressed (www.sciencehistory.org/learn/science-matters/case-of-rare-earth-elements/home). Semiconductors, mobile phones, magnets, TVs, and even satellites are just some of the many modern technologies that can benefit from REE. Research into rare earth elements (REEs) has attracted a lot of attention in the past few decades because of their usefulness in

DOI: 10.1201/9781003618164-8

modern technology and culture. This chapter covers what we know about the factors that contribute to rare earth element pollution. The current effort aims to contribute to the growing body of research on the environmental and health effects of REEs. The rare earth elements (REEs) are complex, with both positive and negative impacts on human health.

8.2 INSIGHT ON RARE EARTHS

In recent years, rare earth elements have gained more attention and study. These diverse, and sometimes foreign, components are found in many of the objects on which we depend daily. The use of rare earth elements (REEs) is becoming increasingly crucial in cutting-edge engineering. Many cutting-edge technologies rely on rare earth elements (REEs), making it imperative that the supply be increased through the discovery of new sources and the development of more efficient extraction methods.

A miner in Ytterby, Sweden, found a mysterious black rock in 1788; this was the first known instance of a rare earth. The geological term for acid-soluble rocks in the 18th century was "rare," therefore, the ore was also called "earth." Yttria was named for the Swedish city where it was discovered in 1794 by chemist Johan Gadolin. Naming new elements was a highly respected and much-contested matter among European scientists in the 19th century. Both cerium and thorium were named after Jöns Jacob Berzelius, who first isolated them in 1803 and 1828, respectively. During Carl Gustaf Mosander's systematic study of a mixture of rare earths that he began in 1839, the elements lanthanum, erbium, and terbium were found and given their present names (Jyothi et al., 2020; Mehmood, 2018; Drobniak and Mastalerz, 2022; Balaram, 2019). The elements neodymium and praseodymium were both named after a student of Robert Bunsen's, Carl Auer von Welsbach.

The fields of chemistry that focus on rare earth elements are geochemistry, marine chemistry, and environmental chemistry (REEs). Due to their unique physical qualities, rare earth elements (REEs) are indispensable components in modern technology, especially in the semiconductor sector (Abigail and Paul, 2011). While their position at the bottom of the periodic table may appear out of place, many "green" and modern comforts rely on these elements. Rare earth metals consist of the lanthanides (from lanthanum to lutetium) plus scandium and yttrium (REEs). The valence electrons of all lanthanides are found in 4f orbitals, making them members of the f-block at the bottom of the periodic table. Non-bonding electrons in lanthanides, protected by 4d and 5p electrons, are responsible for the element's distinctive properties, such as light emission and magnetism. The f-electrons of Eu3+ can temporarily absorb UV light by hopping to a higher energy level, only to emit light as they return to Earth. One reason REEs can be found in lasers, energy-efficient light bulbs, and electronic displays is because of the ability of "excited" f-electrons to emit light. Many contemporary innovations rely heavily on rare earths, and this is explained by the elements' electrical configurations (King, 2021; Voncken, 2016).

The so-called "rare-earth" elements are actually rather common and have various modern technical products, applications (Behrsing et al., 2024), uses, and services (Figures 8.1 and 8.2) that are crucial to our daily lives in the areas of communication, work, education, safety, energy, transportation, etc. Like tin, lead, and zinc, they can be found in great quantities in the planet's crust.

The rare earth elements now play critical roles in many areas of modern technology, including magnets (neodymium and dysprosium) in wind turbines, automobiles, computers, and military guidance systems; lighting phosphors (europium and terbium); exhaust emission catalyst support ceramics and industrial catalysts.

Because of their vital role in producing strong permanent magnets—which are necessary parts of electric vehicle motors, wind turbine generators, and some hybrid car batteries—rare earth elements—particularly neodymium, praseodymium, dysprosium, and terbium—are mainly utilized in the energy and transportation sectors. In addition to being found in automobile equipment,

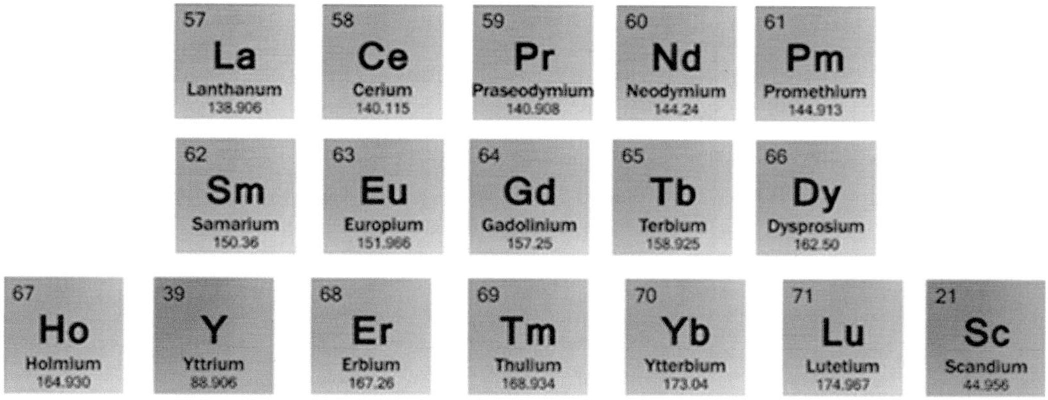

FIGURE 8.1 Rare Earth elements (Behrsing et al., 2024, Open Access.)

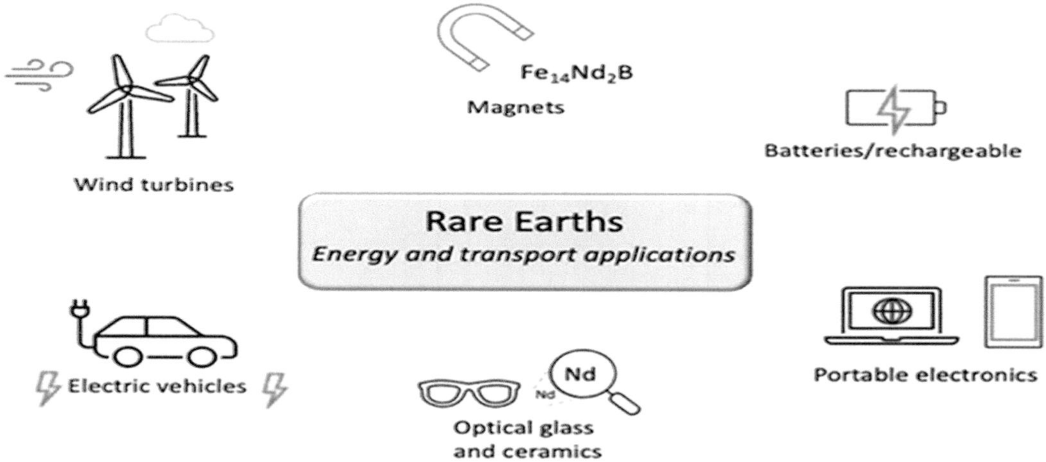

FIGURE 8.2 Uses of Rare Earths in Energy and Transport (Behrsing et al., 2024, Open Access.)

rare earth oxides, metals, and alloys are also utilized in the production of transportation-related components. Both the catalytic converter and the refinement of the gasoline used to power internal combustion engines (ICEs) require rare earth elements. In the refining of petroleum, rare earth elements (REEs) are crucial.

The discovery of these elements in very small mineral occurrences in the nineteenth century prompted the name "rare earths." The discovery of rare earths is always accompanied by the discovery of other rare earths or radioactive elements like uranium and thorium; rarely are they ever found alone or in large amounts (Gschneidner and Pecharsky, 2024). Rare earth elements are notoriously difficult to isolate from one another and their natural settings due to their distinct chemical characteristics (Voncken, 2016; Aide and Nakajima, 2020). These characteristics make it difficult to keep the surroundings clean. Rare earth elements (REEs) have become much more necessary in recent decades, mostly due to their employment as valuable additions in cutting-edge technology. Because of their distinct electrical, magnetic, and optical characteristics as well as their numerous uses in modern industries, including metallurgy, electronics, and optics, rare earth elements (REEs) are becoming increasingly important (Table 8.1). To extract a relatively small amount of rare earth

TABLE 8.1
Rare-earth elements (REEs) are used as components in high-technology devices

REE	Symbol	Application
Scandium	Sc	Aerospace lightweight high-strength Al-Sc alloys, electron beam tubes. Aluminium-scandium alloys for the aerospace industry and, defence industry.
Yttrium	Y	Capacitors, phosphors, microwave filters, glasses, oxygen sensors, radars, lasers, and superconductors. Red phosphors, fluorescent lamps, metals, Y-Fe-garnet resonators, ceramics.
Lanthanum	La	Ceramics, car catalysts, phosphors, hydrogen storage, special optical glass, and accumulators. Battery alloys, glasses, lasers, pigments, and accumulators.
Cerium	Ce	Polishing powders, ceramics, phosphors, glasses, pigments, misch metal, UV filters, and abrasives. Pigments. Catalysis, particularly in auto petroleum refining, metal alloys, Phosphors, corrosion protection, Carbon Arc Lamps, and cigarette lighter flints.
Praseodymium	Pr	Sources for measuring devices, miniature nuclear batteries, phosphors, Permanent magnets, photographic filters, ceramics, glasses, and pigments.
Samarium	Sm	Permanent magnets, microwave filters, nuclear industry, and reactor control rods.
Europium	Eu	Phosphors (red and blue), lasers, and fluorescent glass.
Terbium	Tb	Green phosphor, permanent magnets, and lasers.
Dysprosium	Dy	Phosphors, ceramics, the nuclear industry, and permanent magnets.
Erbium	Er	Ceramics, dyes for glass, optical fibers, lasers, and the nuclear industry.
Ytterbium	Yb	Metallurgy, chemical industry, and lasers.
Thulium	Tm	Electron beam tubes, visualization of images in medicine, and Magnets.
Gadolinium	Gd	Good Visualization of images in medicine, optical and magnetic detection, ceramics, glasses, crystal scintillators, Nuclear fuel bundles, medical imaging, and electronics.

Source: Yan et al., 2020. Creative commons Attribution 3.0 license.

metals, the present manufacturing procedures require a lot of ore and produce a lot of toxic waste. Byproducts of various processing techniques include acid, radioactive water, and poisonous fluorine.

Rare-earth elements (REEs) are used as components in high-technology devices, including smartphones, digital cameras, computer hard disks, fluorescent and light-emitting-diode (LED) lights, flat screen televisions, computer monitors, and electronic displays. Large quantities of some REEs are used in clean energy and defense technologies. Rare earth elements are crucial to the operation of many cutting-edge machines. REEs are used in a wide variety of industries, including electronics, clean energy, aerospace, automotive, and defense. Predictions for REE demand in 2020 indicate that magnet manufacturing will consume 29% of all REEs used (Yang et al., 2017). Permanent magnets are essential to the operation of many electronic devices, including cell phones, televisions, computers, vehicles, wind turbines, and jet aircraft. REEs are used extensively in cutting-edge technology and eco-friendly products due to their luminescent and catalytic properties (Ambaye et al., 2020).

8.3 ENVIRONMENTAL IMPACTS OF RARE EARTH MATERIALS

There is a significant reliance on rare earth elements in green technology (REEs). Rare earth elements are not hard to come by. They form a crucial part of the foundation upon which modern society stands. They are the impetus behind a wide range of technological advances. Some rare earth elements are highly sought after because of their exceptional magnetic, luminescent, and electrical properties, which are used in a wide variety of technological and consumer applications. Rare earth elements are indispensable in today's technological society. This is because rare earth elements are crucial for numerous medical devices. Since these technologies are at the heart of global communication, transportation, medical, energy creation, surveillance, and even war, questions of access and application are at the center of hot geopolitical debates. Rare Earths availability in the world is illustrated in Figure 8.3.

Although they tend to be found in trace amounts throughout the crust, rare earths are not truly uncommon on Earth. As of 2023, China is the world's largest country in terms of rare earth deposits, with an estimated 44 million metric tons of rare earth oxide equivalent (REO) in reserves. With 210,000 MT, the nation was by far the largest producer of rare earths in the world in 2022. To aid in the shift to a future free from climate change, we require rare earth elements. There are 22 million MT of rare earth reserves in Vietnam. Russia and Brazil share the third-largest reserves of rare earth elements in the world. India produced 2,900 MT and has 6.9 million MT of rare earth reserves. However, the mining and manufacturing processes used to get REEs often pose environmental risks. There has been some discussion on the environmental impacts of mining and producing rare earth elements (REEs) because of their increasing importance in green technologies, including wind turbines, electric vehicles, and low-energy lighting (Zapp et al., 2022). The outlook for rare earths is supported by strong supply and demand fundamentals as the world heads into a new economic era with a focus on clean energy and technological advancements. Rare Earths availability in the world is shown in Figure 8.3.

The geology of the deposit, the nature and concentration of the minerals, the extraction methods, the local infrastructure, and the presence of rules aimed at mitigating environmental damage are all relevant factors. Primary deposits are managed by open-pit and underground mining operations, including drilling and blasting for hard rock resources. The secondary deposit managers are in charge of all mining techniques, including dredge mining, manual surface collection, and mechanized dry mining with the use of ordinary earthmoving equipment. Rare earth elements frequently coexist with other elements that are damaging to human health when taken from the ground and processed into fine dusts, such as heavy metals, arsenic, and fluorite. Scientists studying the environment study toxic trace metals such as arsenic, lead, cadmium, mercury, and uranium.

Life cycle assessment (LCA) is a method for determining how harmful a product or piece of technology is to the environment. The LCA method, which takes into consideration the greenhouse gas effect, is used to classify emissions as having an acidification, eutrophication, or toxicity impact (Paul and Campbell, 2011).

There is a substantial quantity of hazardous waste and tailings produced as a result of REE processing due to the presence of radioactive and possibly harmful byproducts. Crushed rocks and fluids used in mills and concentrators make up tailings. At every phase of rare earth element extraction, the surrounding environment is in jeopardy. Mining is the primary means by which rare earth elements can be obtained. Once this is done, it is delivered to a refinery before being dumped. There are problems with mining, processing, and disposal. One of the drawbacks of REE mining is the contamination and depletion of local water supplies. Rare plants and animals are threatened by the depletion of their water source. Thousands of people were compelled to leave their homes and farms

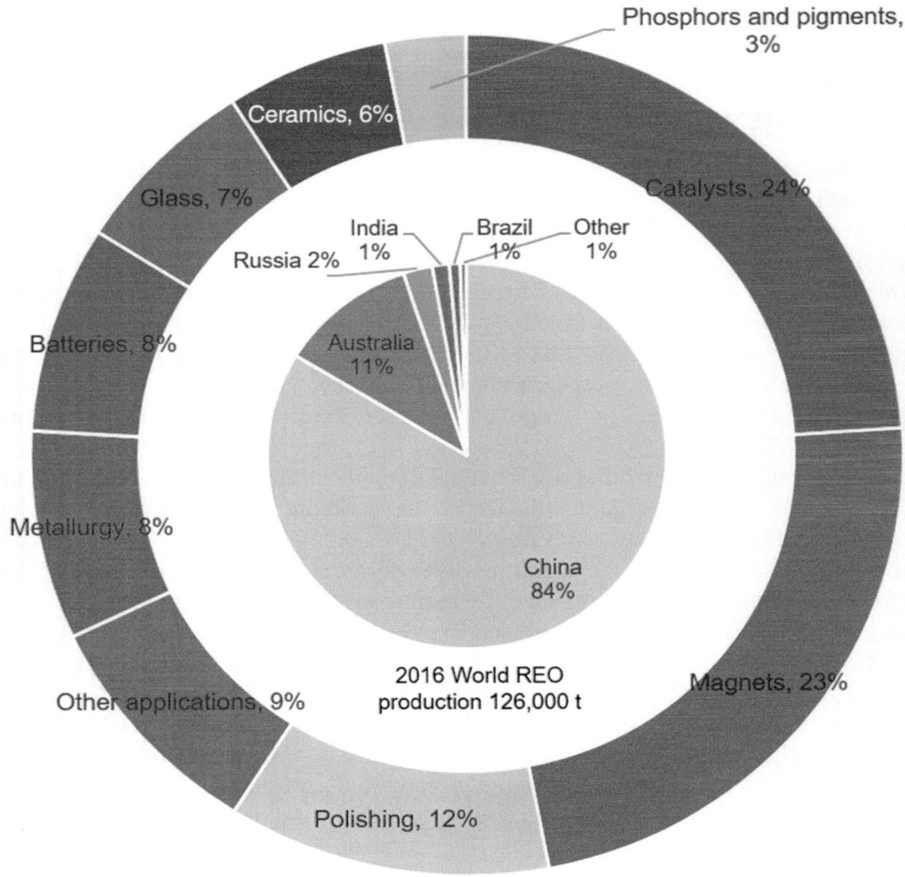

FIGURE 8.3 Global Potential of Rare Earth Resources (Zhou et al. 2017. https://doi.org/10.3390/min7110203. Open access.)

because they were poisoned by rare earth mining waste. Health issues such as painful legs, diabetes, osteoporosis, and heart problems are common among inhabitants (Ali, 2014).

Balaram (2019) offered a thorough evaluation of the many issues connected to rare earth elements. The theoretical framework for evaluating risks is given by Kapustka et al. (2019) and Weber and Reisman (2012). Every form of mining has the potential for unintended repercussions, but the severity of these effects depends on factors like mine management and safety protocols, processing processes, and waste treatment systems. The geology of the area around the REE deposit (and the likely existence of other metals and agents inside it) is also crucial. This includes the local climate, the closeness of streams, lakes, and underground aquifers, and the availability of water. Exposure to REE is one possible source of health problems, while the cumulative effect of mining on the ecosystem is another. Negative effects on nearby flora and animals may result from the mine's rock and dust waste contaminating the land.

The same rocks that contain radioactive components like uranium and thorium also contain rare earth elements. Rare earth element processing generates radioactive waste that persists for very long periods. The existence of radioactive isotopes in rare earth elements necessitates the stringent regulation of ore quality. Radiation concerns typically arise during ore processing, which can lead to thorium production. Most thorium decay produces alpha particles, so focusing on monitoring for

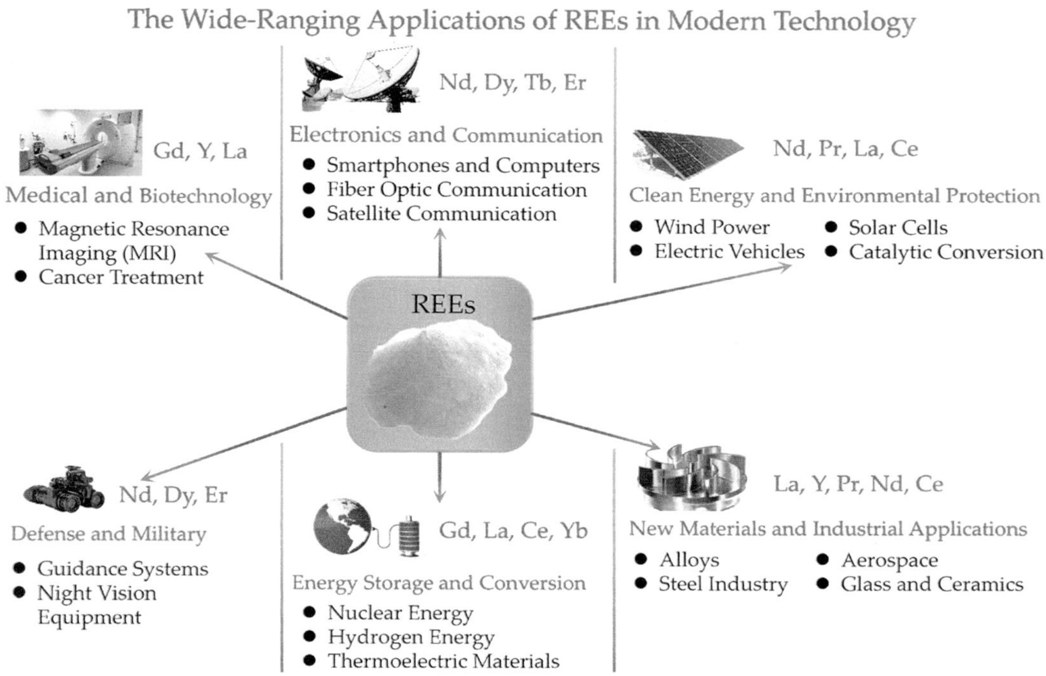

FIGURE 8.4 The many uses of rare earths: Most common application areas are electronics, Glasses, Ceramics, Magnets, and Metal alloys (Liang et al., 2024, https://doi.org/10.3390/molecules29194624. Open access.)

these sources specifically is important. Ingesting alpha radiation is more harmful than being exposed to it because of its short range. Natural rare earth carbonates provide a buffer against the acidity that can result from the various leaching processes. Possible impacts have been identified, including kidney disease, heart disease, occupational lung disease, and an increased chance of acquiring multiple types of cancer (Klinger, 2018; Rim, 2013, 2016).

However, longitudinal epidemiological studies are lacking. Rarely, exposure to REE can cause nephrogenic systemic fibrosis, a condition characterized by an abnormal accumulation of connective tissue in the skin, joints, eyes, and internal organs. A higher chance of having a heart attack is linked to this disorder. Blood levels of a certain protein were much lower in those who had previously resided in a REE-rich region. Children who were exposed to REEs had a significant decrease in IQ (Fan et al., 2005). This has altered the rate at which human red blood cells divide and multiply, perhaps due to changes in neurotransmitters in the brain. DNA replication and repair were both hampered. Environmental contamination caused by REEs has been related to an increased risk of leukemia (Figure 8.4). Many authors have given the applications and uses of rare earth elements (Walters and Lusty, 2011; Hu et al., 2018; Haque et al., 2014). Figure 8.4 presents the uses and applications of rare earths.

The boom in technological advances in recent decades has led to increased demand for rare earth elements (REEs) (also known as rare earth metals) across various industries with wide-ranging industrial applications, including in the clean energy sector, but with some health and environmental, economic, and social footprint concerns. Rare earth elements and their effects on the environment and human health are given in Table 8.2 (Giese, 2018; Hu et al., 2018; Haque et al., 2014).

Among other things, REEs are utilized in the manufacturing of energy systems, permanent magnets, storage batteries, radar (and radiography) equipment, portable electronics (like smartphones), sophisticated television sets, and many types of electric vehicles (both light and heavy). Some of

TABLE 8.2
Rare earth elements and their effects on the environment and human health

La	Lanthanum oxide nanoparticles can be used for MRI, Hybrid engines, metal alloys, catalysis, phosphors, Carbon Arc Lamps, and cigarette lighter flints.
Ce	Cerium-doped lutetium orthosilicate is a scintillator that has been mainly used for PET imaging, a type of test that reveals tissue and organ function, petroleum refining, metal alloys, Phosphors, corrosion protection, Carbon Arc Lamps, and cigarette lighter flints.
Pr	Praseodymium oxide nanoparticles have been used in radiotherapy techniques, Magnets, Optical Fibers, and Carbon-Arc Lamps.
Nd	Neodymium has been used in lasers as crystals and is employed in the treatment of skin cancers, as well as laser hair removal. Hard drives in laptops, headphones, hybrid engines, and Nd-Fe-B magnets.
Sm	A Radioisotope Sm-153 has been used to treat severe pain in patients whose tumors have advanced into bone tissues. Sm-Co magnets, IR absorption in glass.
Eu	Europium presents bio applications due to its optical properties as Nano probes with an emphasis on their heterogeneous/homogeneous bio detection as well as in vitro and in vivo bio imaging. Red colour for TV and computer screens.
Gd	Gadolinium enhances MRI images of tumors, and its magnetic properties are also of use in intravenous radio-contrast agents in MRI scans. Magnets, nuclear magnetic resonance imaging, phosphors.
Tb	A Radioisotope Tb-149 has been used in targeted cancer therapy. Phosphors, particularly for fluorescent lamps, and magnets.
Dy	A radioisotope Dy-165 has been employed in the treatment of rheumatoid knee Effusions.
Ho	Holmium-based solid-state lasers have been used for non-invasive medical procedures for treating cancers and kidney stones.
Er	Erbium-based lasers have been used in medical and dental practice. Red, green phosphors, amplifiers for optical fibre transmission, pink in glass melts, sunglasses.
Tm	A Radioisotope Tm-167 has been used as a power source in portable X-ray devices.
Yb	A radioisotope Yb-176 can be used to produce Lu-177, which is known to be a promising radioisotope for a medical application.
Lu	Lutetium is being researched for its potential uses in targeted radiotherapy, for the advancement of new cancer therapies such as prostate cancer. Ce-doped Lu-glass used in detectors for positron emission tomography—PET.

Sources: Giese, 2018; Hu et al., 2018; Haque et al., 2014. Open access.

their uses are described in Table.8.3. In addition to being crucial as catalysts and ingredients in a variety of industries (such as petroleum refining, specialty steel, the glass and lighting industries, the aerospace and defense industries), REEs are anticipated to receive increased attention for the production of solar panels, electric motors, electric vehicles and their batteries (hybrid, full battery, and fuel cell), wind turbines and their systems, and other energy systems as the problems associated with climate change and the pressure on sustainability responses increase. The versatile nature of rare earth elements is evident in their applications (Table 8.3).

The rare earth industry, along with the associated research, technology, and policy, co-evolves with broader societal, cultural, and political forces operating at several scales. Protecting the environment and public health from the additional contaminants created by mining and processing rare earth elements is crucial to achieving sustainable development. In most soil cases (Aide and Aide, 2012), the whole soil REE concentrations are more directly related to the presence of a few trace minerals, such as apatite and zircon. Ce and Eu were negatively anomalous. REE emissions from human activities will accumulate in groundwater and soil.

Ramos et al. (2016) looked at where rare earth elements (REE) are found in soils (Aide and Aide, 2012) and the Environment. They examined the significance of rare earth elements (REE) for both soils and the plants they sustain. Risks to ecosystems and human health from REE in soils were also demonstrated. The presence of rare earth elements (REE) in soils and their possible transport

TABLE 8.3
The prominent latest applications of the REEs

Name Usage	Usage
Scandium (Sc)	As an alloy additive in aluminium-based products (such as Al-Mg alloys) for strength. In "super-alloys", such as steel. Additionally, in NiMH batteries.
Lanthanum (La)	In optics and lighting (e.g., lasers, LEDs, and fluorescent lamps, though these are being phased out), and more specifically, in carbon lighting. Capacitors, sensors, and semiconductors. Additionally, in NiMH batteries, super-alloys, AL-Mg alloys, and steel. Flat display electronics. X-ray imaging. Petroleum refining, automotive catalysts, and diesel additives. Water treatments. Fuel cells. Polishing colorants.
Cerium (Ce)	The catalyst for lighters. In carbon arc lighting. For glass polishing. In NiMH batteries, super-alloys, AL-Mg alloys, and steel. Flat display electronics. X-ray imaging. In capacitors, sensors, and semiconductors. Petroleum refining, automotive catalysts (converters), and diesel additives. Water treatments. Fuel cells. Polishing colorants.
Praseodymium (Pr)	In alloys for aircraft engines. In motors and generators, more widely. For NdFeB magnets instead of Nd. In NiMH batteries, super-alloys, AL-Mg alloys, and steel. Magnetic resonance imaging (MRI) contrasting agents. In capacitors, sensors, and semiconductors. In flat display electronics. In addition, for X-ray and MRI imaging. In HD drives, microphones, and speakers. For petroleum refining, automotive catalysts, and additives. For water treatments. In fuel cells. For polishing colorants.
Neodymium (Nd)	Production of magnets incorporated into electronics, such as computers and cell phones. The basic element for the NdFeB magnets. It is widely used in many vehicle components, such as motors, brakes, speakers, doors, windows, and air conditioners. In NiMH batteries, super-alloys, AL-Mg alloys, and steel. Magnetic resonance imaging (MRI) contrasting agents. Also in HD drives, microphones, and speakers.
Europium (Eu)	The most reactive REE and not so widely used. It may be used in lasers. Flat display electronics. X-ray imaging. Capacitors, sensors, and semiconductors. In addition, in NiMH batteries, super-alloys, AL-Mg alloys, and steel. Flat display electronics. Petroleum refining, automotive catalysts, and diesel additives. Water treatments. Fuel cells. Polishing colorants. Fluorescent powders are used in lighting equipment, though this is being phased out by LED.
Yttrium (Y)	As additive alloys (mostly with aluminium and magnesium) are used to increase the strength of materials. It can be used to produce YVO_4 europium and Y_2O_3 europium phosphors, and gives the red color in color television sets. Flat display electronics. X-ray imaging. Fluorescent powders are used in lighting equipment, though this is being phased out by LED.
Gadolinium (Gd)	Not widely used, but may be used in alloy production. NiMH batteries, super-alloys, AL-Mg alloys, and steel. Flat display electronics. In HD drives, microphones, and speakers. In capacitors, sensors, and semiconductors. For X-ray and MRI imaging contrast agents. For petroleum refining, automotive catalysts, and diesel additives. Water treatments. Fuel cells. Polishing colorants.
Terbium (Te)	Calcium fluoride, calcium tungstate, and strontium molybdate are all used in the making of solid-state devices. Flat display electronics. X-ray and MRI imaging contrasting agents. In HD drives, microphones, and speakers.
Dysprosium (Dy)	The basic element in NdFeB magnets is used for a range of vehicle components. May feature in nuclear reactors. It may be combined with other REEs in laser material. For magnetic resonance imaging (MRI) contrasting agents. In HD drives, microphones, and speakers.
Holmium (Ho)	In laser production. In capacitors, sensors, and semiconductors. In NiMH batteries, super-alloys, AL-Mg alloys, and steel. Flat display electronics. X-ray imaging. Petroleum refining, automotive catalysts, diesel additives. Water treatments. Fuel cells. Polishing colorants.
Ytterbium (Yb)	It may be used in memory devices and lasers. Furthermore, as an industrial catalyst.
Thulium (Tm)	The least abundant REE. It may be used in X-ray machines. Has the potential as an energy source.

Source: Filho et al., 2023. Open access.

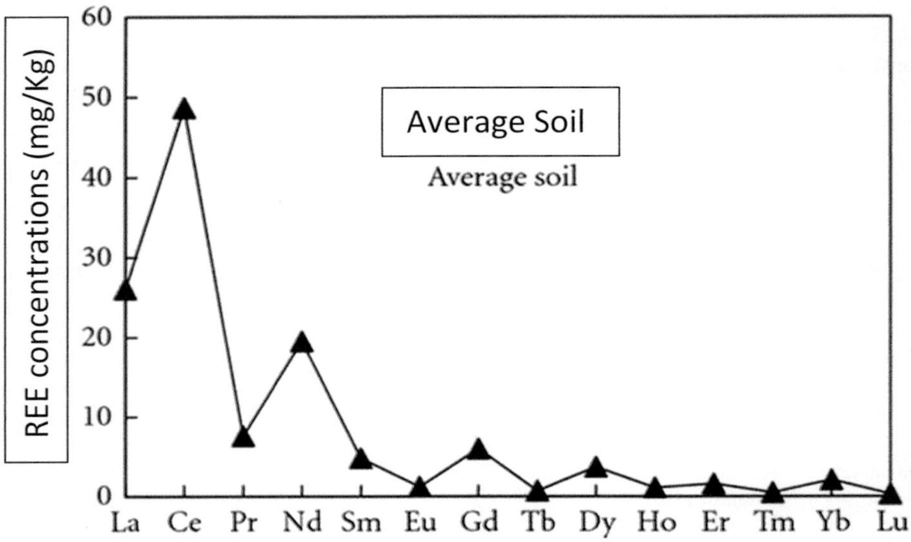

FIGURE 8.5 REE concentration values for an average soil (Aide and Aide, 2012. https://creativecommons.org/licenses/by/3.0/.)

up the food web is discussed by Ramos et al. (2016). Bone, liver, and lungs are three of the most common storage organs. The list of potential adverse effects is long (Li et al., 2013; Wyttenbach et al., 1994), and includes problems with the heart, liver, blood, and kidneys as well as the intestine, bones, nervous system, lungs, and cytogenetics. REE accumulation in plants has been improved by the development of more sensitive technologies, such as inductively coupled plasma-mass spectrometry (ICP-MS) (Trimmel et al., 2023). Figure 8.5 shows REE concentration versus the average soil relationship. Concentration values were reported in Pendias Kabata (2000).

8.4 RARE EARTH ELEMENT EXPOSURE IN HUMANS

Rare Earths (REs) are essential in many fields and are sometimes referred to as industrial vitamins. Human health may be affected by occupational and environmental rare earth element (REE) exposures due to the widespread usage of REEs in a variety of technologies (Pagano et al., 2019). Over the past few decades, researchers have amassed a wealth of experimental evidence on the toxicity of REEs, giving crucial context for understanding the risks associated with REE exposures. Researchers have looked at the health of people living near REE mines from a geographical perspective, reporting on bioaccumulation and links between residential exposure to REEs and negative health outcomes. Increased levels of several REEs in exposed people have been linked to tobacco use and indoor smoking in a recent line of investigations. Damage to the respiratory system was observed in investigations of workers exposed to REEs in the workplace. Bioaccumulation of REEs in scalp hair, elevated REE levels in urine, and faulty gene expression have all been linked to working with rare earth elements (REEs). Other occupations that expose workers to REEs include: a) those in the film industry, where they are exposed to REE aerosol; b) those involved in the processing of electronic waste; and c) those involved in the repair and maintenance of diesel engines, where they are exposed to exhaust microparticulate matter (containing nanoCeO_2 as a catalytic additive). Multiple pathogenic effects in animals exposed via respiratory or systemic pathways to diesel exhaust microparticulate have been documented in animal models (Pagano et al., 2015a, b).

A series of studies evaluating scalp hair REE accumulation as a function of distance from mining sites and with different accumulation extent according to gender and age have been conducted in regions of China involved in REE mining and processing, raising health concerns for children who are highly exposed to REEs. Several research groups have published accounts of these experiments (Wei et al., 2013; Wang et al., 2018). Two studies by Wang et al. (2017, 2018) found that REE hair levels and exposure to REE-containing indoor air pollution were both associated with an increased risk of hypertension in Shanxi Province housewives. The effects of REE on human health were discussed by Wang et al. (2024).

Rare earth elements (REEs) affect people's daily lives in profound ways since they are present in so many modern conveniences, including automobiles, computers, cell phones, televisions, and energy-efficient fluorescent light bulbs. Rare earth elements (REEs) have been the subject of scant environmental research (Ali, 2014). The impact of REEs on the environment is typically overlooked in favor of that of more harmful elements like arsenic, cadmium, mercury, and lead. For instance, additional hazardous elements contained in REE-bearing ore and in waste streams pose the greatest risks when mining REE ore. Naturally occurring radioisotopes of elements like thorium and uranium are present in REE processing waste. The closing of the Mountain Pass mine in California in the 1990s was partially attributable to the high price of cleaning up radioactive wastewater released during the processing of rare earth element (REE) ore.

In the early 1980s, scientists began to notice that mining and processing REEs had a negative effect on the ecosystem. San Pedro Shelf sediment cores were analyzed, and the results showed that there were 890-6900 times as many light REEs (La, Ce, Nd, and Sm) as there were in crustal material. The wastewater outfall that discharged the REEs into the basin supplied water to 3.5 million residents, tens of thousands of commercial and industrial establishments, and most significantly, seventeen oil refineries. Most wastewater is released into rivers, which can lead to REE contaminants being washed ashore or out to sea during floods. On average, the Rhine River dumps 1.5 metric tons of man-made La into the North Sea annually (www.envchemgroup.com/rare-earth-elements.html).

Environmental dangers associated with mining rare earth elements (REEs) are distinct from those associated with other types of mining due to the radioactive pollution and REE toxicity that result from the process. Co-occurrence of REE deposits with radioactive elements like U (uranium) and Th (thorium) and their decay products (like radium) is the primary cause of radioactive pollution during REE production (https://cdnsciencepub.com/doi/10.1139/er-2020-0115).

The area around a REE mining site typically has a greater concentration of radioactive materials. In the Bayan Obo mining zone, for instance, the mean content of airborne Th in total suspended particulate ranged from 820 Bqm^3 in mining regions to 39,720 Bqm^3 in smelting areas, which is significantly higher than the 0.5 Bqm^3 global background level (Wang et al., 2014). Therefore, there is an increased danger of radioactive element exposure in biota located close to REE mining. Due to the negligible levels found in the environment naturally and the benefits of REEs in agricultural production, the direct toxicity of REEs is typically neglected despite radioactive pollution's longstanding status as a public health priority. Due to increased REE mining and production (Petra et al., 2022), there has been a dramatic rise in REE concentration in some surface settings. For instance, the Sidaosha River (which flows through Baotou City, where the Bayan Obo REE deposit is located) has REE concentrations that are 30,000 gL^{-1} more than the average value of REEs of other rivers in Northern China in terms of both suspended particles and surface sediments (Liang et al., 2014).

The energy required to obtain 1 t of an individual REE varies from 38 to 48 GJ, except for scandium (148 GJt 1) and yttrium (75 GJt 1) (Peiró and Méndez, 2013). In Bayan Obo, Mongolia, for example, the extraction of 1 t of REE requires 4.41 t of sulfuric acid, 12.32 t of sodium chloride, and 1.64 t of sodium (Peiró and Méndez, 2013). After mining companies released contaminated wastewater into streams near the mines, a survey of nearly one hundred locals living near the mines in Kachin Special Region 1 in 2018, found that locals had not used watercourses near the mines for four years.

The mining and processing of REEs can alter their geochemical behavior and environmental fate, turning them into toxicants (Šmuc et al., 2012; Arienzo et al., 2022). First, the production process leaves behind goods and tailings that include residual REEs, which add to the REE composition of surface habitats. The second concern is that mining operations may alter the physicochemical properties of the surrounding soil and water. Increased bioavailability and environmental mobility of REE components can occur when they are solubilized by acidity but are otherwise stable. High levels of rare earth elements (REEs) have been found in soils, vegetables, and other crops, as well as human bodies, in areas close to mining and processing facilities (Li et al., 2014). While other types of mining may also cause the aforementioned environmental problems, the radioactive contamination and toxicity linked to REEs production are drawing increased international attention as unique environmental dangers. Deforestation, soil erosion, water contamination, destruction of wildlife habitats, and changes in landscape structure are just some of the negative environmental effects of REE extraction from ore (Machácek, 2019).

The effects of REEs vary depending on factors such as the kind of rock surrounding the REE deposit, the presence of other metals or agents inside the rocks, the local climate, and the proximity of streams and lakes to the mine. Vegetation in mining locations is also destroyed as a result of REE mining due to the increased footprint of mining operations, which hurts agricultural land and water quality (Lindagato et al., 2023).

Some of the negative effects of REEs on environmental health include bioaccumulation, radiation exposure, species invasion, and biodiversity loss. Toxic consequences on the soil microfauna community have resulted from the accumulation of REEs in soils and their usage in agriculture as microelement fertilizers. Because of this, REEs are increasingly found in ecosystems and throughout the food chain. The active migration of pollutants also causes damage to the surrounding environment, including degraded landscapes, crops, soils, water sources, and the standard of living for people in the area (Charalampides et al., 2016).

Chemicals and acids are used in REE processing, and they can be dangerous if not handled properly. Soil and groundwater contamination can result from the improper disposal of spent chemicals and trash during REE processing at hazardous waste sites. Furthermore, e-waste, which often contains rare earth metals and other harmful compounds, might have indirect environmental and health implications from the usage of REEs in technology. There are environmental and public health risks associated with improper e-waste disposal, according to research by Migaszewski et al. (2016).

Globalization and increased industrialization are major factors in the growing demand for rare earth metals (REEs). The medical, renewable energy, electrical, agricultural, and military sectors are just a few of the many uses for these metals around the world. Since REEs are predicted to play a significant role in the world's future, we can anticipate a rise in contaminated areas over the next few decades as production ramps up. As a result, it has serious consequences for the health of the planet, its inhabitants, and the animals that share it (Oladipo et al., 2023). There are few studies looking at REE production, deposits, and associated health implications in the African context, despite the growing interest in REEs beyond China in recent years. Sustainable processing methods, waste handling systems, and proper mine management have all been discovered and proposed on a worldwide basis, but the nature and scope of adopting these management protocols on the African continent have been less evident.

The widespread usage of these elements in a variety of industrial and consumer items has led to an increase in human and animal exposures in the last decade due to food and water contamination. There is growing evidence that ingesting or being exposed to contaminated environments containing REEs poses a health risk. For instance, REE miners have been linked to an increased risk of chronic interstitial lung disease and pneumoconiosis. Oladipo et al. (2023) investigated the toxicity of REEs at the crossroads of human, animal, and environmental health. Toxic effects on the immune system, on the development of cancer, and on the nervous system are only a few of the many that fall under this category.

Studies of mine employees and other regular users of REE, whose exposure is often far higher than what the general population would experience, provide nearly all the information about the negative consequences of REE (www.carexcanada.ca/Timis Mining REE.pdf). When REE levels are this high, there is a greater chance of a heart attack. Pneumoconiosis is a lung condition caused by exposure to mining dust and metal vapors. Nephrogenic systemic fibrosis is an uncommon illness characterized by the accumulation of connective tissue throughout the body, including the skin, joints, eyes, and internal organs. Blood protein levels were abnormal in those who had lived in an area with high REE concentrations. The IQs of children who were exposed to REE were significantly lower. This is likely caused by a chemical imbalance in the brain, which influences alterations in the generation of human red blood cells. It also interfered with DNA replication and repair. REEs in environmental contamination were linked to leukemia (Qin et al., 2002).

There are always human rights and environmental concerns associated with mining raw resources. Many cell phone manufacturers source their gallium and indium from China and South Korea. Less than 10 grammes of a phone's total weight is made up of such components. However, the sale of these grammes supports a global mining sector that is responsible for radioactive waste disposal, water contamination, and the forced relocation of indigenous peoples. No one, not even those working in the sector, would argue that extracting lithium doesn't have a major impact on the natural world. However, lithium isn't the only raw mineral with negative consequences. Large areas of China have been ravaged by the quest for rare earth elements (www.dw.com/en/toxic-and-radioactive-the-damage-from-mining-rare-elements/a-57148185), with just one ton of rare earth elements producing two thousand tons of toxic waste. To extract rare earths from other ores, local businesses have taken to spraying acid over the pollution created by rare-earth-element mining regions. Afterwards, the land is typically abandoned. They can no longer be used for farming. The Environmental Performance Index assesses 180 countries on conservation, air quality, waste management, and emissions; states like Madagascar, Rwanda, and the DRC fare poorly. Therefore, environmentalists are especially worried about the mining of very dangerous elements in these nations, such as beryllium, tantalum, and cobalt. Villages in China pay a high price for rare-earth mining. This information is almost ten years old. Farms and towns in the area where the precious minerals are processed are being contaminated by pollution.

There has been a rise in popularity for recycling electronic and industrial scrap, such as printed circuit boards, scrap vehicle magnets, and hard drives, to recover rare earth elements (REE). However, recycling poses risks to the environment and the health of the people involved in the element recovery process. The difficulty of recycling is exacerbated by the fact that REEs are often only used in trace amounts. Thus, it is estimated that only around 20% of e-waste gets recycled and recorded, while the remaining 80% is just dumped in landfills. No function or necessity for REE is known to exist in biological systems (Rim et al., 2013). Ce, Eu, Gd, La, Nd, Pr, and Sm are considered "light" REE, while Dy, Er, Ho, Lu, Tb, Tm, Y, and Yb are considered "heavy" REE (www.reehandbook.com/intro.html). The heavy REE tend to resemble calcium and deposit in the bone, while the light REE are primarily found in the liver.

When inhaled, REEs cause acute irritative bronchitis and, over time, pneumoconiosis and progressive lung fibrosis, especially among workers (www.nrc.gov/docs/ML1513/ML15134A344.pdf). Pneumoconiosis and lung fibrosis are most common in those who work with carbon arc lamps for lengthy periods of time, such as photoengravers and film projectionists (Porru et al., 2001). In many but not all of these instances, the existence of a radioactive component is unrelated to the elevated REE concentrations found in the lungs. Exposure to both radioactive elements and silica dust complicates the assessment of such cases. Occupational exposure limits (OELs) have not been established for any of the REEs except yttrium, despite the well-documented effects of REEs on the lung. Elevated levels of REE in the fingernails have been linked to inhalation exposure.

Conjunctivitis, corneal damage, and scarring have all been linked to REE exposure in the eyes (Harbison, 1998).

8.5 EVALUATION OF RARE EARTH ELEMENT EXPOSURE IN LOCAL CEREALS

Multiple exposure routes, mainly through food consumption, have been found to be responsible for the uptake of REE from soil and water by humans (Zhuang et al., 2017). The consequences of REE accumulation on organisms are still incomplete and inconsistent because REEs are not necessary components. There is still a lack of knowledge on the toxicological mechanisms and associated environmental risk. Long-term exposure to REE may be associated with health problems, such as alterations in the brain and bone, even though there have been no reports of human poisoning through the food chain. This is because REEs are stored in the blood, brain, and bone after entering the human body. In addition to this, the effect of REE on children's neurodevelopment is gaining attention. Children's studies (Zaichick et al., 2011; Chen, 2005) suggested a link between REE and lower IQ and cognitive function.

Wheat had 150 percent greater REE contents than maize in both mining and control zones. Higher REE concentrations in maize were found in the mining region compared to the control region ($P = 0.03$). There was no statistically significant change in REE concentrations between the mining region and the control area for either wheat or legume. Wheat has the highest REE concentration out of all the cereals, followed by legumes, and then corn. According to research on rare earth element absorption and distribution (Zhuang et al., 2017), REE concentrations in maize roots were significantly greater than in the rest of the plant. More research revealed that REEs were tightly linked to the root cell walls. This could explain why maize has lower quantities of rare earth elements (REEs) than wheat and legumes.

Multiple routes of exposure, including food consumption, exist when rare earth elements (REEs) in soil and water are destabilized and enter the human body. Unintentionally, their increased usage has led to their incorporation into FCM (Food and Food contact materials) (www.digicomply.com/blog/rare-metals-risk-assessment). Concerns have been expressed concerning the possible effect of chronic exposure to low levels of REEs on human health, even though there have been no reports of human poisoning via the food chain. Long-term exposure to REEs may cause health concerns, including alterations to the brain and bones, and there is evidence that they accumulate in the human body after entering the bloodstream. Rare earth elements (REEs) have been detected in the rib bone of a healthy human being, suggesting that they accumulate there throughout a lifetime.

Rare earth elements (REEs) are widely used as fertilizers and feed additives, as described by Li et al. (2012). The accumulation of REEs in the food chain has hurt human health, and their concentration has increased in environmental and biological samples. Because of their toxicity and rising prevalence, the quantification of trace REEs has become increasingly important. Li et al. (2012) utilize inductively coupled plasma mass spectrometry and walnut shell as a green adsorbent for online preconcentration and detection of rare earth elements (REEs) in food and agricultural goods (ICP-MS). The walnut shell-packed microcolumn has a high adsorption efficiency and a high tolerated capacity for coexisting ions because of its porous surface and numerous COO- groups. Li et al. (2012) noted that at optimum conditions, the adsorption efficiency of the walnut shell packed microcolumn reached 99.9 percent for 15 REEs, and that the tolerable concentration ratios for 37 types of coexisting interfering ions ranged from 2,000 to 80,000,000. Using a loading amount of 4.7 mL for the sample, they found that the enhancement factors for 15 REEs ranged from 79 to 102. The sensitivity ranged from 2 to 34 pg g^{-1}. For a 2.5 ng L^{-1} REEs solution, the relative standard deviation for 11 replicate pre-concentrations was between 0.5 and 2.0%. The proposed method has been effectively used for the selective identification of REEs in 18 locally gathered food and agricultural products, as well as 4 recognized reference materials from the environment and biology.

Glass polishers, photoengravers, and movie projectionists are just a few examples of occupations that expose workers to REEs, and in 2015, Pagano et al. highlighted some of these exposures and presented a few case reports on respiratory health impacts. There were no case-control or cohort studies found that looked at REE exposures in the workplace. Biomonitoring of communities living in REE mining sites has revealed the buildup of REE due to environmental exposure. NMR contrast agents containing gadolinium have been cited as a potential source of iatrogenic REE exposure (Bernstein et al., 2012). The liver, lungs, and blood are only some of the organs that have been proven to be negatively impacted by REE poisoning in animal studies. On the other hand, zootechnical operations often mention REE as a safe and promising device, presumably pointing to a hermetic effect known for both REE and other xenobiotics.

8.6 THE EFFECTS OF RARE EARTH ELEMENTS (REES) ON MARINE LIFE

Heavy metal pollution has significant impacts on aquatic fauna and flora. It accumulates in marine organisms, both plants and animals, which are then consumed by humans. This can lead to various health problems, such as organ damage and the development of cancer. Additionally, this pollution causes biological magnification, where the toxicity concentration gradually increases as aquatic organisms continuously accumulate metals. The sources of these pollutants come from anthropogenic activities such as agricultural waste, household wastewater, docking and embarking activities, landfill erosion, sewage from industrial sources, and a few natural approaches that have become heavily polluted with trash and metals.

Malhotra et al. (2020) provide a comprehensive analysis of the toxic effects of rare earth elements (REEs) on aquatic organisms. The possibility of REE pollution in water bodies may grow as a result of the disposal of obsolete equipment containing REEs, the usage of REEs containing phosphate fertilizers, mining, and dispersion from indigenous rocks. Therefore, the contamination can cause their dispersal into the natural environment. Malhotra et al. (2020) concentrated on the bioavailability, accumulation, interaction, and toxicity parameters of REEs with aquatic species. Malhotra et al. (2020) noted that significant knowledge gaps persist on the health implications, but that more research is needed. There are both positive and negative outcomes associated with REE exposure in terms of health. Research has shown that REEs are stimulatory or protective at low doses but toxic at large doses, suggesting a hermetic concentration-related pattern. Malhotra et al. (2020) pointed out that additional in-depth investigations on the impacts of REEs are needed because there is still so little data on the topic. Research results should be broken down by specific endpoints, including growth inhibition, embryo toxicity, cytogenetic impacts, and organ toxicity. Aquatic biota, according to these researchers, can serve as a useful model for examining how REEs affect human health. Pagano's (2016) book, Rare Earth Elements in Human and Environmental Health, explores the many facets of REEs, including both their possible advantages and severe health impacts. The adverse impacts of REEs on human and environmental health raise a growing concern not only in the scientific community but also among a number of stakeholders. Malhotra et al. (2020) concentrated on the bioavailability, accumulation, interaction, and toxicity parameters of REEs with aquatic species. The occurrence and prolonged presence of HMs in aquatic ecosystems is increasing rapidly, which is obvious everywhere in the world, and their lead paves the way for a major concern on a global scale.

8.7 THE RECYCLING OF RARE EARTH ELEMENTS AND THEIR POTENTIAL TOXIC EFFECTS

As REE production rises in the coming decades, more and more locations are likely to become poisoned (Aik et al., 2020). When REE levels in living things rise, the toxicity limits of those things might be crossed, leading to harm (e.g., growth inhibition or illness). The toxicity of REEs is garnering more attention as more potential exposure sources come to light, including mine tailings, wastewater, dust, e-waste, REE-containing items, and medical uses. REEs, unlike monazite and bastnasite, have mines in a relatively small number of countries while having relatively large mineral deposits. Dust, metal, radioactive material, organics, and wastewater generated during mining, purification, and separation contribute to bioaccumulation, human toxicity, and environmental contamination (Shin et al., 2019).

Due to the employment of potentially dangerous chemicals or the discharge or production of other potentially harmful substances, the REE recycling procedures may necessitate extreme care. As the REE sector grows, so do concerns about worker safety and health. Workers who were regularly exposed to REEs developed lung fibrosis and pneumoconiosis. Supporting these concerns are reports of the in vivo and in vitro toxicity of REEs. As with REE mining and processing, REE recycling can pose risks to human health and the environment due to its similarities to these activities (Pagano et al., 2015a; Waring and Watling, 1990).

Using concentrated sulfuric acid (96 percent) during the high-temperature calcination process results in the generation of 9,600 to 12,000 m^3 of waste gas, including hydrofluoric acid, sulfur dioxide, and concentrated dust, and 75 m^3 of acidic wastewater, as reported by Hurst (2010). The oral and inhalation lethal dose 50 (LD50) in experimental animals was greater than roughly 2,000 mg/kg or 5,000 mg/m^3 (4 h) for most REE compounds (Damian, 2014). However, in a 13-week inhalation toxicity study of Ce oxide in rats, the lowest detected adverse impact level was 50.5 mg/m^3 due to increased alveolar epithelial hyperplasia, indicating an RfC of 0.0009 mg/m^3.

Due to pulmonary alveolar proteinosis and inflammatory alterations, the LD50 for La and Nd oxides in rats after 4 weeks of inhalation toxicity testing was determined to be 0.5 mg/m^3. Ce and Nd revealed signs of lung toxicity and possible fibrosis in in vitro investigations utilizing macrophages (Palmer et al., 1987). In the Draize eye irritation test with rabbits, both Dy and Nd chlorides caused conjunctivitis, and in the skin sensitization test, both chlorides caused severe irritation on the scratched skin of rabbits. Acute irritant bronchitis was induced by short-term inhalation exposure to REEs, while pneumoconiosis and progressive lung fibrosis were induced by long-term exposure to REEs in the context of occupational safety and health concerns. There is some debate on whether radioactive pollutants or REEs are to blame for the development of interstitial pulmonary fibrosis (Shin et al., 2019).

Hydrochloric acid is commonly used in the rare-earth recycling process, which is a dangerous chemical. It also generates considerable thermal energy consumption. Even if some metal is recovered, it can be a negligible amount. For example, there may be less than an ounce (a few grams) of rare-earth metals in a computer's hard drive. A few products may have only one part per thousand. There are no large reserves of rare earth elements. To extract rare earths, miners must first excavate massive volumes of ore, which must then go through a series of physical and chemical procedures to concentrate the minerals. It's a messy, energy-intensive process that frequently produces some radioactive waste that needs to be contained and eventually disposed of safely. Ingesting excessive amounts of rare earth elements is harmful to health. The frightening reality is that we may be unknowingly ingesting these poisonous minerals regularly. The concept of bioaccumulation is key here. Toxins can seep out of improperly discarded rare earth metals, poisoning nearby vegetation, animals, and humans. Due to their indigestibility, heavy metals tend to build up in organisms' tissues and then be bioaccumulated and passed on to higher trophic levels. Our bodies will become increasingly contaminated with these substances over time, which is detrimental to our health (https://glescrap.com/blog/rare-earth-production-why-are-rare-earth-metals-important-to-recycle/). The impact of Rare Earth Elements on human organs and systems is shown below (Figure 8.6) (Wang et al., 2024).

Excess REEs have adverse impacts on organ function (Yin et al., 2021). Trivalent cationic REEs are the dominant toxic chemical species. Therefore, drivers affecting the transfer of REE fractions to this more soluble/free form (e.g., pH) are also factors for REE toxicity by increasing bioavailability.

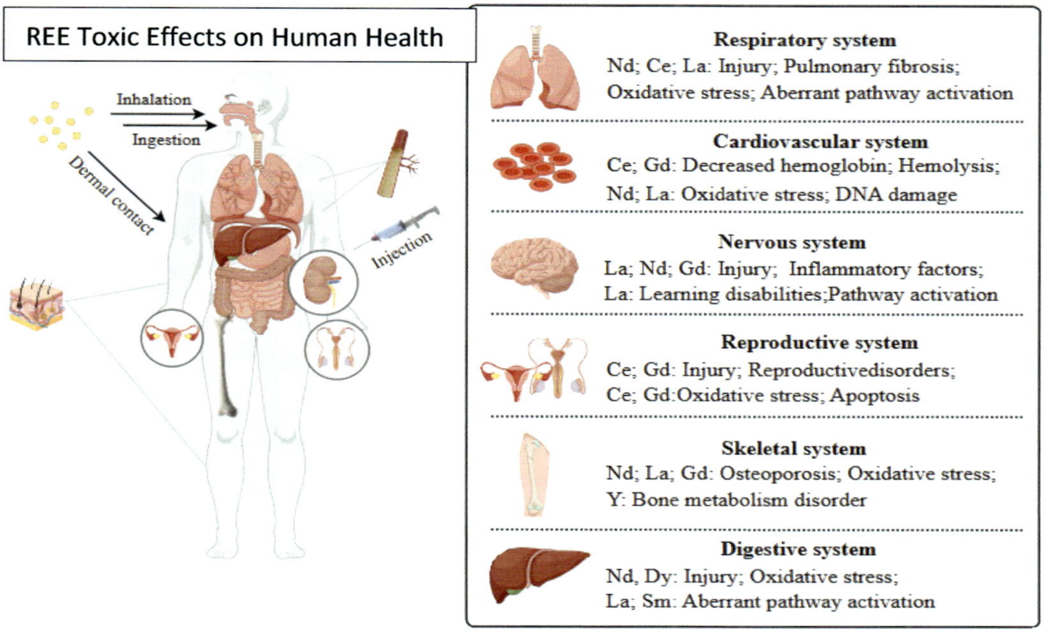

FIGURE 8.6 Toxic Effects of Rare Earth Elements on Human Health (Wang et al., 2024. https://doi.org/10.3390/toxics12050317.)

Leaching with an aqueous electrolyte solution (sodium chloride or ammonium sulphate) through an ion exchange procedure extracts trivalent cationic REEs from clay minerals. The greatest ecological and health threats come from REEs as emerging pollutants and radioactive elements (U and Th). Canada has some of the world's richest REE reserves, and it's widely anticipated that it will become a top producer of REEs in the next decades.

8.8 RARE EARTH METALS AND THEIR EFFECTS ON HEALTH

Rim et al. (2013) and Rim (2016) provided a high-level overview of OHS (Occupational Health and Safety Concerns). These researchers have provided a thorough overview of the literature concerning RE accumulation and toxicity in humans, focusing on the possible threat to environmental and occupational health posed by RE fertilizers used in agricultural production.

Construction and use of access roads, leakage of drilling fluids, and increased sediments can all affect the water chemistry and cause acid to drain from the rock and into nearby streams and lakes. The river or Rare Earth Elements (www.carexcanada.ca/Timis Mining REE.pdf) in the area can suffer damage from acid runoff. Wells and municipal water systems that rely on deep subsurface water may be contaminated by mining operations. Water is often utilized in the REE extraction process, and this water must be either disposed of in a wastewater stream or treated for reuse. The efficient management of wastewater is essential to any mining operation. It is important to carefully control any radioactive compounds found in the REE ore to prevent them from contaminating nearby water supplies. There are strict limits on how much metal and other ingredients (radioactive materials, sediments, chemicals, etc.) can be present in discharged wastewater before it can be dumped into nearby waterways.

According to Pang et al. (2021), solid fuel combustion is a major contributor to the environmental release of rare earth elements (REEs), which may have harmful impacts on human cardiovascular health. Pang et al. (2021) research attempted to determine if there were any consistent exposure biomarkers for REE intake, and if so, what effect that intake would have on blood pressure. The dose-response association between REE administration and BP variation was studied using a linear mixed-effects model. Pang et al. (2021) discovered that there were substantial dose-response connections between the REE concentration in fur samples and the administration dose, but not in blood samples. It implied that hair samples are a more effective matrix than blood samples for identifying a population's exposure level to REEs. Because REEs did not have an immediate detrimental effect on blood pressure in the rat model (Pang et al., 2021), the researchers concluded that hair samples are a more appropriate sample type to reliably measure exposure to REEs than blood samples.

While rare earth elements (REEs) such as yttrium, scandium, and lanthanides are already widely used in health and medical applications, their future in diagnostic imaging analysis could increase their presence. REEs are essential in many bio-medical and food-related fields, offering creative solutions and filling knowledge gaps in natural systems (Reddy and Pranav, 2024; Reddy, 2024). Many different types of human exposure to REEs result from human activity. Particles in the atmosphere are subject to REE-enrichment from human-made sources and the effects of high winds. Large levels of REEs are inhaled, ingested, and absorbed via the skin by workers in recycling plants (www.frontiersin.org/articles/10.3389/fenvs.2022.948041/full). Since REEs are utilized as feed additives to improve animal health and productivity without harming their safety, and in pesticides, herbicides, and fertilizers to boost crop yield and quality, they can also be conveyed to the human body through the food chain. In contrast to animals, however, plants accumulate REE not just in the soil but also in their roots and shoots. Some of the occupational health and safety issues with RE are represented in Table 8.4.

The lanthanides, or rare earth elements, have been of interest since the 19th century (Kilby, 2014), indicating atomic numbers from 57 to 71. The ions of rare earth elements are employed in medicine because their biological properties are comparable to those of calcium ions. As an antiemetic medication, cerium oxalate was the first therapeutic use of rare earth elements. Cerium is one of the most abundant rare earth elements. In the middle of the 19th century, scientists first detailed how to make active rare earth ion complexes, and they continued to do so until the middle of the 20th century. Salts of rare earth metal ions were used to cure tuberculosis, lanthanum chloride was used for its anti-atherosclerotic actions, and lanthanum carbonate was used to treat hyperphosphatemia, all of which date back to the early 20th century (www.frontiersin.org/articles/10.3389/fmed.2022.946100/full). In addition to their potential use as antibiotics, lanthanide coordination compounds have shown to be useful in the diagnosis and treatment of cancer. Both gadopentetic acid and gadoteric acid are gadolinium complexes commonly used as MRI contrast agents for the diagnosis of cancer. Cancer imaging and therapy have both made use of lanthanide radioisotopes (like 177 Lu) (Larson et al., 2015; He et al., 2021). In addition to Yb and Nd, Sm is also useful lanthanides for in vivo luminescence imaging. Nanoparticles, nanodrums, and nanocrystals made of lanthanide oxide are also intriguing candidates for use as imaging agents and cancer therapeutics.

Patients with tongue cancer had higher serum praseodymium (Pr), dysprosium (Dy), and lanthanum (La), and lower serum cerium (Ce) and scandium (Sc) levels than the control group, despite eating fewer fish, seafood, fruit, green leafy vegetables, and non-green leafy vegetables. Some REEs and dietary types were found to have an interactive effect. La and Thorium (Th) components in green vegetables are responsible for their protective effect against tongue cancer (P 0.05, the mediated proportion was 14.933 percent and 25.280 percent, respectively) (Wang et al., 2023) in part. The Sc components in seafood (P 0.05, the mediated proportion was 26.120%) and the Pr, Dy, and Th in non-green leafy vegetables (P 0.05, the mediated proportions were 0.408 percent, 12.010 percent, and 8.969 percent, respectively) contribute to their effect on the incidence of tongue cancer.

Constant accumulation of rare earth elements can occur when they move from the soil to the food chain. Furthermore, cereals, fresh marine goods, fresh vegetables, fresh meats, and eggs all have

TABLE 8.4
REE-associated toxicity

	Element	Section Studied	Toxicity Outcome
Respiratory System	Y	Endotracheal	Dyspnea and pulmonary edema, pleural effusions
	Ce	Environmental exposure, skin contact	Extra pulmonary translocation, interstitial lung disease, pulmonary fibrosis, pneumoconiosis, cytotoxicity
	La	Environmental exposure	Phosphate deposition, pulmonary fibrosis
	Ce	Occupational exposure, environmental exposure	Cytotoxicity, Genotoxicity, lung cancer, inflammation, granulomas, mobilization
	Dy	Endotracheal instillation	Lung injury, oxidative stress, inflammatory response
	Sm	Endotracheal	Lung injury, inflammatory response, pulmonary fibrosis
	Th	Environmental exposure, skin contact	Dyspnea, pneumoconiosis, lung cancer
Nervous System	Gd	Iatrogenic exposure	Deposits in the brain, brain damage
	Nd	Environmental exposure, food chain	Fetal neural tube defects
	La	Environmental exposure, skin contact, food chain	Learning and memory impairment, decreased spatial discrimination, cytotoxicity, memory disorders
Cardiovascular System	Ce	Environmental exposure,	The hemoglobin level is reduced, anemia
	Gd	Endotracheal instillation	Cytotoxicity, hematopoietic destruction
	La	Occupational exposure	Deposition in blood vessels
	Nd	Environmental exposure, skin contact, food chain	DNA damage, cytotoxicity, abnormal cardiovascular and cerebrovascular development
Reproductive System	Ce	Environmental exposure, oral administration	Oxidative stress, placental dysfunction, fetal abortion, growth restriction
	Gd	Iatrogenic exposure	Inflammatory or invasive skin diseases, stillbirth, neonatal death
Skeleton	Gd	Iatrogenic exposure	Bone deposits, osteoporosis
	Y	Iatrogenic exposure	Bone deposits
	Nd	Occupational exposure, environmental exposure	Disorders of bone metabolism, decreased bone mineral density
	La	Environmental exposure, food chain	Abnormal metabolism of calcium and phosphorus, decreased bone mineral density

Source: Wang et al. (2024) Under creativecommons.org/licenses/by/4.0/.

different total REE concentrations (Turra, 2018). There is substantial evidence linking eating habits to the development of cancerous tumors (especially oral cancer).

Fewer reports have shown positive benefits of REEs in boosting plant development, primarily in field trials or in their use as feed additives in livestock, compared to the large body of literature associating REE exposure to deleterious effects in plants (> 3000 citations in PubMed). Rice plant growth was quantified in the presence of REE. Martinez et al. (2018) reported that a significant decrease in rice plant growth was observed upon the addition of increasing concentrations of REE. Outside of China, there is scant data on the prevalence or concentration of REE-containing fertilizers in the soil. Studies of REE toxicity in plant models were carried out on several crop and native species, showing decreased seed germination, root elongation, and mitotic activity for REE levels. Plant models have been noted in reports of REE-associated toxicity, with endpoints including growth suppression, chromosomal aberrations, and redox abnormalities (Tommasi et al., 2020).

Anthropogenic REE primarily come from the following: pharmaceuticals, fertilizers, animal feeds, electronic trash, and recycling factories. Hydrological and wind-driven mechanisms are responsible for the mobilization and environmental movement of REEs. Reduced plant growth, function, and nutritional quality; genotoxicity and neurotoxicity in animals; trophic bioaccumulation; chronic and acute toxicity in soil organisms; these are all examples of ecotoxicological impacts. Consumption of polluted water and food, inhalation, and direct ingestion during medical administration all contribute to human exposure to REEs. Human hair, nail, and biofluid samples have all tested positive for REEs. REEs produce nephrogenic systemic fibrosis and severe damage to nephrological systems in humans, as well as neurodegenerative disorders, fibrotic tissue injury, oxidative stress, pneumoconiosis, cytotoxicity, anti-testicular effects, and male sterility due to Gd-based contrast agents. The epidemiological data relating REEs in the environment to human health issues are poor, except REEs in medical equipment.

8.9 CONCLUSION

Among the many contemporary technologies that can profit from REE are semiconductors, cell phones, TVs, magnets, and even satellites. Many advanced machineries depend on rare earth elements (REEs) to function. Numerous industries, including electronics, clean energy, automotive, aerospace, and defense, employ rare earth elements (REEs). However, there are frequently environmental hazards associated with the mining and manufacturing procedures used to get REEs. Rare earth elements can be found in the same rocks that contain radioactive elements like uranium and thorium. When rare earth elements (REEs) in soil and water become unstable and enter the human body, there are several ways to be exposed, including through food consumption. The processing of rare earth elements produces radioactive waste that lingers for a very long time. Because REE processing produces radioactive and potentially toxic byproducts, a significant amount of hazardous waste and tailings are created. Due to radioactive waste and REE toxicity, the environmental risks connected with mining rare earth elements (REEs) differ from those involved in other forms of mining. The main organs in the body where REEs are stored are the liver, lungs, and bones. Exposure has been associated with a wide range of health issues, including those affecting the intestine, bones, neurological system, lungs, cytogenetics, heart, liver, blood, and kidneys.

Over the past ten years, human and animal exposures have increased as a result of food and water contamination brought on by the extensive use of these elements in a wide range of industrial and consumer products. When REEs are used as microelement fertilizers in agriculture, they accumulate in soils and have toxic effects on the population of soil microfauna. Bioaccumulation, radiation exposure, species invasion, and biodiversity loss are further detrimental effects of REEs on the environment. A thorough grasp of the biochemical processes and destiny of rare earth elements (REEs) in marine habitats—one of the last sinks of pollutants—as well as mitigation strategies for REE contamination, is necessary for the safe expansion of high-tech industries that use these elements. When rare earth elements migrate from the soil to the food chain, they may accumulate continuously. Additionally, the overall REE concentrations varied among cereals, fresh meats, fresh eggs, fresh veggies, and fresh marine goods. To determine the possible health risk, researchers in the mining zone compared the amounts of rare earth elements in the tap water and urine of local children to those in the control area. The environmental effects of extracting REs from ore include deforestation, soil erosion, water contamination, wildlife habitat damage, and changes in landscape structure.

Both natural and man-made activities release REEs into the environment, where they can have a negative impact. For the cleanup of REE, several plants have been employed, including *Taraxacum officinale, Phragmites australis*, and *Phytolacca americana*. Recovering REE from marsh plants can be accomplished using a variety of extraction techniques, including electro precipitation, leaching, and organic complexations. The extracted REE can be used as a fertilizer and to promote the growth of seeds or plants.

REFERENCES

Abigail, W. and L. Paul. 2011. *Rare Earth Elements* (p. 54). British Geological Survey. (Mineral Commodity Profile) (Unpublished). https://nora.nerc.ac.uk/id/eprint/17448/

Aide, M. and T. Nakajima. (eds.) 2020. *Rare Earth Elements and Their Minerals*. Intech Open. https://doi.org/10.5772/intechopen.77602

Aide, M.T. and C. Aide. 2012. Rare Earth Elements: Their importance in understanding soil genesis. *Int. Scholarly Res. Network.* 2012. Article ID 783876: 11. https://doi.org/10.5402/2012/783876

Aik, D., H. Amlund, J.J. Sloth and M. Hansen. 2020. Review of potentially toxic rare earth elements, thallium and tellurium in plant-based foods. *EFSA J.* 18(S1): e181101.

Ali, S.H. 2014. Social and environmental impact of the rare earth industries. *Resources*. 3: 123–134. https://doi.org/10.3390/resources3010123 4

Ambaye, T.G., M. Vaccari, F.D. Castro , S. Prasad and S. Rtimi. 2020. Emerging technologies for the recovery of rare earth elements (REEs) from the end-of-life electronic wastes: A review on progress, challenges, and perspectives. *Environ. Sci. Pollut. Res.* 27: 36052–36074. https://doi.org/10.1007/s11356-020-09630-2

Arienzo, M., L. Ferrara, M. Trifuoggi and M. Toscanesi. 2022. Advances in the fate of rare earth elements, REE, in transitional environments: Coasts and estuaries. *Water* 14: 401. https://doi.org/10.3390/w14030401

Balaram, V. 2019. Rare earth elements—A review of applications, occurrence, exploration, analysis, recycling, and environmental impact. *Geoscience Front.* 10(4): 1285–1303. https://doi.org/10.1016/j.gsf.2018.12.005

Behrsing, T., V.L. Blair, F. Jaroschik, G.B. Deacon and P.C. Junk. 2024. Rare earths—the answer to everything. *Molecules*. 29: 688. https://doi.org/10.3390/molecules2903068

Bernstein, E.J., Schmidt-Lauber and J. Kay. 2012. Nephrogenic systemic fibrosis: A systemic fibrosing disease resulting from gadolinium exposure. *Best Pract. Res. Clin. Rheumatic*. 26(4): 489–503.

Charalampides, G., K. Vitali, V. Karayiannis and A. Baklavaridis. 2016. Environmental defects and economic impact on global market of rare earth metals. In Charalambides (ed.) *IOP Conference Series: Materials Science and Engineering* (Vol. 161, No, 1, p. 012069). IOP Publishing. IOP Conf. Series: Materials Science and Engineering 161 (2016) 012069. doi:10.1088/1757-899X/161/1/012069

Chen, Z.Y. 2005. Accumulation and toxicity of rare earth elements in brain and their potential effects on health. *Rural Eco-Environ.* 21: 72–73.

Damian, P. 2014. *Standalone Report 14 – A Literature Review of the Health and Ecological Effects of the Rare Earth Elements* (p. 47). United States Nuclear Regulatory Commission.

Drobnjak, A. and M. Mastalerz. 2022. Rare earth elements—A brief overview: Indiana geological and water survey. *Indiana J. Earth Sci.* 4. https://doi.org/10.14434/ijes. v4i1.33628

Fan, G.Q., H.L. Zheng and Z.K. Yuan. 2005. Effects of thulium exposure on IQ of children. *J. Environ. Health.* 22(4): 256–257.

Feng, J. and S.Q. Xu. 2000. Biological effect of rare earths in RE-High Background Region. *J. Rare Earths.* 18(4): 356.

Filho, W.L., R. Kotter, P.G. Özuyar, I.R. Abubakar, J.H.P.P. Eustachio and N.R. Matandirotya. 2023. Understanding rare earth elements as critical raw materials. *Sustainability.* 15: 1919. https://doi.org/10.3390/su15031919

Giese, E.C. 2018. Rare earth elements: Therapeutic and diagnostic applications in modern medicine. *Clin. Med. Rep.* 2(1): 1–2. https://doi.org/10.15761/CMR.1000139

Gschneidner, K.A. Jr. and V.K. Pecharsky. 2024. Rare-earth elements. In: *Encyclopedia Britannica*. www.britannica.com/science/rare-earth-element .

Haque, N., A. Hughes, S. Lim and C. Vernon. 2014. Rare earth elements: Overview of mining, mineralogy, uses, sustainability and environmental impact. *Resources.* 3: 614–635. https://doi.org/10.3390/resources3040614

Harbison, R.D. (ed.) 1998. *Hamilton & Hardy's Industrial Toxicology* (p. 682). 5th ed. Mosby.

He, B., J. Wang, J. Lin, J. Chen, Z. Zhuang, Y. Hong, L. Yan, L. Lin, B. Shi, Y. Qiu, L. Pan, X. Zheng, F. Liu and F. Chen. 2021. Association between rare earth element cerium and the risk of oral cancer: A case-control study in Southeast China. *Front. Public Health.* 9: 647120. https://doi.org/10.3389/fpubh.2021.647120

Hu, Y., J. Florek, D. Lariviere, F.G. Fontaine and F. Kleitz. 2018. Recent advances in the separation of rare earth elements using mesoporous hybrid materials. *Chem. Rec.* 18: 1261–1276.

Hurst, C. 2010. *China's Rare Earth Elements Industry: What Can the West Learn?* Institute for the Analysis of Global Security (IAGS). WWW.IAGS.ORG.

Jyothi, R.K., T. Thenepalli, J.W. Ahn, P.K. Parhi, K.W. Chung and L.Y. Lee. 2020. Review of rare earth elements recovery from secondary resources for clean energy technologies: Grand opportunities to create wealth from waste. *J. Cleaner Prod.* 267: 122048.

Kabata-Pendias A. 2000. *Trace Elements in Soils and Plants.* 3rd ed. CRC Press.

Kapustka, K., D. Klimecka and G. Ziegmann. 2019. The management and potential risks reduction in the processes of rare earths elements. *CzOTO.* 1(1): 77–84.

King, H.M. 2021. REE – *Rare Earth Elements and Their Uses.* https://geology.com/articles/rare-earth-elements/

Kilby, C. 2014. China's rare earth trade: Health and the environment. *The China Quarterly.* 218: 540–550. https://doi.org/10.1017/S0305741014000320

Klinger, J.M. 2018. Rare earth elements: Development, sustainability and policy issues. *The Extractive Ind. Soc.* https://doi.org/10.1016/j.exi.2017

Larson, S.M., J.A. Carrasquillo, N.K.V. Cheung and O.W. Press. 2015. Radioimmune therapy of human tumors. *Nat. Rev. Cancer.* 15: 347–360. https://doi.org/10.1038/nrc3925

Li, X., Z. Chen and Y. Zhang. 2013. A human health risk assessment of rare earth elements in soil and vegetables from a mining area in Fujian Province, Southeast China. *Chemosphere.* 93(6): 1240–1246. https://doi.org/10.1016/j.chemosphere.2013.06.085

Li, Y., J-l. Yang and Y. Jiang. 2012. Trace Rare earth element detection in food and agricultural products based on flow injection walnut shell packed microcolumn preconcentration coupled with inductively coupled plasma mass spectrometry. *J. Agric. Food Chem.* 60(12): 3033–3041. https://doi.org/10.1021/jf2049646

Li, Z., Z. Ma, T. Jan van der Kuijp, Z. Yuan and L. Huang. 2014. A review of soil heavy metal pollution from mines in China: Pollution and health risk assessment. *Sci. Total Environ.* 468: 843–853. https://doi.org/10.1016/j.scitotenv.2013.08.090

Liang, T., K. Li and L. Wang. 2014. State of rare earth elements in different environmental components in mining areas of China. *Environ. Monit. Assess.* 186(3): 1499–1513.

Liang, B., Gu, J., Zeng, X., Yuan, W., Rao, M., Xiao, B., Hu, H. 2024. A review of the occurrence and recovery of rare earth elements from electronic waste. *Molecules.* 29: 4624. https://doi.org/10.3390/molecules29194624

Lindagato, P., Y. Li, J. Machá˘cek, G. Yang, I. Mungwarakarama, A. Ndahimana and H.P.K Ntwali. 2023. Lithium metal: The key to green transportation. *Appl. Sci.* 13: 405.

Machá˘cek, J. 2019. Typology of environmental impacts of Artisanal and small-scale mining in African Great Lakes Region. *Sustainability.* 11: 3027.

Malhotra, N., H.S. Hsu, S.T. Liang, M.J.M. Roldan, L.S. Lee, T.R. Ger and C.D. Hsiao. 2020. An updated review of toxicity effect of the rare earth elements (REEs) on Aquatic Organisms. *Animals* (Basel). 10(9): 1663. https://doi.org/10.3390/ani10091663

Martinez, R.E., O. Pourret, M.P. Faucon and C. Dian. 2018. Effect of rare earth elements on rice plant growth. *Chem. Geol.* 489: 28–37. https://doi.org/10.1016/j.chemgeo.2018.05.012

McDonald, J.W., A.J. Ghio, C.E. Sheehan, P.F. Bernhardt and V.L. Roggli. 1995. Rare earth (cerium oxide) pneumoconiosis: Analytical scanning electron microscopy and literature review. *Mod. Pathol.* 8: 859–865.

Mehmood, M. 2018. Rare earth elements- A review. *J. Ecol. Nat. Resour.* 2(2): 000128.

Migaszewski, Z.M., A. Gałuszka and S. Dołęgowska. 2016. Rare earth and trace element signatures for assessing an impact of rock mining and processing on the environment: Wisniewski case study, south-central Poland. *Environ. Sci. Pollut. Res.* 23: 24943–24959.

Oladipo, H.J., Y.A. Tajudeen, E.O. Taiwo, A.O. Muili, R.O. Yusuf, S.A. Jimoh, Kouadio, I.O. Olagundoye, O.M. Egbewande and Y.I. Sodiq. 2023. Global environmental health impacts of rare earth metals: Insights for research and policy making in Africa. *Challenges.* 14: 20. https://doi.org/10.3390/challe14020020

Pagano, G., F. Aliberti, M. Guida, R. Oral, A. Siciliano, M. Trifuoggi and F. Tommasi. 2015a. Rare earth elements in human and animal health: State of art and research priorities. *Environ. Res.* 142: 215–220. https://doi.org/10.1016/j.envres.2015.06.039

Pagano, G., M. Guida, F. Tommasi and R.Oral. 2015b. Health effects and toxicity mechanisms of rare earth elements—Knowledge gaps and research prospects. *Ecotoxicol. Environ. Safety.* 115: 40–48. https://doi.org/10.1016/j.ecoenv.2015.01.030

Pagano, G. (ed.) 2016. *Rare Earth Elements in Human and Environmental Health: At the Cross Roads Between Toxicity and Safety.* Pan Stanford Publishing Pte. Ltd. Stanford Publishing Penthouse Level, Suntec.

Pagano, G., P.J. Thomas, A. Di Nunzio and M. Trifuoggi. 2019. Human exposures to rare earth elements: Present knowledge and research prospects. *Environ Res.* 171: 493–500. https://doi.org/10.1016/j.envres.2019.02.004. Epub

Palmer, R.J., J.L. Butenhoff and J.B. Stevens. 1987. Cytotoxicity of the rare earth metals cerium, lanthanum, and neodymium in vitro: Comparisons with cadmium in a pulmonary macrophage primary culture system. *Environ. Res.* 43(1): 142–156.

Pang, Y., J. Jiang, K. Li, L. Yan, Y. Feng, J. Wang, X. Cao, Z. Li and B. Wang. 2021. Effects of rare earth elements on blood pressure and their exposure biomarkers: Evidence from animal experiments. *Int. J. Environ. Res. Public Health.* 18(18): 9836. https://doi.org/10.3390/ijerph18189836

Paul, J. and G. Campbell. 2011. Investigating rare earth element mine development in epa region 8 and potential environmental impacts (908R11003). U.S. Environmental Protection Agency. www.epa.gov/region8/mining/Report/On/Rare/Earth/Elements.pdf. https://reviewboard.ca/upload/project_document/EA1011-001_Investigating_Rare_Earth_Element_Mine_Development_in_EPA_Region_8_and_Potential_Environmental_Impacts.PDF

Peiró, L.T. and G.V. Méndez. 2013. Material and energy requirement for rare earth production. *J. Minerals, Metals and Materials Society (JOM).* 65: 1327–1340. https://doi.org/10.1007/s11837-013-0719-8

Petra, Z., S. Andrea, M. Josefine and K. Wilhelm. 2022. Environmental impacts of rare earth production. *MRS Bull.* 47: 267–275. https://doi.org/10.1557/s43577-022-00286-6

Porru, S., D. Placidi, C. Quarta, E. Sabbioni, R. Pietra and S. Fortaner. 2001. The potential role of rare earths in the pathogenesis of interstitial lung disease: A case report of movie projectionist as investigated by neutron activation analysis. *J. Trace Elements in Med. Biol.* 14: 232–236.

Qin J., X. Chen and Z. Li. 2002. Rare earth element' influence on human health. *Guangdong Trace Elem. Sci.* 9(6): 1–17.

Ramos, S.J., G.S. Dinali and C. Oliveira. 2016. Rare earth elements in the soil environment. *Curr. Pollution Rep.* 2: 28–50. https://doi.org/10.1007/s40726-016-0026-4

Reddy, R.R. 2024. Prospective potential applications and emerging tendencies in rare earth materials. *Rare Earth: A tribute to the late Mr. Rare Earth, Professor Karl Gschneidner.* 164: 177–210.

Reddy, R.R. and A. Pranav. 2024. Rare Earth Element (REE) insights for health and diagnostic imaging. *Rare Earth: A tribute to the late Mr. Rare Earth, Professor Karl Gschneidner.* 164: 230–256.

Rim, K.T. 2016. Effects of rare earth elements on the environment and human health: A literature review. *Toxicol. Environ. Health Sci.* 8: 189–200. https://doi.org/10.1007/s13530-016-0276-y

Rim, K.T., K.H. Koo and J.S. Park. 2013. Toxicological evaluations of rare earths and their health impacts to workers: A literature review. *Safety and Health at Work.* 4:12–26. https://doi.org/10.5491/SHAW.2013.4.1.12

Shin, S.H., H.O. Kim and K.T. Rim. 2019. Worker safety in the rare earth elements recycling process from the review of toxicity and issues. *Safety Health Work.* 10(4): 409–419. https://doi.org/10.1016/j.shaw.2019.08.005

Šmuc, N.R., T. Dolenec, T. Serafimovski, M. Dolenec and P. Vrhovnik. 2012. Geochemical characteristics of rare earth elements (REEs) in the paddy soil and rice (*Oryza sativa* L.) system of Kočani Field, Republic of Macedonia. *Ganoderma*, 183–184: 111.

Tommasi, F., P.J. Thomas, G. Pagano, G.A. Perono, R. Oral, D.M. Lyons, M. Toscanesi and M. Trifuoggi. 2020. Review of rare earth elements as fertilizers and feed additives: A knowledge gap analysis. *Arch. Environ. Contam. Toxicol.* 81(4): 531–540. https://doi.org/10.1007/s00244-020-00773-4

Trimmel, S., T.C. Meisel, S.T. Lancaster, T. Prohaska and J. Irrgeher. 2023. Determination of 48 elements in 7 plant CRMs by ICP-MS/MS with a focus on technology-critical elements. *Anal. Bioanal. Chem.* 415: 1159–1172. https://doi.org/10.1007/s00216-022-04497-3.

Turra, C. 2018. Sustainability of rare earth elements chain: From production to food—A review. *Int. J. Environ. Health Res.* 28: 23–42. https://doi.org/10.1080/09603123.2017.1415307

Voncken, J.H.L. 2016. The rare earth elements—an introduction: Springer. In: *Briefs in Earth Sciences.* Springer. https://link.springer.com/book/10.1007/978-3-319-26809-5

Walters, A. and P. Lusty. 2011. *Rare Earth Elements* (p. 54). British Geological Survey. https://nora.nerc.ac.uk/id/eprint/17448/

Wang, B., Y. Zhu, Y. Pang, J. Xie, Y. Hao, H. Yan, Z. Li and R. Ye. 2018. Indoor air pollution affects hypertension risk in rural women in Northern China by interfering with the uptake of metal elements: A preliminary cross-sectional study. *Environ. Pollut.* 240: 267–272. https://doi.org/10.1016/j.envpol.2018.04.097

Wang, B., L. Yan, W. Huo, Q. Lu, Z. Cheng, J. Zhang and Z. Li. 2017 Rare earth elements and hypertension risk among housewives: A pilot study in Shanxi Province, *China. Environ. Pollut.* 220(Pt B): 837–842.

Wang, X., X. Shan, S. Zhang and B. Wen. 2014. Distribution of rare earth elements among chloroplast components of hyperaccumulator *Dicranopteris dichotoma. Anal. Bioanal. Chem.* 376(6): 913–917.

Wang, N., F. Liu, Y. Chen, M. Xie, B. Gao, Y. Qiu, L. Lin, B. Shi, F. Chen and B. He. 2023. The role of rare earth elements and dietary intake in tongue cancer: A mediation analysis in southeast China. *Front Public Health.* 11:1058013. https://doi.org/10.3389/fpubh.2023.1058013

Wang, W., Y. Yang, D. Wang and L. Huang. 2024. Toxic effects of rare earth elements on human health: A review. *Toxics*. 12: 317. https://doi.org/10.3390/toxics12050317

Waring, P.M and R.J. Watling. 1990. Rare earth deposits in a deceased movie projectionist. A new case of rare earth pneumoconiosis? *Med. J. Aust.* 153: 726–730.

Weber, R. J and D.J. Reisman. 2012. *Rare Earth Elements: A Review of Production, Processing, Recycling, and Associated Environmental Issues*. US EPA.

Wei, B., Y. Li, H. Li, J. Yu, B. Ye and T. Liang. 2013. Rare earth elements in human hair from a mining area of China. *Ecotoxicology. Environ. Saf.* 96: 118–123.

Wyttenbach, A., P. Schleppi, J. Bucher, V. Furrer and L. Tobler. 1994. The accumulation of the Rare Earth Elements and of Scandium in successive: Wheedle Age classes of Norway Spruce. *Biol. Trace Element Res.* 41:13–29.

Yan, D., S. Ro, O. Sunam and S. Kim. 2020. On the global rare earth elements utilization and its supply-demand in the future. *Earth Environ. Sci.* 508: 012084. https://doi.org/10.1088/1755-1315/508/1/012084

Yang.Y, A. Walton, K. Sheridan, K. Guth, R. Gauß, O. Gutfleisch, M. Buchert, B.-M. Steenari, T. Van Gerven, P.T. Jones and K. Binnemans. 2017. REE recovery from end-of-life NdFeB permanent magnet scrap: A critical review. *J. Sustain. Metall.* 3: 122–149.

Yin, C.X., C. Martineau, I. Demers, N. Basiliko and N.J. Fenton. 2021. The potential environmental risks associated with the development of rare earth element production in Canada. *Environ. Rev.* 29: 354–377. https://doi.org/10.1139/er-2020-0115

Zaichick, S., V. Zaichick, V. Karandashev and S. Nosenko. 2011. Accumulation of rare earth elements in human bone within the lifespan. *Metallomics.* 3: 186–194.

Zapp, P., A. Schreiber, J. Marx, and W. Kuckshinrichs. 2022, Environmental impacts of rare earth production, *MRS Bulletin*, 47, 1–9. mrs.org/bulletin

Zhuang, M., L. Wang and G. Wu. 2017. Health risk assessment of rare earth elements in cereals from mining area in Shandong, China. *Sci. Rep.* 7: 9772. https://doi.org/10.1038/s41598-017-10256-7

Zhou, B., Z. Li and C. Chen. 2017. Global potential of rare earth resources and rare earth demand from clean technologies. *Minerals* 7: 203. https://doi.org/10.3390/min7110203

9 Microplastics' Dangers to Humans and Ecosystems

9.1 INTRODUCTION

There are microplastics in everything we eat and drink; plastic is not a fantastic way to live. Up to 5 g of plastic is found in commonly consumed foods and drinks every week. Life has become plastic for us. Air, ocean, water, and food are all contaminated with microplastics. Human health is not in immediate danger from microplastic contamination, but action must be taken to prevent future risks. The most common source of microplastics in the average person's diet is drinking water, both tap and bottled water, as well as groundwater and surface water. The average person consumes around 1,769 pieces of microplastic weekly, according to a 2019 World Wide Fund for Nature (WWF) analysis. Researchers found that 94% of American tap water and 72% of European tap water had microplastics. Four plastic fibers per 500 ml were found in tap water in India (https://theprint.in/opinion/we-are-consuming-up-to-5-gms-of-plastic-every-week-in-our-water-apples-fish-beer/730790/).

The creation of new commercial products and the breakdown of bigger plastics both contribute to the widespread proliferation of microplastics. Microplastics are a contaminant that can have negative effects on ecosystems and wildlife. Microplastics are entangled with living creatures. A water sample taken from the Pacific Ocean near Hawaii reveals the pervasive presence of microplastics in the diets and behaviors of marine organisms. Microfibers shed from garments and other textiles, like fishing nets and microscopic particles manufactured for commercial use, like cosmetics, are primary microplastics. The term "secondary microplastics" refers to the smaller particles that form when bigger plastic objects, such as water bottles, break down. Sunlight and ocean waves, in particular, are blamed for this deterioration. As with plastics of all sizes, the problem with microplastics is that they do not easily decompose into innocuous molecules. Decomposing plastics can take hundreds to thousands of years, during which time they wreak havoc on ecosystems. Microplastics appear as small, colorful plastic fragments in the sand on beaches. Microplastic pollution is a major problem in the water, and marine animals frequently ingest it. While littering is certainly to blame for some of this pollution, it's also important to note that storms, water runoff, and winds all contribute to the problem by washing plastic debris, both large pieces and smaller microplastics, into the oceans. Microplastic effects and pathways on the environment and human health are highlighted by many Scientists. The majority of the plastic debris floating around came from single-use goods like straws that were only used once and thrown away.

People have contributed to a global increase in plastics production over the past seven decades, and that growth has spread throughout the environment to the point that we now truly do inhabit a plastic world. These synthetic polymers are not only valid indicators of the modern and contemporary time, mainly after the middle of the 20th century, but they are also environmental contaminants and vectors for the transfer of many kinds of chemicals. Marine water, freshwater, agroecosystems, the

DOI: 10.1201/9781003618164-9

atmosphere, food and drinking water, biota, and other remote areas have all been found to have microplastic particles of varying shapes, polymers, sizes, and concentrations (Campanale et al., 2019, 2020). They can be as flimsy as a veil and blown by the wind over hundreds of miles, or they can be as solid and heavy as a rock. Because of their extensive global distribution, many scientists view them as a crucial geological marker of the Anthropocene (Prata, 2018). When plastics are viewed as recent and reliable indicators of earth deposits, they can be employed as stratigraphic markers in the archaeological field.

Anthropogenic trash, mostly composed of plastic, has been increasing rapidly in both aquatic and terrestrial ecosystems over the past few decades. Over 280 million metric tons of plastic are produced yearly around the world (www.sciencedirect.com/science/article/abs/pii/S0269749119302039). Mass production of plastics began in the 1950s. Every year, between 4.8 and 12.7 million metric tons of improperly handled plastic garbage flow into the oceans from coastal countries. Microplastics are now found in every oceanic environment (Ahmad et al., 2020); (Osman et al., 2023), from the deepest parts of the ocean to the highest (Tibetan Plateau) plateau on Earth. Governments on various continents have restricted plastic use (particularly single-use) and are promoting recycling of plastics to reduce microplastic pollution. While scientists have devoted a great deal of time and energy to studying microplastics in the marine environment, little is known about the nature and amount of microplastic pollution in continental habitats such as rivers, lakes, soil, and air. Therefore, investigating microplastics' consequences is crucial. Figure 9.1 exhibits microplastic effects and pathways on the environment and human health (Ghosh et al., 2023).

There are two types of microplastics: primary microplastics, which are the raw materials used in household and personal care products, and secondary microplastics, which are the particles of raw plastic that have degraded due to environmental factors such as physical, chemical, and biological

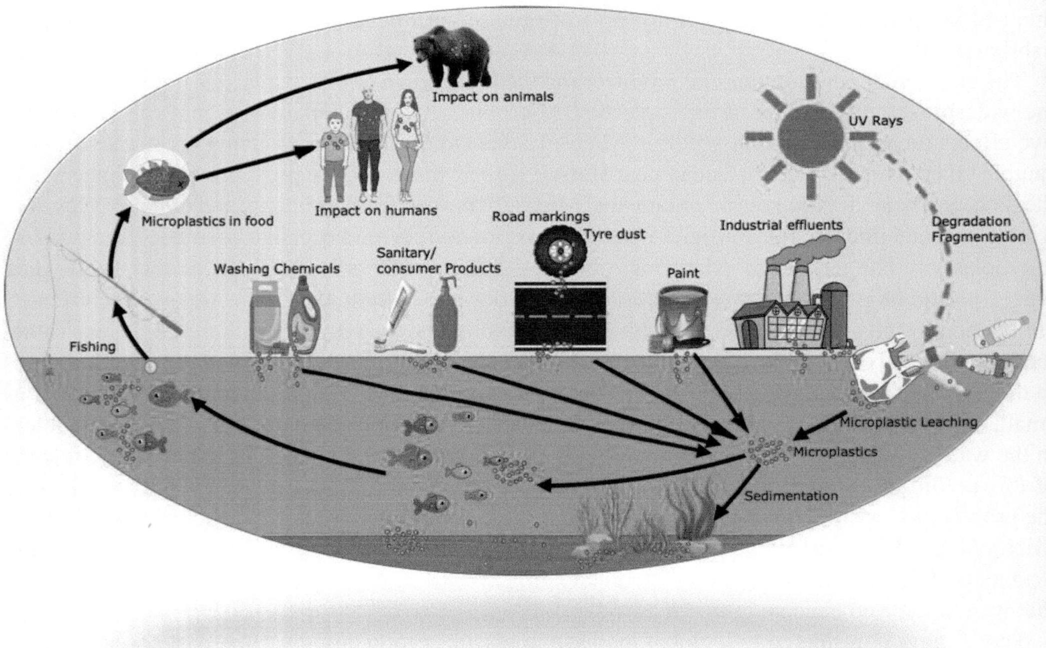

FIGURE 9.1 Microplastics generation, transportation, and ingestion in the environment affect the whole ecosystem (Ghosh et al., 2023. https://doi.org/10.3390/su151410821. Open access.)

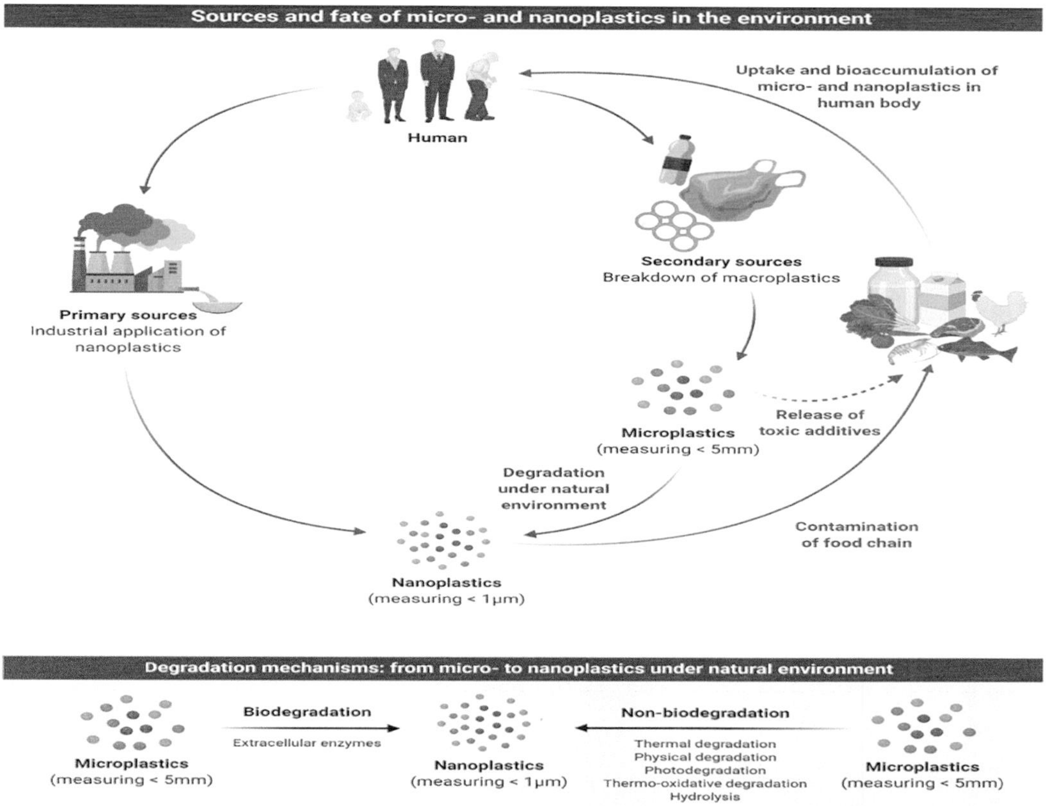

FIGURE 9.2 Microplastic sources, fate, and degradation mechanisms of the microplastics (Yee et al., 2021. Open access.)

processes. Microplastics are a major cause for concern among environmentalists and biologists because of their polymeric nature, which allows them to persist for a long time and facilitates their movement between different environments. Polyethylene terephthalate (PET), polyurethane (PU), polystyrene (PS), polyvinyl chloride (PVC), polypropylene (PP), polyesters, polyethylene (PE), and polyamide are all commonly used raw polymers (PA, nylon). Microplastics have become a global problem due to improper plastic waste handling. Microplastics sources, transformation, and transport (Yee et al., 2021) are shown in Figures 9.2 and 9.3. The occurrence of microplastics in the aquatic environment has become a growing concern globally. Microplastics pose a hazard to the ecological system, and their presence, particularly in the water, hurts human health and the ecosystem (Padervand et al., 2020).

MPs pollution (Ziani et al., 2023) has gradually spread from aquatic environments to the rest of the environment and has caused adverse effects. Therefore, it is important to develop removal, recycling, and degradation technologies for MPs and prevent the potential ecological risks (Pan et al., 2023). Figure 9.3 explains the microplastics in water and aquatic systems (Issac and Kandasubramanian, 2021).

Microplastics have been detected in marine organisms from plankton to whales, in commercial seafood, and even in drinking water. Microplastics and microscopic plastics are fragments of any type of plastic that are being produced today as plastic waste originating from anthropogenic activities. Such microplastics are discharged into the environment, and they enter back into the human body through different means. Singh et al. (2021) and Karthik et al. (2018) evaluated the presence of microplastics in various water sources as well as the ecological and human health impacts of microplastics. Various types of plastics and their effect on marine organisms (Chatterjee and Sharma, 2019) are shown in Figure 9.4.

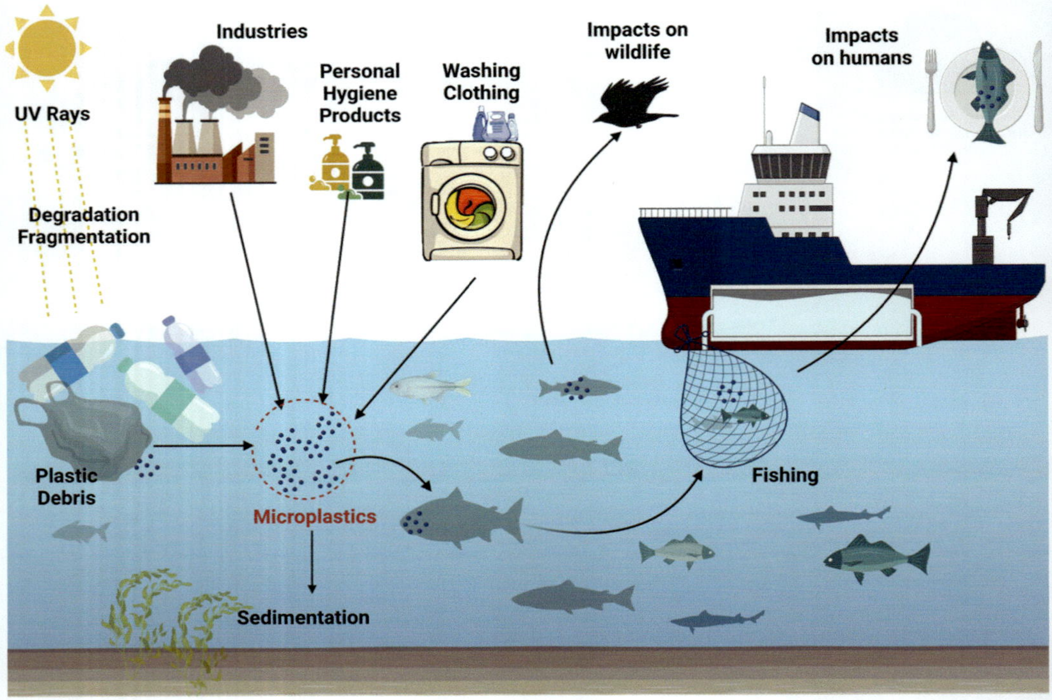

FIGURE 9.3 Microplastics in the aquatic environment (Ziani et al., 2023. https://doi.org/10.3390/nu15030617.)

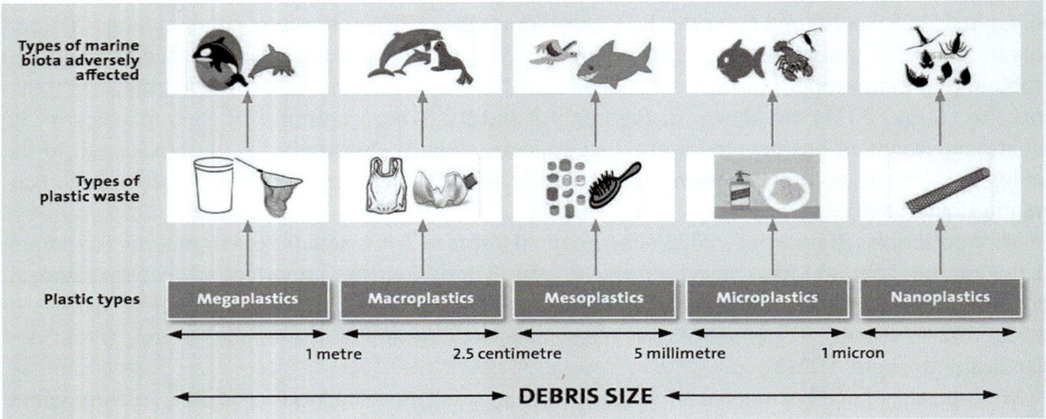

FIGURE 9.4 Pictorial representation of different types of plastics and their effect on marine organisms. Debris sizes of the plastics are also given (Chatterjee and Sharma 2019, Creative Commons – Attribution 4.0 International – CC BY 4.0).

9.2 STUDIES OF MICROSCOPIC PLASTICS

Microplastics were studied by Gela and Aragaw (2022), who examined sediment and agglomerated sewage water samples taken from four big ditches in Bahir Dar, Ethiopia. Microplastics are more prevalent in sediment than they are in water, according to the authors' findings. These findings and data help investigate the effects of microplastics in Tana Lake and the Blue Nile River.

In a freshwater adult male African catfish study, Sayed et al. (2021) examined the negative consequences of microplastic intake on reproductive success in *Clarias gariepinus*. The authors have investigated the possible preventive effects of three distinct antioxidant supplements—lycopene, citric acid, and *Chlorella*—against reproductive damage caused by microplastics in fish. The findings indicate that ingesting microplastics may cause stress on reproduction. To determine if *Chlorella*, citric acid, and lycopene supplementation can reduce the neurotoxicity of polyethylene (PE) type microplastics on the brain tissue of African catfish (*Clarias gariepinus*), Hamed et al. (2022) conducted experiments. It was observed that catfish brain enzymes and histology are altered after ingestion of microplastics.

Daphnia magna mortality was examined, as were changes in the diet of free-living and host-associated microorganisms when exposed to or free from microplastics and Chromium (VI) (Varg et al., 2022). This research added new knowledge on how pollutants affect human mortality and the microbiome. According to Jiang et al. (2020), when plastic waste degrades, it creates microplastic (MP) and nanoplastic (NP) particles, with MP being less than 5 mm in diameter and NP being between 1 and 100 or 1'000 nm in diameter, respectively.

There was a workshop on microplastics in Washington, DC, and during that time, a more inclusive working definition was developed. Microplastics are small bits of plastic, greater than 1 μm. According to Müller (2021), macroplastics are defined as particles larger than 2.5 cm, mesoplastics as >1 mm, microplastics greater than 1 μm, and nanoplastics less than 1 μm (Figure 9.5). Plastics originally manufactured at sizes less than 5 mm are also regarded as microplastics, which are termed as primary microplastics. The term microplastic was first used to describe microscopic fragments of plastic debris (~20 μm in diameter) in a 2004 publication.

Plastic debris is often subdivided into four categories: nano-, micro-, meso-, and macroplastics. Nanoplastics, which have particles smaller than 100 μm in their largest dimension, have historically been included with microplastics (particles < 5 mm in size) but are generally considered a separate category today. The plastic litter we typically see along roads and in streams falls into the category of mesoplastics (5 mm to 2.5 cm) and macroplastics (>2.5 cm).

Microplastics (MPs henceforth) refer to any plastic particles, beads, granules, or fibers that are less than 1 millimeter in size, as defined by Silva et al. (2019). Thompson first put up the idea of MPs in a 2004 issue of Science. Primary and secondary microplastics (MPs) have been identified. Primary MPs include microfibers, beads, and plastic particles on garments, all of which are less than 5 mm in size before entering the environment. When large plastic items degrade in the environment due to natural weathering, secondary MPs are created. UV light, oxidation, and the mechanical grinding of plastic products are common environmental sources of MPs. Researchers are beginning to shift their attention from microplastics (MPs) to nanoparticles (NPs) as their study progresses. Particles of plastic smaller than 1 micrometer in size are typically considered nanoplastics.

Qin et al. (2020) performed a bibliometric analysis of global MPs research conducted from 2014 to 2019. Fen et al. (2020) also created a bibliometric profile by searching the Web of Science Core Collection database for the topic "microplastic or (micro near/1 plastic)." As plastic particles decrease in size, it also becomes easier for them to be transferred into the environment, resulting in a greater health effect on organisms. Therefore, it is necessary to study plastic particles with smaller sizes, such as nano-level particles. MPs can be specifically manufactured (Cole et al., 2016) or they are the consequence of mechanical, biological, or photo degradation and subsequent fragmentation of large plastic items (Athey et al., 2020). One of the major concerns about MPs is that

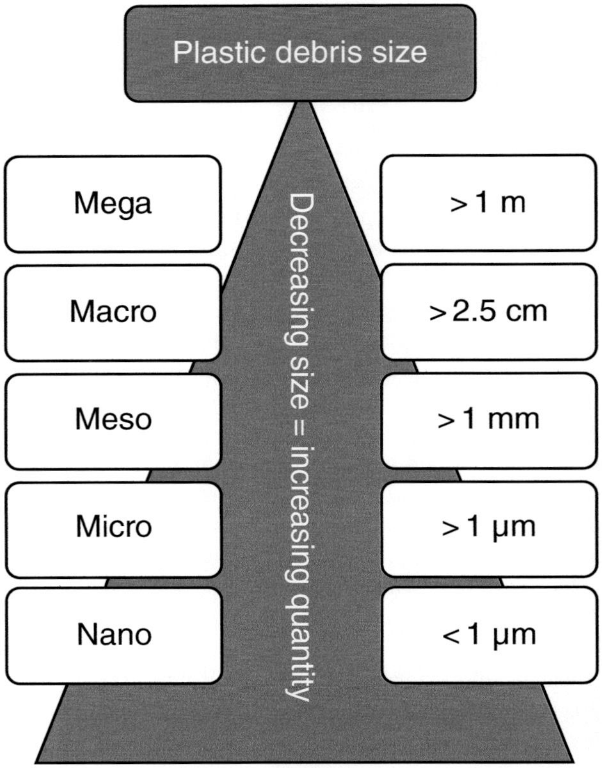

FIGURE 9.5 Classification of plastic debris according to their size (Müller 2021. https://creativecommons.org/licenses/by/4.0/.)

they are highly bioavailable in marine environments, freshwater, and terrestrial ecosystems (Wagner and Lambert, 2018). A large variety of organisms have been reported to ingest and accumulate MPs, from zooplankton to fish (Mattsson et al., 2015). MPs can hurt organisms, compromising the individual fitness by reducing feeding, energetic reserves, growth rate, fecundity, and survival. Moreover, the bioavailability of MPs and their possible interactions with different pollutants in the environment (Rochman et al., 2014) might open the possibility of synergic toxic action of MPs with other pollutants such as toxic metals. The synergic effects of MPs and metal ions on organism fitness and its associated microbes remain mostly unexplored, constituting one of the new frontiers in toxicology.

Peng et al. (2022) explained that it is of particular relevance that once plastic litters enter the environment, they can persist for a long time and will be broken down into small debris (alias mesoplastics), microplastics (MPs, <5 mm), and even nanoplastics (NPs, <100 nm) via physical, chemical and biological processes before reaching complete decomposition. The MPs and NPs have attracted public concern due to their small size, worldwide distribution, and potential ecotoxicological effects. As a result, the studies have been increasing in the past few years. Qin et al. (2020) undertook a bibliometric study of international MPs studies published between 2014 and 2019. The smaller the plastic particles are, the more easily they are transported into the environment and have a detrimental effect on living things. This calls for research into even smaller pieces of plastic, such as nanoparticles. MPs can be made on purpose in factories (Cole et al., 2016), or they can be the result of the breakdown of larger plastic objects by mechanical, biological, or photodegradation (Athey et al., 2020). MPs' high bioavailability in marine, freshwater, and terrestrial ecosystems is a serious cause for alarm (Wagner and Lambert, 2018). Many different kinds of animals, from zooplankton to fish, have been found to consume and build up MPs in their bodies (Mattsson et al., 2015). As a result of being exposed to MPs, an organism's fitness may suffer, including its ability to feed, store energy, grow, reproduce, and survive. Synergistic toxic action of MPs with other pollutants, such as toxic metals, may be feasible due to their bioavailability and potential interactions with various environmental contaminants (Rochman et al., 2014). Toxicologists have only just begun to scratch the surface of the synergistic impact that MPs and metal ions have on organism fitness and the microorganisms associated with it.

According to Peng et al. (2022), once plastic litters enter the environment, they can persist for long periods and will be broken down into small debris (alias mesoplastics), microplastics (MPs, 5 mm), and even nanoplastics (NPs, 100 nm) via physical, chemical, and biological processes before reaching complete decomposition. Due to their diminutive stature, global spread, and possible eco-toxicological impacts, MPs and NPs have garnered public attention. This has led to an uptick in research activity in recent years. To date, estimates of the global yearly MP inputs into the ocean vary from several thousand to a few million tons (www.frontiersin.org/journals/marine-science/articles/10.3389/fmars.2022.947309/full).

9.3 MICROPLASTICS IN OUR FOOD MATERIALS

Recent evidence indicates that humans constantly inhale and ingest microplastics through contaminated seafood, including fish and shellfish. Additionally, microplastics have been found in tap water, bottled water, and even commonly consumed beverages, such as beer and salt. More than 80% of land-generated trash ends up in waterways, and much of that trash, whether it's trash or sewage, is plastic. Marine life and, consequently, humans who eat seafood are exposed to and ingesting tiny plastic particles that do not biodegrade but instead keep breaking down under the influence of water and sunlight.

Microplastics can throw the entire food web and ecology out of whack if they make their way into the food chain. When marine life ingests microplastics from the ocean, they absorb chemical poisons like DDT and BPA from manufacturing effluents and other sources. In addition to cancer, the poisons may also damage the liver and kidneys. Microplastics can be found in a wide variety of consumer products, including plastic bags, garments, fabrics, toothpaste, exfoliants, and more. Microplastics are a particularly pernicious form of pollution because they are so small, yet they are also ingested by marine organisms.

9.4 THE ROLE MICROPLASTICS PLAY IN OUR DIETS

It's not shocking that these microplastics are entering our bodies, given how commonplace they are nowadays. The water we drink, the fish we eat, the salt we use, and the beer we drink all include trace amounts of plastic. While drinking tap water is a significant contributor, switching to bottled water may more than double your daily microplastic intake. Drinking bottled water may contribute an additional 90,000 microplastic particles to the annual ingested total of 74,000 to 121,000. Shellfish are the second-most-common source of microplastics. Shellfish have particularly high amounts of microplastics since they are ingested whole, but they are now discovered in a wide range of sea species, which can often mistake plastic for food or simply ingest it through the water. However, microplastics aren't solely a water pollution problem. Improper composting is a major source of microplastics in our soil and food supply, according to studies. Plastics are entering our compost, which is then used as fertilizer for our food supply, largely as a result of sloppy composting on an individual basis. When these plastics decompose, they leave behind microplastics that end up everywhere.

Riverine fauna in general is at risk from microplastic pollution. We know that even microscopic species like zooplankton can consume them. Microplastics pose a threat to marine life since they can be swallowed and cause gastrointestinal obstruction or mislead creatures into thinking they aren't hungry (Masó et al., 2016; Browne et al., 2011; Van Cauwenberghe et al., 2013). Plastic pollution in terrestrial, aquatic, and atmospheric ecosystems is shown in Figure 9.6.

Debris in the ocean, particularly plastic, is a major environmental concern worldwide. Half of the projected ten million tons of plastic that enter the water each year will float at first. But only about 0.3 million tons of plastic are actually visible in the open ocean (Cozar et al., 2014). Microplastics are particularly problematic because many harmful substances can bind to their surfaces, and ingestion of these microplastics could subject creatures to toxic levels (www.nottingham.ac.uk/connectonline/research/2018/the-big-problem-of-microplastics.aspx).

Greater amounts of these plastic particles are entering our food chain as more microplastics fill our marine environment (https://phys.org/news/2021-05-ocean-plastic-boost-wave-induced-size.html, www.nature.com/articles/d41586-021-01143-3) and are ingested by the organisms that inhabit our waterways. Evidence of microplastics has been identified in people for the first time, according to recent studies conducted, for example, by the Austrian Environment Agency and the University of Vienna. Once in the marine environment, microplastics are subject to a variety of physicochemical processes, including biofouling, leaching, and the integration of secondary pollutants. Microplastics vary in quantity, size, and shape; as a result, different parts of the marine ecosystem (including the benthos) have variable amounts of plastic debris available to marine life. The causes and harmful consequences of microplastics on marine biota were discussed by Chatterjee and Sharma (2019). Figure 9.7 explains the sources and deleterious effects of microplastics on marine biota and control measures.

9.5 WHERE DO MICROPLASTICS EXIST?

9.5.1 Microplastics in Bottled Water

Using Nile red dye to fluoresce particles in the water, researchers analyzed 259 bottles purchased from 19 different places in 9 different nations; the dye prefers to cling to the surface of plastic but not most natural materials. They discovered that 93% of the bottles of water they examined contained microplastics. Researchers concluded that there are approximately 10.4 microscopic plastic particles per liter of bottled water. Also, there were approximately 315 microplastic bottle fragments identified per liter of water studied (https://youmatter.world/en/microplastics-types-effects-humans-nature/).

9.5.2 Microplastics in Tea Bags

Many shoppers are becoming informed about the dangers of plastic pollution and are demanding that the usage of disposable plastics be curtailed. However, some producers are developing plastic alternatives to paper packaging, such as plastic tea bags. Approximately 11.6 billion microplastics and 3.1 billion nanoplastics are released into a single cup of tea when steeping.

Microplastics were detected in 24 different brands of German beer by chemical analysis. These microplastics included fibers, pieces, and granules. Of these, 12 were standard Pilseners, 5 were wheat beers, and 7 were non-alcoholic versions of the popular beverage. Microscopic plastic was discovered in all 24 beer samples. Although most of the fibers were clear, some were colored blue, black, or green. The granular and fragmentary particles were mostly translucent or white, with some green and yellow inclusions. Beer contains microplastics, which, while not alarming in and of themselves, show that the human environment is contaminated by micro-sized synthetic polymers to a significantly greater extent than previously thought (https://youmatter.world/en/microplastics-types-effects-humans-nature/).

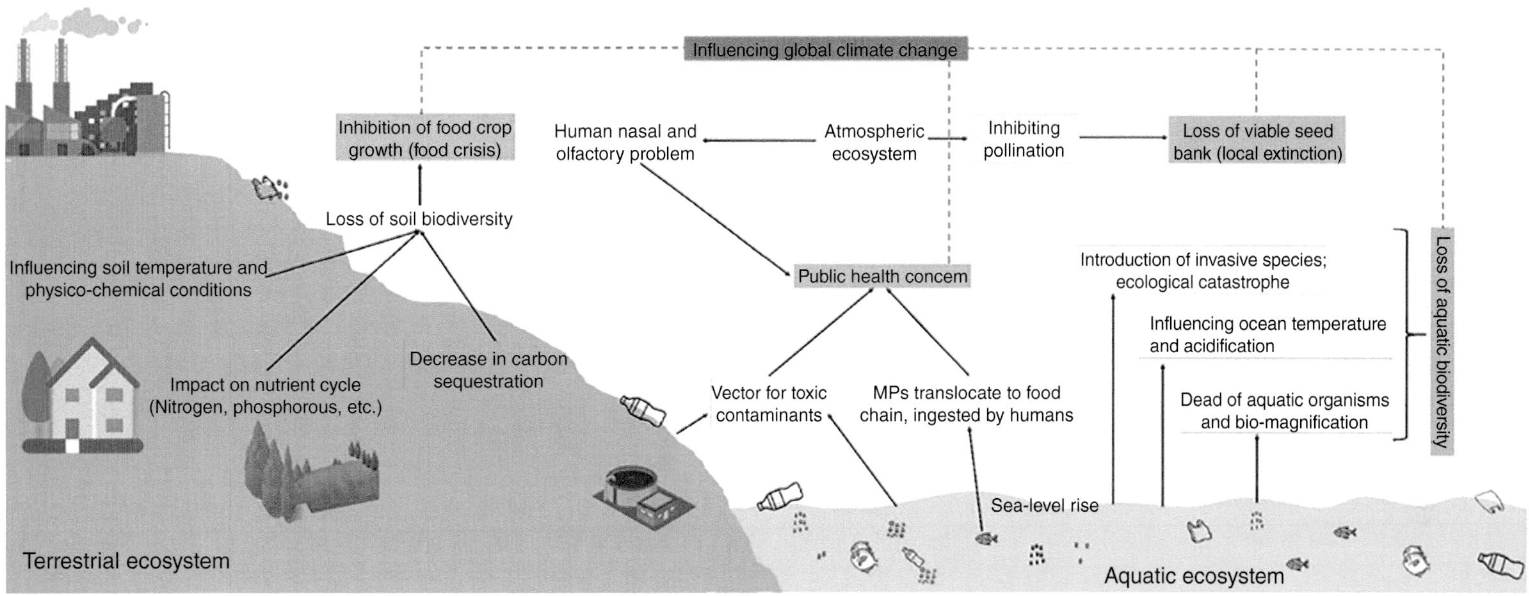

FIGURE 9.6 MPs and NPs affect various ecosystem services and climate change on terrestrial, aquatic, and atmospheric ecosystems (Kumar et al., 2021. Open access.)

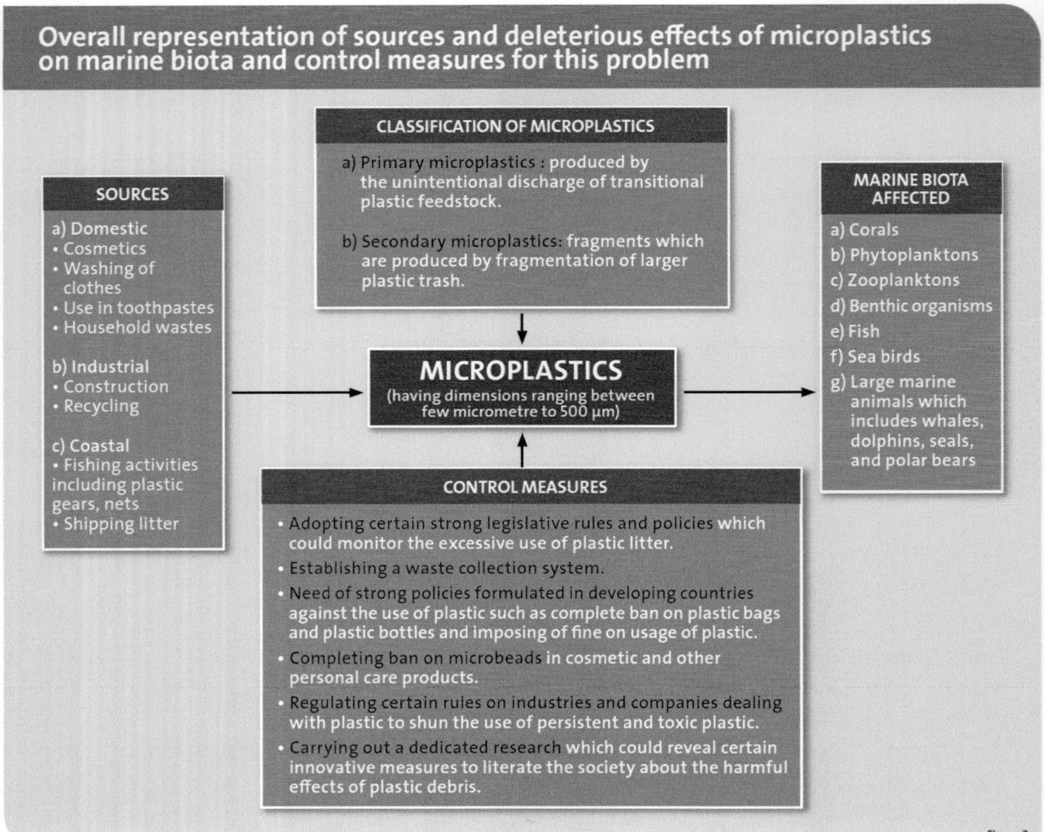

FIGURE 9.7 Sources and deleterious effects of microplastics on marine biota and control measures for this problem (Chatterjee and Sharma 2019. https://creativecommons.org/licenses/by/4.0/.)

9.5.3 Rainwater Containing Microplastics

Parashar and Haitet (2023) reported finding plastic in 90% of rainwater samples across all 6 sites, i.e., urban and peri-urban areas of the East Indian region, India. Most of the plastic components were filaments, which could be seen only under a microscope and came in a rainbow of hues. Magnification also revealed fragments and beads made of plastic. The researchers conclude in the discussion that additional work is required to examine potential environmental consequences and improve sampling, identification, and quantification of plastic deposition. The sources of MPs in atmospheric fallout are thought to be multifaceted, including the breakdown of larger plastic debris (Abbasi, 2021). Globally, intense rain events (such as monsoon events) could represent a significant means of MPs transportation in the atmosphere and an important source of MPs to the continents and oceans.

9.6 MICROPLASTICS IN ANIMALS, FISH, FOOD, AND SOIL

Microplastics are so microscopic that many marine animals accidentally consume them. Over 800 animal species have been exposed to plastic by ingestion or entanglement, according to a 2016 UN report (69% more than in 1977). About a quarter of these 800 species have been observed swallowing microplastics in the wild. Particles of plastic are frequently found in the digestive tracts of animals across a wide range of trophic levels, including marine mammals, fish, invertebrates, and fish-eating birds. Micro and nanoplastics are particularly difficult to eliminate from an animal's system because they are so small that they can enter the bloodstream and spread throughout the body.

The influences of MNPs (micro (nano) plastics) on the soil systems and plant crops have also been overlooked, despite that MNPs can accumulate in the plant root system and generate detrimental impacts to the terrestrial environments (Allouzi et al., 2021). MNPs impact the soil-plant system significantly because they accumulate in the soil and interact with PTEs and organic pollutants, concentrating them in the soil. MNPs can affect the growth of plants and their chlorophyll content. Besides, MNPs can interact with soil organic debris and stay in soil for hundreds of years, which may affect the physio-chemical properties of the soil and pollute the groundwater. The crop yield of agricultural soil contaminated with MNPs will be affected negatively due to low pH levels and a lower number of productive earthworms.

Evidence of human exposure to microplastics via seafood is described in a paper titled "Microplastics in Seafood and the Implications for Human Health" by Smith et al. (2018). Microplastic ingestion has been recorded in planktonic species, larvae at the base of the food chain, tiny and big invertebrates, and fish of varying sizes, as described in the literature review. Microplastics are most likely to enter the human diet through the consumption of bivalves and small fish that have been ingested whole. However, is there evidence that microplastics pose a threat to human health?

9.7 MICROPLASTICS IN WATER

Tap water from around the world is contaminated with microplastics (www.downtoearth.org.in/news/water/even-bottled-water-can-have-microplastics-58676). The world produces around 350 million tons of plastic waste each year. Ritchie (2023) presented that 1.5 million tons out of the 1.7 million that enter the oceans remain close to the shoreline. Figure 9.8 shows Plastic waste ends up in the ocean (https://ourworldindata.org/how-much-plastic-waste-ends-up-in-the-ocean).

Microplastics have been detected in seawater, sewage, freshwater, food, air, and drinking water (both bottled and tap water). Microplastics typically enter freshwater ecosystems by surface run-off and wastewater effluent (both treated and untreated), combined sewer overflows, industrial effluent,

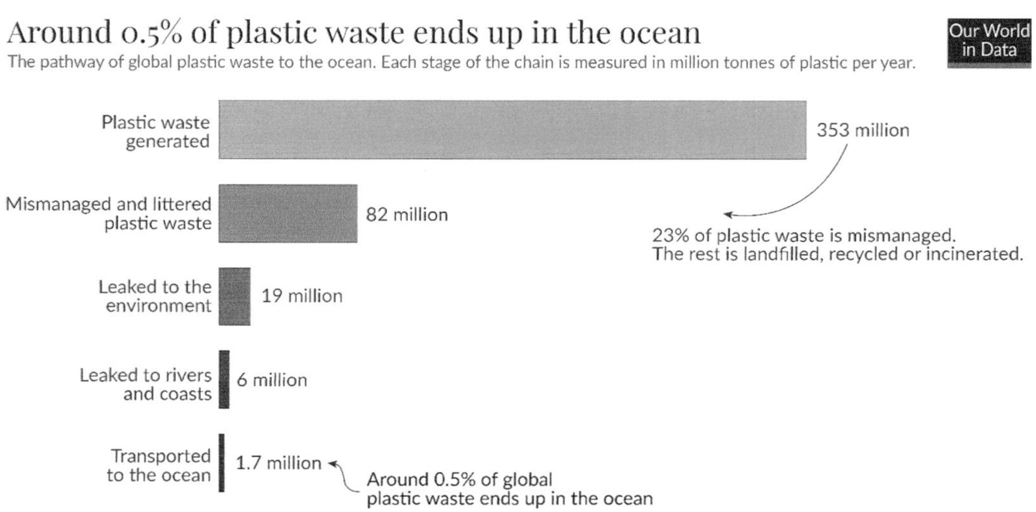

FIGURE 9.8 Plastic waste ends up in the ocean (Ritchie 2023. https://ourworldindata.org/how-much-plastic-waste-ends-up-in-the-ocean. Reuse this work freely: https://creativecommons.org/licenses/by/4.0/.)

degraded plastic debris, and air deposition. However, data that could quantify the inputs and their antecedents is lacking. Microplastics in drinking water may also arise in the methods used to produce both tap water and bottled water, according to the limited data currently available (WHO, 2019).

The frequency of microplastic particles by polymer type in fresh water was shown to be related to the density and production volume of plastics. A wide range of sizes and shapes was found. Only nine studies have looked into the presence of microplastics in drinking water, with fragments and fibers being the most common types. The most widely recognized polymers were PET (polyethylene terephthalate) and PP (polypropylene). The smallest particles detected were typically determined by the size of the mesh used in the sample process, which varied substantially between studies of freshwater and tests of drinking water. Particle concentrations in fresh water ranged from very low (0 particles/L) to very high (103 particles/L), while drinking water filtration systems use very fine mesh sizes to catch even small particles, studies have still detected microplastic concentrations in drinking water ranging from 0 to 104 particles per liter, with average values typically between 0.001 and 1,000 particles per liter, highlighting that even with smaller mesh sizes, microplastics can still be present in drinking water. Fine mesh with an aperture size ranging from 100 to 300 microns is highly effective in capturing these tiny contaminants, making it ideal for applications requiring precise filtration. Although 1μm is the smallest particle size detected so far, this discovery is constrained by the present techniques (WHO, 2019). The mesh sizes used in freshwater studies are often an order of magnitude greater than those used in drinking water studies because they focus on larger particles. Therefore, studies on potable water cannot be directly compared to studies on freshwater.

In addition to polluting the ocean, microplastics have been found in drinking water (WHO, 2019; Danopoulos et al., 2020). Microplastics (MPs) are omnipresent in the environment, including the human food chain; a likely important contributor to human exposure is drinking water. Plastic fibers, which are extremely small and difficult to detect, were found in 83% of the more than 150 tap water samples gathered from cities and towns across five continents. After the United States and Lebanon, India ranked third for having the highest percentage of microplastic contamination in drinking water (82.4 percent). Only Delhi provided samples from India. Microscopic plastic fibers were found in 14 of 17 samples of Delhi's tap water. It also claimed that microplastics were present in American bottled water. Even bottled water can have microplastics (www.downtoearth.org.in/news/water/even-bottled-water-can-have-microplastics-58676), which has already been established by sources. Plastic does not biodegrade; thus, it fragments into smaller pieces when exposed to water, sunlight, and waves. As its size decreases, so does its safety.

Although there is not yet agreement on a definitive size range for microplastics (MPs), these particles typically vary from 1 mm to 5 mm in size. MPs have been found in both saltwater and freshwater (lakes, rivers, reservoirs, and groundwater), while most studies have focused on marine MPs (Eerkes-Medrano et al., 2019). With the ongoing increase in plastic manufacture, consumption, and waste, MP contamination of aquatic habitats is also predicted to rise. The introduction of MPs into the food network presents a new threat to consumer health. A newly recognized (emerging) hazard poses an emerging concern to food safety when it has the potential to cause widespread exposure.

9.8 PLASTICS AFFECT THE AQUATIC ECOSYSTEM

When animals become entangled in plastic nets or rings, they might suffer physical limitations that may ultimately lead to death. Plastics can also be mistaken by animals for edible items. Things like plastic bags and thin film plastics evoke images of jellyfish and drifting kelp. Microplastic pollution in aquatic environments and its impacts on food chains are depicted in Figure 9.9.

Many fish and other aquatic animals mistake microplastics for the plankton they rely on for food (Chaukura et al., 2021). There are numerous ways through which microplastics and associated contaminants can be incorporated by the aquatic biota. The damage that plastics cause to marine life.

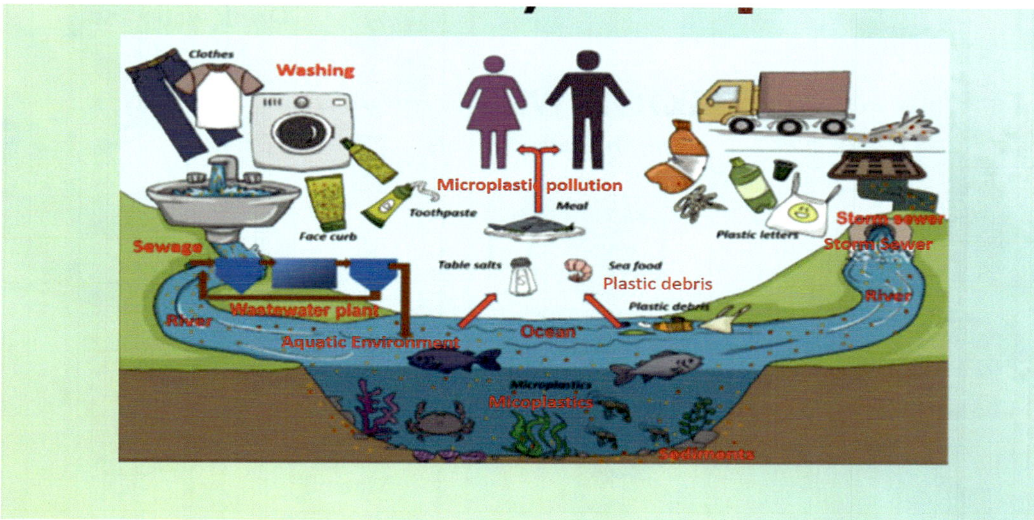

FIGURE 9.9 Microplastic pollution in aquatic environments and impacts on food chains (Westphalen and Rasoul 2018. http://creativecommons.org/licenses/by/3.0.)

Plastics are difficult for animals' digestive systems to process, so they tend to become stuck there. This might trick the animal into thinking it has eaten when it hasn't. In addition, even the smallest pieces of plastic can have sharp edges that can puncture organs.

Bacteria, algae, crustaceans, and bivalves are all at risk, as well as higher levels of the marine food chain. Bacteria's biofilm formation and general growth can be hindered by some nanoplastics. Many food webs rely on algae for oxygen production and nutrient provisioning; as such, changes to algae's typical growth and behavior can have far-reaching consequences for many different organisms (Bergami et al., 2017). Some species of microalgae have been demonstrated to be significantly affected by even short periods of exposure to nanoplastics (96 hours) (Venâncio et al., 2019). Exposure to nanoplastics has been linked to a number of negative effects in crustaceans, including nanoplastic accumulation in the stomach (Bergami et al., 2017), slower swimming speed (Gambardella et al., 2017), and disruption of embryo development.

Furthermore, microplastics may contain persistent organic pollutants (POPs) that are not easily broken down by the environment. Surface POPs on microplastics can have harmful effects on the immune, neurological, and reproductive systems. Animals and other aquatic species may bioaccumulate them, and they could then travel up the food chain. The harmful effects of microplastics on soil and living beings are also highlighted in the literature. Microplastics in the food web and others are given in Figure 9.16. Microplastics often have toxic heavy metals, including lead, arsenic, and copper attached to their surface (Vo and Pham, 2021). Animals with compromised immune systems may be more easily infected by sickness due to other foreign substances on or in the plastics.

The presence of microplastics in a fish's digestive tract indicates that the fish has ingested small plastic particles from the surrounding water, which can be a significant environmental concern as it can impact the fish's health and potentially move up the food chain through consumption by larger predators. Twenty-five percent or more of the fish samples taken from the South Pacific had microplastics. According to this study (www.sciencelearn.org.nz/resources/2809-how-harmful-are-microplastics), larger marine creatures that ingest microplastics typically excrete them. But studies suggest that microplastics can stay in the digestive tract for long periods, potentially abrading and damaging interior tissues. Nanoplastics can enter the bloodstream and circulate throughout the

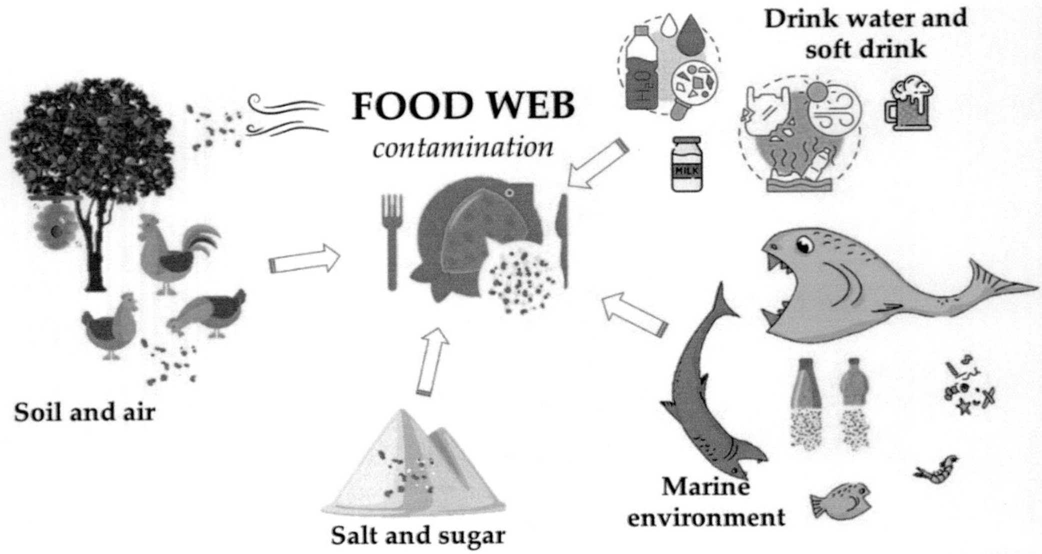

FIGURE 9.10 Scheme of food web contamination due to MP pollution (Pironti et al., 2021. https://doi.org/10.3390/toxics 9090224.)

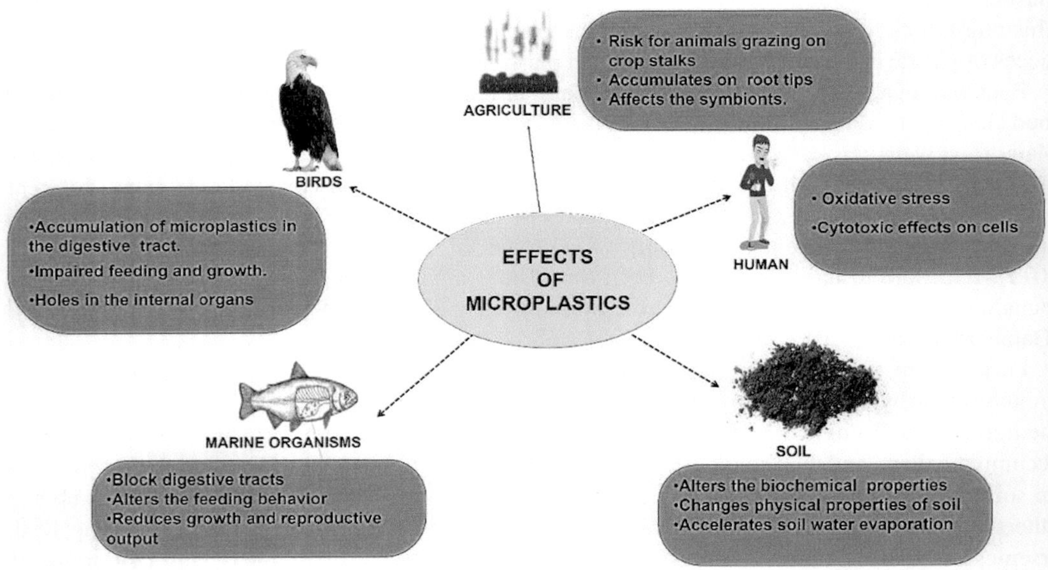

FIGURE 9.11 The harmful effects of microplastics on soil and living beings (Loganthan and Kizhakedathil 2023. DOI:10.33263/BRIAC132.126. https://creativecommons.org/licenses/by/4.0/.)

body, wreaking havoc in organs like the liver and lungs. The potential health effects of microplastic ingestion require additional study. Food web contamination and the effects of microplastics on soil and living beings are shown in Figures 9.10 and 9.11.

Weber et al. (2018) demonstrated that an exposure of the freshwater amphipod *Gammarus pulex* to PET MP over 24 h resulted in a high body burden predominantly consisting of particles <53 μm. However, a long-term exposure over 48 d did not significantly affect feeding activity, energy

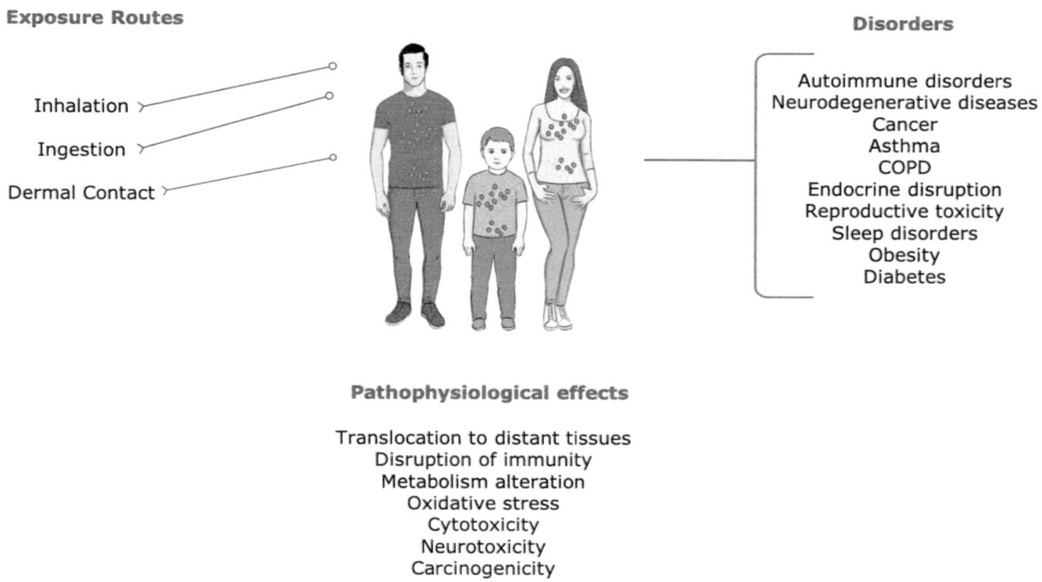

FIGURE 9.12 Microplastics severely impact human health (Ghosh et al., 2023. Open access.)

reserves, molt, or mortality. Ingestion of microplastics can hurt the health of smaller creatures, such as marine and freshwater invertebrates, leading to decreased reproduction and growth. When animals swallow microplastics instead of digestible food, they suffer physical damage, such as lacerations and inflammatory responses, and they also limit their feeding habit. Because they filter large quantities of water while feeding, filter feeders like mussels and oysters are particularly at risk of ingesting microplastics (Sharma and Chatterjee, 2017). The effects are different for different species, different plastics, and higher microplastic concentrations (Figure 9.12).

Through multiple routes, microplastics can invade the human body and cause multiple changes via a downstream cascade of events that cause the body to experience pathophysiological effects, which can lead to a multitude of disorders, including sleep disorders, obesity, cancer, Parkinson's disease, and others.

The negative impacts of plastic pollution on marine biota have made it a subject of contemporary concern. Microplastics have a great potential for bioaccumulation because of their small size. As they are consumed by various marine organisms such as corals, planktons, fish, seabirds, and marine mammals, they are passed up the food chain (Andrady, 2011; Lin, 2016). In addition, plastic polymers absorb numerous hazardous chemicals and pollutants because of their chemical additives and stabilizers. Thus, microplastics operate as a vector for the spread of these hazardous pollutants. The danger posed by microplastics was not recognized as a hazard until very recently. Microplastics are currently found in everything from drinking water to table salt.

9.9 MICROPLASTICS ON THE BEACHES

Now, scientists have mapped almost 3 million pieces of microplastic found on more than 300 beaches around Australia. It is estimated that between 4.8 and 12.7 metric t of plastic trash enter the ocean ecosystem annually, and several studies have identified microplastics as a pervasive pollutant of marine debris. The easy absorption of these micro pollutants by marine organisms, such as bivalves for human consumption, pelagic and demersal fish, or seafood, is of special concern. One of the significant environmental implications of growing plastic use is the proliferation of "microplastics," or plastic trash smaller than 5 millimeters in diameter that is typically invisible to the naked eye but poses a serious threat. The presence of nearby hotels and sewage disposals (effluents) as a potential source of plastic litter is mostly to blame for the increased concentrations of microplastics on the beaches. Beach cleaning processes are very similar to sifting machines. The top layer of sand is gathered and sieved through holes of 10 millimeters in diameter; this is done to remove litter such as cigarette ends, plastic trash, and seaweed left behind by beachgoers. However, the openings are too small for the microplastics to fit through, so they remain on the beach. Although microplastics are too small to be seen by the naked eye, they have a significant impact. Animals that consume microplastics may experience behavioral abnormalities, DNA damage, and other negative effects, according to research. Microplastics are everywhere; scientists have even discovered them in the Mariana Trench, the deepest part of the ocean on Earth. Every week, we consume the equivalent of a credit card in plastic.

There are serious risks to human health and the environment from microplastic pollution. Microplastics are produced on beaches due to the strong UV irradiation and physical abrasion of waves on bigger plastic items. Fragmentation is slowed drastically once a body of water is entered because of the lower temperature and lack of UV light. Microplastics are fragments of plastic that are too small to be seen by the naked eye, but are nonetheless pervasive in the marine environment.

Ten percent to twenty five percent of plastic debris originates from marine sources, including maritime shipping, fishing, and dumping, whereas the majority originates from land-based activities including littering, tourism, and inadequate waste management (Khuyen et al., 2021). There are approximately 5 trillion plastic pieces, or 268,940 tons, in the world's oceans; of this amount, MPs make up 92.4%. MPs have been shown to have negative effects on organisms at a variety of levels, including through the induction of oxidative stress and histological changes, the modification of feeding preferences and behaviors, the suppression of incubation and hatching of eggs, and the reduction of growth and survival.

Raman microscopy was used in the MPs analysis to identify specific particles. Microplastics with a sum of dimensions greater than 25 mm were counted, and their shape, color, and size were examined in microscopic detail. The Raman spectrometer used a charge-coupled device detector and a single-beam laser with a 532 nm wavelength. Each particle was studied using a grating with 900 lines per mm and a wavelength range of 50-3,600 cm-1 over 15 seconds (Khuyen et al., 2021). Microplastics were sorted by form, polymer type, and colour after their original polymer was determined using Raman spectroscopy.

Swordfish, mahi-mahi, and triggerfish (which reportedly tastes like crab) larvae were discovered to contain plastic, as were larvae of other popular seafood. Tuna eat a variety of fish, including flying fish, so it's possible that plastic from these fish could end up in your tuna salad. Microfibers such as polyester, nylon, polyethylene terephthalate, rayon, and artificial cellulose made up almost all of the plastic fragments found inside the fish. They also shared the "mainly blue or translucent in colour" of the fish larvae's preferred meal, zooplankton, making them an obvious food choice.

Plastic waste has become a universally significant environmental problem. Figure 9.13 shows that the ingestion of food and water contaminated with microplastics is the main route of human exposure (Alberghini et al., 2023).

Plastic has become the most common and pervasive kind of marine pollution due to its widespread availability in both marine and terrestrial ecosystems (de Souza et al., 2018). Ingestion, entanglement, and destruction of crucial habitats like breeding beaches and larger ecosystem effects are all potential threats to sea turtles from plastic pollution. Due to the direct influence of physical force, plastic particles are typically found in greater quantities on windward (WW) beaches of these islands. de Souza et al. (2018) have noticed an increase in the number of tiny plastic discs or rings (1/8 inch, or 3 millimeters) washing up on our coasts during the past few years (https://seagrant.oregonstate.edu/visitor-center/found-beach/plastic-discs). They come in a wide variety of hues, though

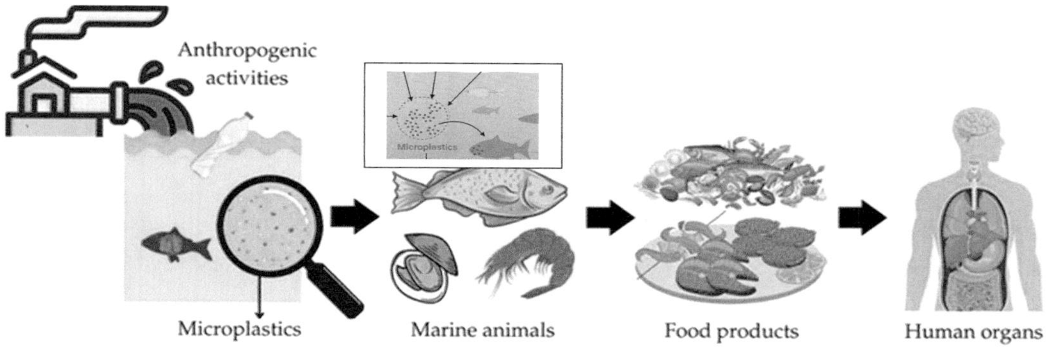

FIGURE 9.13 An illustration of how human activity leads to the entry of microplastics into the food chain, making a path to our food and, ultimately, into our organs (Alberghini et al., 2023. Open access.)

the most common ones are white, grey, and tan. This is the packaging that polyethylene, polypropylene, and polystyrene suppliers use to send their goods to factories.

Many of how plastics are harmful to wildlife are only now being understood by researchers. Birds, fish, and other larger animals can suffer internal damage from ingesting plastics; also, animals may stop eating because they mistakenly believe they are full after consuming plastics. Slidy plastic cups are fatal for hermit crabs who try to use them as shells. Once consumed, the chemicals in microplastics bioaccumulate in tissues and organs, causing hazards like organ failure and reproductive stress at the bottom of food chains.

9.10 HUMAN EXPOSURE TO MICROPLASTICS

Concern is growing about the pervasive presence of microplastics in the environment and people's bodies. Seafood and processed foods, including sugar, beer, salt, bottled water, milk, honey, and plastic teabags, have all been discovered to contain microplastics (Hernandez et al., 2019). A growing number of tissues and human fluids have also been shown to contain microplastics, which may be a direct result of these discoveries. The lungs (Amato-Lourenço et al., 2021), breast milk, placenta, and feces are also sources of infection. Almost every part of the body is covered. The "inevitability" of being exposed to these particles has been argued for.

There are many unanswered questions about the dangers of ingesting microplastics to human health. How high are the average concentrations of exposure from both inhaled and ingested substances? Can microplastics build up inside a living organism? How do bio-persistent microplastics fare in terms of toxicology? Current analytical approaches may have certain gaps when it comes to accurately estimating concentrations in humans. A more complete picture of toxicological effects and potential health impacts can be gleaned from a more thorough understanding of exposure levels.

The toxicity of plastics depends on both the kind of plastic and any possible additional chemicals. Polyvinyl chloride, one of the most widely produced polymers in the world, is also the most carcinogenic (Patil et al., 2022). Bisphenol A and phthalates are just two of the many chemicals that have been linked to adverse health effects, including disruptions in fertility and the endocrine system and the development of asthma in children (Lumio et al., 2021). Due to concerns about their health consequences, these two compounds have been mostly removed from infant items. However, the safety of the dozens of other additives used in plastics remains largely unknown. Figure 9.14 enlightens the Microplastics as a trigger of combined physical or chemical-like effects.

Microplastic pollution has the essential nature of "combined physical and chemical effects" due to its non-natural and polymer-based structure as well as its content rich in xenobiotics and poorly

FIGURE 9.14 Microplastics as a trigger of combined physical or chemical-like effects (Dhiman et al., 2023. Open access.)

soluble bio-persistent hydrophobic particles. There is ecological concern over the widespread use of plastics and the rising concentrations of endocrine active compounds in the environment because endocrine systems were relatively well preserved during vertebrate evolution, and endocrine disruptors may have far-reaching direct consequences for animal health (de Souza Machado et al., 2018). There is no reason to think that exposed biota won't be negatively affected by the several plastic components that are known to have the ability to disrupt endocrine systems. The surface/volume ratio exponentially increases as larger plastic particles split into smaller fragments. Because many additives are physically attached to a polymeric structure but not chemically, they can virtually always leach off the polymer surface, increasing the potential for leaking estrogenically active chemicals.

Upstream environmental degradation happens when activities like mining, fracking, plastics manufacturing, and chemical additives manufacturing release large quantities of hazardous compounds into the air and contaminate the surrounding area. The disposal process is fraught with problems as well, as both landfills and incineration of plastic trash release harmful compounds and micro- and nano-plastics into the air. Local and global biodiversity are subsequently impacted by all of these factors (www.unep.org/resources/pollution-solution-global-assessment-marine-litter-and-plastic-pollution).

Large chunks of plastic that have broken free and floated to the ocean's surface are what most people picture when they hear the term "plastic pollution," but this is only one of the problems. In the ocean, plastic can be discovered at any level. Some of it floats, some of it drifts below the surface, and some of it sinks to the bottom of the ocean (Figure 9.15). The smallest particles of plastic, microplastics smaller than 5mm, are very challenging to monitor.

New studies show that animals' digestive systems also play a role in transporting microplastics, in addition to currents and weather. Many of us have seen images of plastic found in the stomachs of marine animals like seabirds, turtles, and whales, so we know that animals in the water are consuming plastic. Microplastics tend to flow easily through an animal, while larger pieces of plastic might become lodged in an animal's stomach and make eating difficult. Microplastics, which are found in the sea as microbeads and microfibers, are consumed by marine benthic biota. Large amounts of microplastic particles are absorbed by marine mammals, most likely from direct ingestion from saltwater or sediment, as well as trophic transfer, which occurs when prey eat plastic. The gastrointestinal tracts, stomachs, and feces of pinnipeds and cetaceans have all been reported to contain microplastics. Because of their micron-sized particle size, tiny plastic fragments are persistent in the marine ecosystem. A variety of marine biota, including corals, phytoplankton, zooplankton, sea urchins, lobsters, fish, and others, mistake them for food and ingest them, which leads to their transfer to higher trophic levels. Numerous marine species end up with tiny microplastics in their stomachs. Fish can readily access these tiny particles due to certain characteristics of microplastics, including their minuscule size, appealing colors, and high buoyancy. Because they mistake these pieces for plankton or other naturally occurring prey, fish consume microplastics. Tiny microplastics end up in the guts of all sorts of marine life (Figure 9.16).

Microplastics can affect microbial communities in aquatic sediments, changing the composition of species and altering their nitrogen cycling activity. Oceans, coasts, mountains, towns, and rural areas are not immune to plastic and microplastic pollution. Mount Everest, the Mariana Trench, and the Arctic have all been shown to have traces of plastic pollution (www.genevaenvironmentnetwork.org/fr/ressources/nouvelles/plastic-pollution-around-the-world/). At least 85% of all marine litter is made up of plastics, making them the largest, most destructive, and most persistent type of marine litter. In China, Zhao et al. (2016) detected microplastics in the stomachs of 94% of dead terrestrial birds with a wide range of foraging habits. It has also been observed that microplastics from agricultural activities can be found in the stomachs of freshwater continental birds. Because microplastics were often much smaller than the birds' typical diet, accidental ingestion or trophic transfer appear to be the most likely explanations for their consumption (Zhao et al., 2016). Microplastic

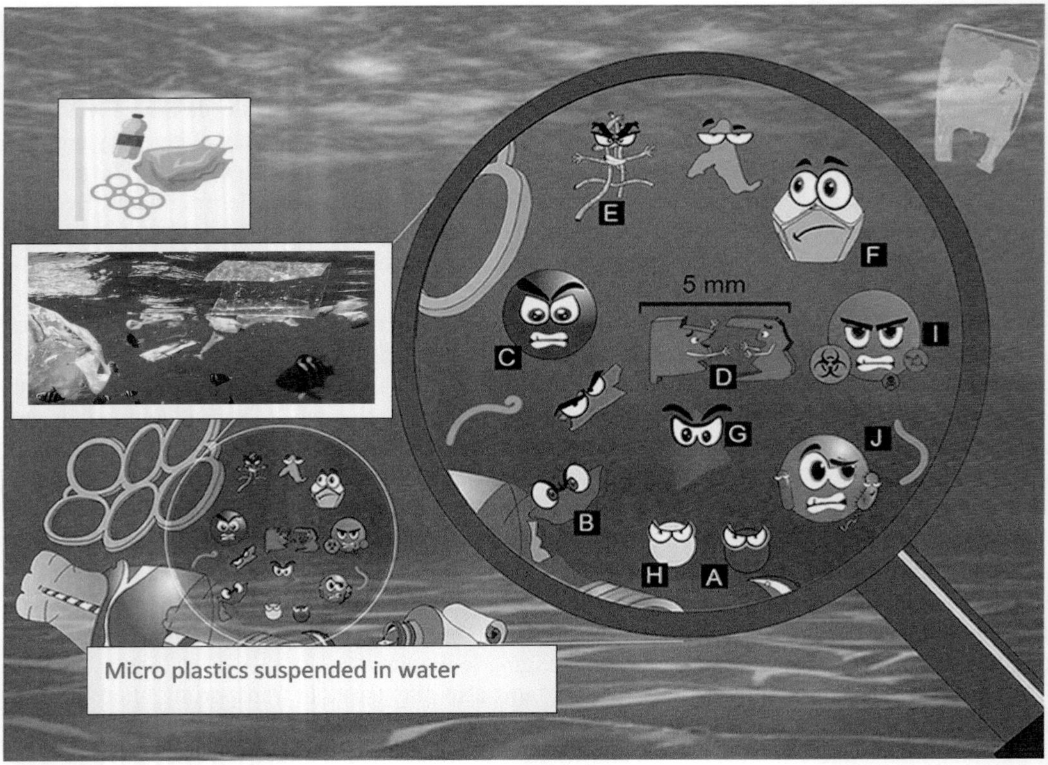

FIGURE 9.15 The Microplastics suspended in water. The types of microplastics and related parts of them are highlighted in the text: (A) primary microplastics, (B) secondary microplastics, (C) pellets, (D) fragments, (E) fibers, (F) foams, (G) films, or (H) granules. Some microplastics can act as fomites, transporting (I) hazardous chemicals or (J) pathogens. (Lujan-Vega et al., 2021. Front. Young Minds. 9:606974. doi:10.3389/frym.2021.606974 https://creativecommons.org/licenses/by/4.0/.)

concentrations have also been shown to be rising in soils (0.9 particles/g), earthworm casts (14 particles/g), and chicken feces (129 particles/g) in the first quantitative investigation of trophic transmission of microplastics. Particles (0.1 m to 150 m, varied compositions) have been shown to enter the bodies of people, dogs, rabbits, and small rodents via the intestinal lymphatic cells.

In particular, seabirds often mistake pieces of trash for food. One hundred and fifty different kinds of marine birds have been observed eating plastic. Half of the seabirds examined in a recent study contained microplastics in their poop, and similar results were discovered in a study of birds off the coast of Canada's Labrador Peninsula. Because of where the birds are defecating, this could be a problem. After foraging in the northern Atlantic, these birds will fly back to their rocky island homes for the night. Swarms of birds numbering in the tens of thousands often defecate near one another (https://blog.marinedebris.noaa.gov/smelly-science-tracking-movements-microplastics). These birds are essentially acting as concentrators, gathering microplastics from vast areas of the water and depositing them in a single location. In their quest for food, these birds may be spreading plastic trash across their nesting grounds.

Birds aren't the only ones consuming plastic and then excreting it. Small marine organisms called zooplankton may be transporting plastic debris when they defecate. California's Monterey Bay mentioned that zooplankton (little critters) could be transporting microplastics from near the ocean's surface down to the briny depths of the ocean floor (https://blog.marinedebris.noaa.gov/

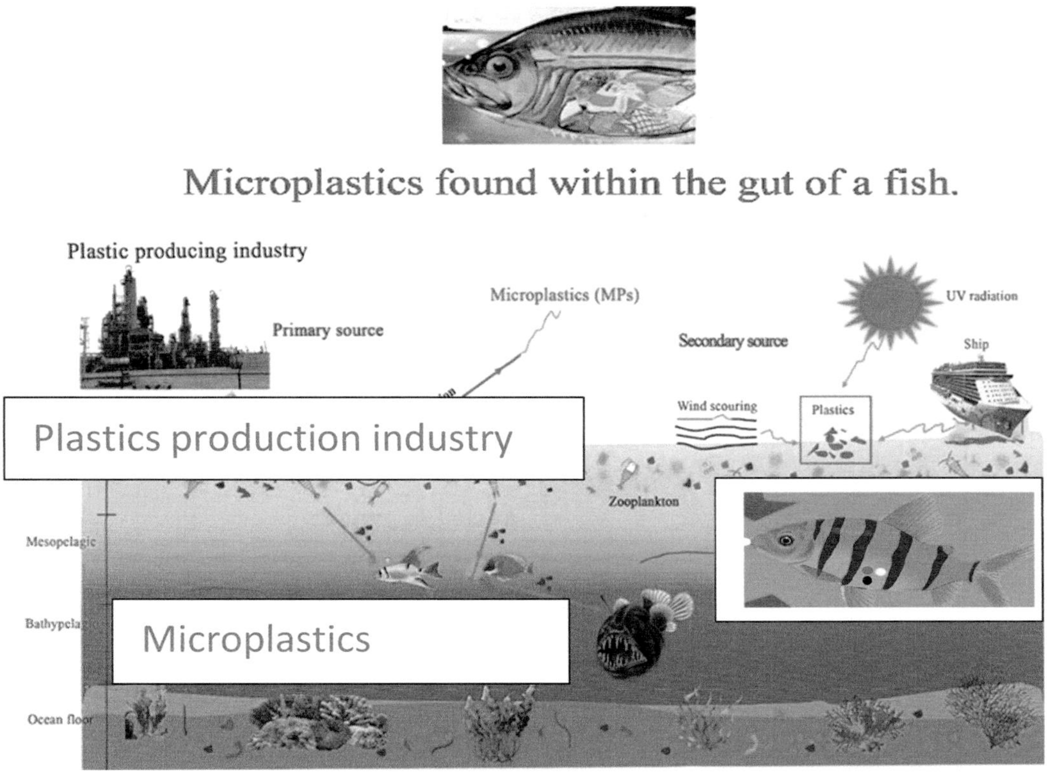

FIGURE 9.16 Tiny microplastics (micron-sized particles) end up in the guts of all sorts of marine life. (Bhuyan 2022, Https://doi.org/10.3389/fenvs.2022.827289. https://creativecommons.org/licenses/by/4.0.)

smelly-science-tracking-movements-microplastics). Small and practically transparent, zooplankton bodies made it possible for scientists to observe vividly colored plastics inside. The microplastics' oceanic migration may also be affected by zooplankton lifestyles. When free of microplastics, the waste produced by the zooplankton known as larvaceans tends to sink quickly, an estimated one thousand feet each day. Microplastics' impact on bowel movements is still a topic of investigation. Although the microplastics themselves did not sink, the microplastic-rich fecal pellets easily settled to the bottom of the zooplankton's holding container. Therefore, these tiny creatures may be bringing microplastics from the ocean's surface all the way to the sandy bottom. Figure 9.17 depicts the impacts of plastic waste on the marine environment.

One way that micro- and nanoplastics can enter the human body is through the digestive system. Following oral consumption, particles undergo a series of processes that have an impact on their interactions, including coming into contact with digestive fluids, interacting with intestinal cells, being absorbed and transported through the intestine and liver, and being excreted (Figure 9.19). Human tissues and organs frequently contain microplastics, which have unique properties and entrance points as well as a variety of testing methods.

Some of the findings from the research on organs and tissues by Paul et al. (2020) are applicable to exposure assessment because we know what parts of animals we eat. Fish tend to store microplastics in organs that aren't often ingested by humans, including their gills, liver, and gut. A European high consumer could ingest approximately 11'000 microplastic particles each year just eating shellfish, according to some research. Microplastics have been found in other types of food as well, according to research (Paul et al., 2020). Among the first edibles to be tested for microplastics

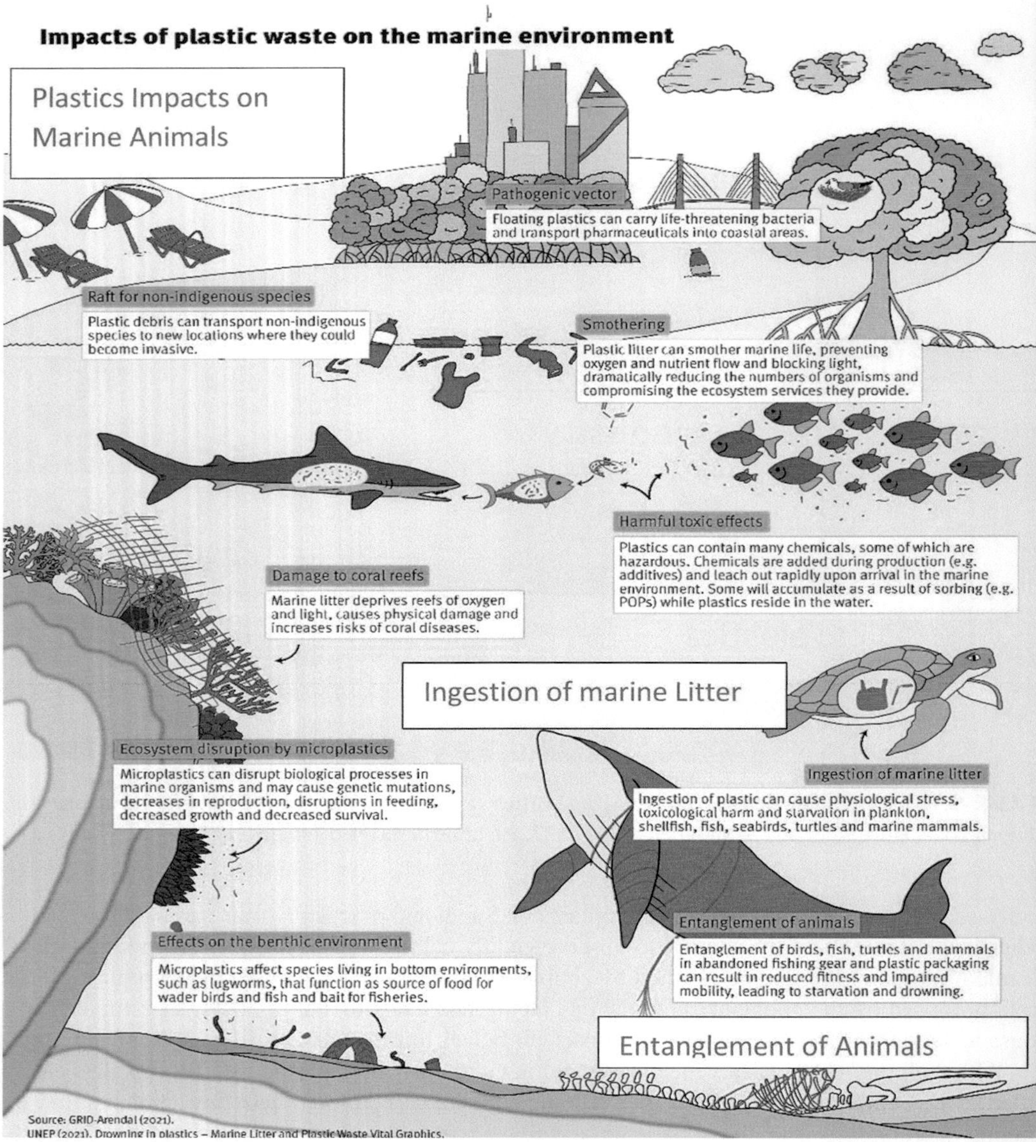

FIGURE 9.17 Impacts of plastic waste on the marine environment (www.grida.no/resources/14899Cartographer: GRID-Arendal.)

contamination were honey and beer. Without the benefit of more sophisticated analytical methods, this earliest research relied on qualitative microscope measurements. The analytical methods used to analyze micro- and nanoplastics and their toxic effects (Adelugba et al., 2024) are shown in Figure 9.19.

Microplastics in salt were the subject of another study. In 2017, data from 14 nations revealed the presence of plastic particles in drinking water for the first time. The United States had the greatest mean concentrations of 0.1-5 mm particles (9.24 11.8 particles per L), while Germany had the

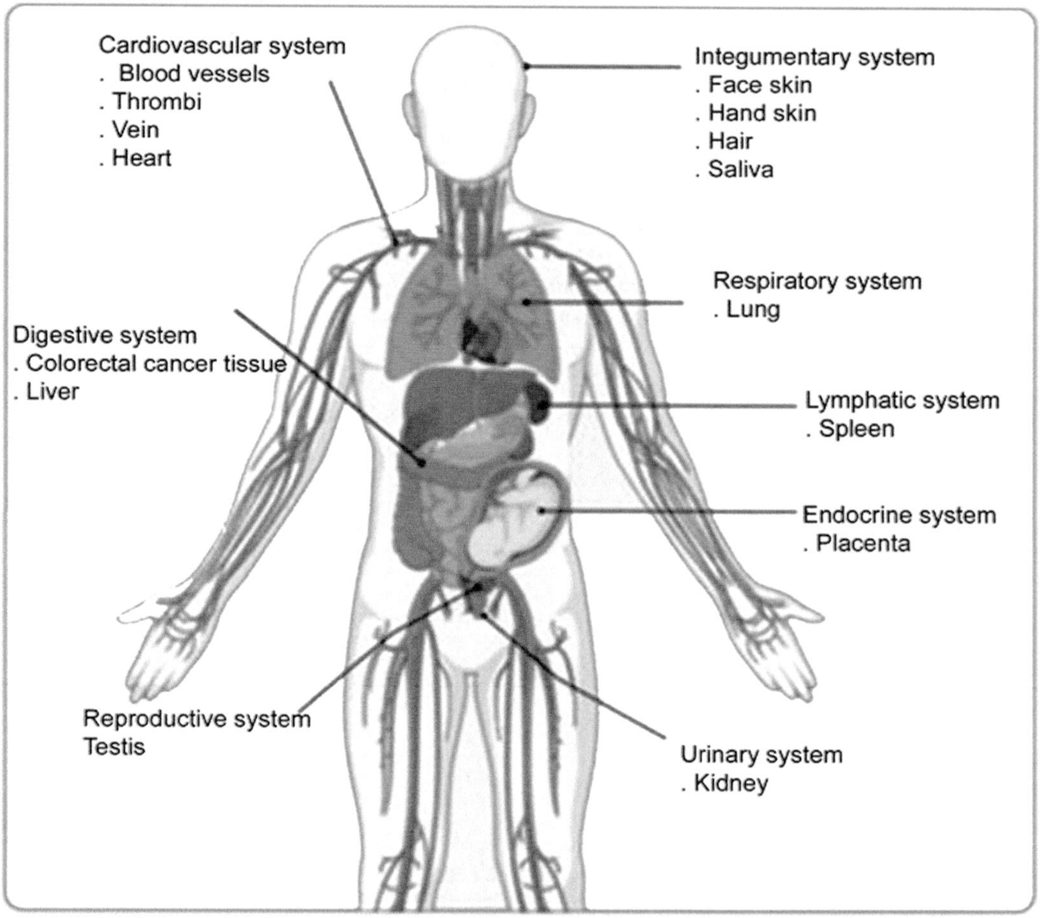

FIGURE 9.18 Detection of microplastics in human tissues and organs (Roslan et al., 2024. https://creativecommons.org/licenses/by/4.0/.)

lowest (0.91 1.29 particles per L). Fibers made up almost 98% of the particles. Additionally, some research on bottled mineral waters has been done so far. Size limits ranged from 1μm to 5mm in an effort to find microplastics. Probably as a result of packaging deterioration, PET was the most common polymer found in the plastic bottles, but microplastic particles were also found in the mineral waters stored in glass containers. Bottled water and groundwater have both been called up as potential sources of human exposure to microplastics in recent years (Weber et al., 2018).

Paul et al. (2020) reached the conclusion that consumers' exposure to microplastics through their mouths is inevitable. It appears that smaller and smaller particles, in particular, can get past the gastrointestinal barrier after being consumed. Rarely have consequences related to cellular toxicity been discovered, and even then, usually only in overload settings and using relatively nonspecific endpoints. There has been relatively little research conducted thus far. Critical examination is precluded by the fact that in vitro research with model particles frequently used extremely high doses, while in vivo investigations gave unimpressive results. The integration of mechanistic data into risk assessment is hindered since neither dose-response correlations nor molecular starting

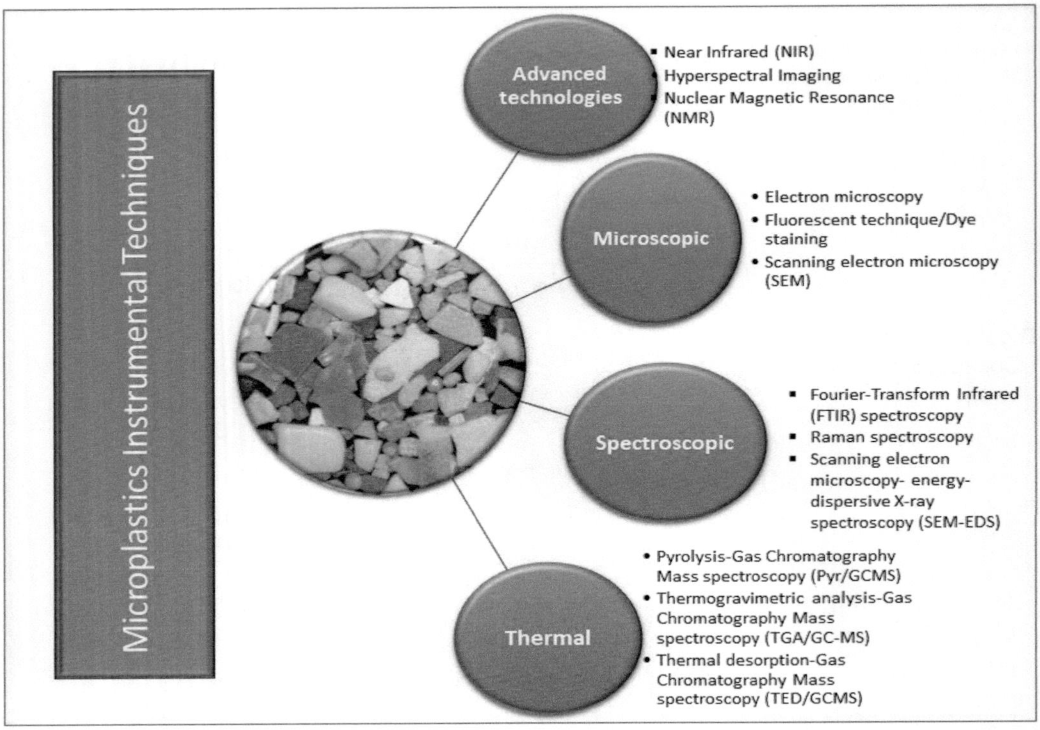

FIGURE 9.19 The analytical methods used to analyze Micro- and nanoplastics and their toxic effects (Adelugba et al., 2024. Open access.)

events can be declared. There is also a lack of chronic and long-term research. Efforts to fill data gaps in the risk assessment of micro- and nanoplastics in human diets must be maintained.

9.11 IMPACT OF MICROPLASTICS

There is a rising tide of microplastics in the seas. The United Nations estimates put the number of microplastic particles in the oceans at 51 trillion, or 500 times more than the number of stars in our galaxy. Marine creatures may consume microplastics from the ocean. The plastic builds up and eventually enters people via the food chain. They have been discovered in a wide variety of beverages and foods. To no one's surprise, scientists have also found plastic debris in human feces. Although the long-term effects of swallowing plastic on humans are not yet established, many types of plastic contain additives that could be damaging to their health, such as stabilizers or flame retardants. Fish with behavioral problems are probably suffering from brain damage brought on by contact with plastic.

As we've seen, numerous marine animal species are ingesting microplastics due to the pervasive plastic contamination. It is critical to investigate microplastics detected in seafood and fish to understand the possible effects on human health, as these microplastics are linked to chemicals (some from manufacturing and others absorbed from the surrounding environment).

To what extent, then, do microplastics endanger human health? Due to information gaps in the study of microplastics, it is difficult to assess the actual amount of microplastics humans may be exposed to via food and the precise nature of their impact. Microplastics have been linked to health concerns by some groups. When it comes to human health, the World Health Organization (WHO) has stated, "microplastics are increasingly identified in drinking water, although there is no evidence so far that this poses a concern to humans." The human body has an excretory mechanism that rids itself of microplastics; it is estimated that >90% of micro- and nano-plastics consumed end up in the feces.

Other research, however, suggests that certain microplastics can transfer living cells, "such as M cells or dendritic cells, to the lymphatic and/or circulatory system, collect in secondary organs, and damage the immune system and cell health." Tissue inflammation, cell growth, and necrosis are all possible outcomes of ingesting microplastics, and they may also damage immune cells. More is known about how toxicants are distributed and stored in the body than about the actual physical effects of microplastic accumulation. Nonetheless, preliminary studies, like those conducted by Wright and Kelly (2017), have revealed some potentially worrying effects of microplastics on human health. These include an uptick in inflammation, particle toxicity that increases with plastic size, the transmission of chemical pollutants that have been adsorbed, and alteration of the gut microbiota.

There is no doubt that further study is needed to fully comprehend and lessen threats to human health. However, banking on waste treatment techniques and plants, as well as adopting circular economy practices that limit or even replace plastic production, is essential to lessen the damage caused by plastic across all human habitats. Potential exposure routes and adverse effects of nanoplastics on humans are depicted in Figure 9.21. Microplastic contamination poses a serious risk (Lai et al., 2022) to ecosystems and biodiversity; it could have adverse effects on organisms, enter the food chain, and eventually reach humans through ingestion, inhalation, and skin contact, leading to severe health problems.

The term "microbiome" describes the combined genetic material, or community, of microorganisms inhabiting a particular environment. Environmental plastics can concentrate in the digestive tracts of organisms from diverse trophic levels. Figure 9.21 shows the Plastics and the microbiome impacts and solutions. Researchers continue to explore diverse microbiomes, including those of soil, marine, freshwater, atmosphere, and subsurface environments.

Microplastics (MPs) are globally known for their negative impacts on different ecosystems and living organisms. MPs are easily taken up by the ecosystem in a variety of organisms due to their small size, and cause immunological, neurological, and respiratory diseases in the impacted organism. Fish is seen as a viable protein source for humans. People at the bottom of the food chain ate a wide variety of fish and crustaceans that included microplastics (MPs). Therefore, there is growing concern that this kind of intake may pose a risk to human health in recent years (Dey et al., 2021). Fish, crabs, and bivalves are only some of the aquatic creatures that have MPs in their tissues and intestines (Lin et al., 2022). The bioavailability of MPs to aquatic organisms is caused by a number of parameters, including form, color, size, and density. MPs were shown to collect on the tissues of the mussels, and the latex-based spheres were observed on the gill's surface in various regions of the rainbow trout fish epidermal cell surface under the skin phagocytes, as demonstrated by the study. Since plastic particles cling to and can penetrate epithelial cells, it follows that epithelial cells play a crucial role in the entry of plastic into fish. Su et al. (2018) found that MPs accumulate in the gills and gut but are not present in the liver or muscles of fish.

PVC and PS are two types of plastic that are known to release carcinogenic monomers when disposed of in landfills (Wang et al., 2016). It's also possible for various PVC additives to build up in human blood. In contrast, PS-based nanoplastics have been shown to harm gastric cancer cell shape, cell viability, and inflammation in humans, all of which can lead to increased gene expression. Plastic microparticles (MPs) are dangerous to breathe in because they can serve as carriers for other pollutants, harmful microorganisms, and chemical parasites. Endocytosis is a fundamental way by which MPs cause damage in the human respiratory and digestive systems.

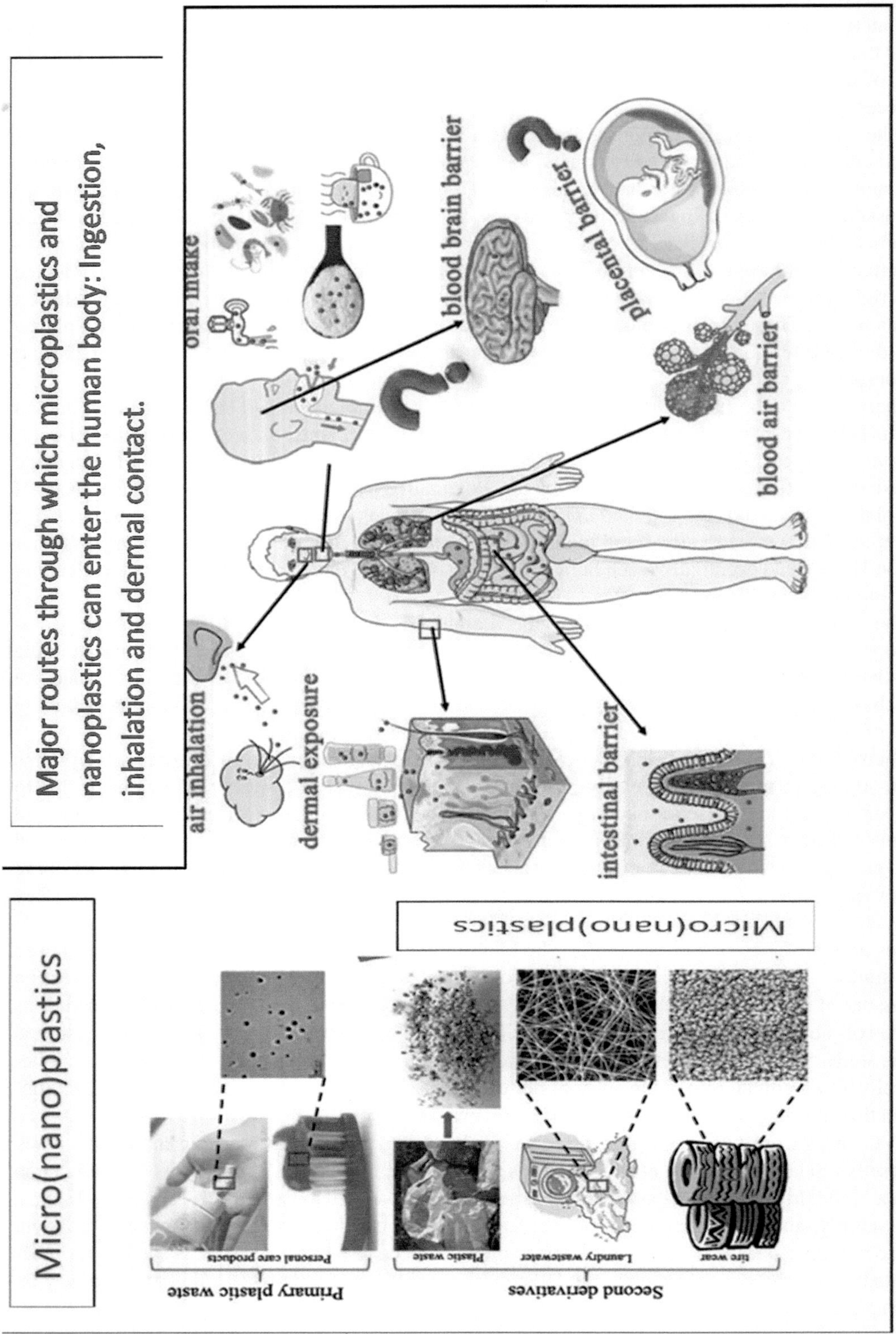

FIGURE 9.20 Potential exposure routes and adverse effects of nanoplastics on humans (Lai et al., 2022, Open access.)

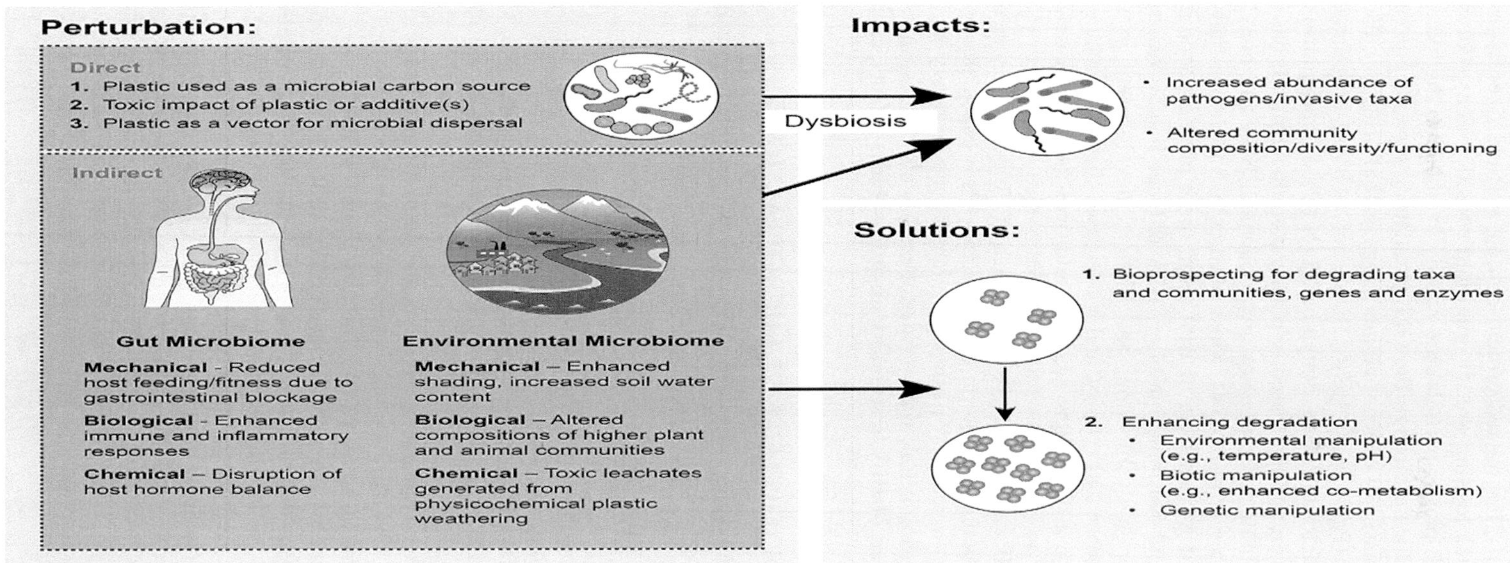

FIGURE 9.21 Plastics and the microbiome (soil, marine, freshwater, atmosphere, and subsurface environments): impacts and solutions (Lear et al., 2021. Open access.)

In addition, the gastrointestinal tract (GIT) circulatory system can transport ingested microplastics to other organs; the related protein corona, particle size, hydrophobicity, surface functionalization, and surface loading are all important considerations. Polystyrene (PS)-based nanoplastics and other small plastic particles are readily available in the blood and other organs due to their ability to translocate through the GIT circulatory system. The presence of various synthetic fibers in the lung tissue of cancer patients suggests that MPs may also have a role in the development of lung cancer (Pauly et al., 1998). Bisphenol A (BPA), brominated flame retardants (BFR), phthalates (PBB), triclosan (TCP), nonylphenol (NYP), and organotin compounds (OTC) are some of the most common additives used in plastic manufacturing. It has been hypothesized that when organisms containing microplastics are consumed, BPA and nonylphenol may leach to biological cells (Koelmans et al., 2014). Most plastics use the potentially harmful antioxidant or plasticizer bisphenol A. Leaching of BPA from these plastic goods make them unfit as food containers. As a thyroid hormone inhibitor, BPA reduces insulin resistance, alters reproductive function, causes brain dysfunction, and disrupts development in the womb. It also increases cardiovascular disease and obesity, and it decreases the performance of pancreatic beta cells (Srivastava and Godara, 2017). There are other additives used as plasticizers that have been linked to health issues such as sterility, low birth weight babies, and cancer (Gómez and Gallart-Ayala, 2018). As a result, MPs are linked to a wide range of illnesses because of the carcinogenic effects of their many ingredients. Long-term exposure to MPs is especially problematic because of their propensity to act as a vector for other pollutants, such as heavy metals and infectious microorganisms. MPs' pollution is linked to a wide range of health issues, such as gastrointestinal (GI) dysfunction, lung cancer, obesity, respiratory disorders, birth defects, cardiovascular illness, viral diseases, and asthma.

9.12 HOW TO LESSEN YOUR USE OF MICROPLASTICS

Microplastics have been found throughout the human body. We compiled some resources so you can better understand how microplastics and plastic additives are entering our bodies, and some tips to limit your exposure.

Here are some suggestions for reducing your exposure to microplastics:

Cut back on the quantity of plastic you use in your daily life.
Food should never be heated in plastic.
Clean often, limit your intake of processed meals, and avoid using plastic to keep food.
Steer clear of canned food that has plastic liners. Choose your seafood carefully.

You can't avoid using plastic altogether. We consume up to 5 gms of plastic every week in our water, apples, fish, and beer, and there are only a few things we can do about it (https://theprint.in/opinion/we-are-consuming-up-to-5-gms-of-plastic-every-week/730790/).

Microplastics can be found in both tap water and bottled water. To avoid double sources of contamination, use tap water over bottled water. Using a high-quality water purifier is your best bet. Plastic utensils are unsafe to use in a microwave. Heat meals in glass containers.

Glass or steel containers are best for preserving food, but you can also use aluminium foil or plastics designated as "recyclable," "biodegradable," or "environmentally friendly." Try to eat as many home-cooked meals as you can. Stay away from food that has been wrapped or packaged in plastic, and don't carelessly throw away plastic. Beaches, storm drains, and other bodies of water should be kept clean. Always do everything you can to help your neighborhood, city, state, and country maintain a clean environment.

9.13 PLASTIC WASTE MANAGEMENT

Plastic waste management is the collection, processing, and disposal of plastic waste (Evode et al., 2021). It is a complex issue that involves a variety of stakeholders, including governments, businesses, and individuals. Reduce your use of plastic. One of the best ways to reduce plastic waste is to reduce your use of plastic in the first place. This means choosing products that are made from sustainable materials, such as paper or glass, and avoiding products that are packaged in plastic. Recycle plastic whenever possible (www.insightsonindia.com/environment/environment-pollution-and-control/was). Recycling plastic helps to reduce the amount of plastic that ends up in our landfills and oceans.

The problem of plastic waste in the environment is one of the most urgent problems of our time. Most of the plastics we use don't break down or dissolve easily, so they're piling up in the drains (Figure 9.22), lakes, and oceans, and will take centuries to decompose, causing serious problems for marine life, human health, and the marine ecosystem (www.conserve-energy-future.com/causes-effects-solutions-of-plastic-pollution.php). Numerous efforts have been made to curb plastic waste and minimize pollution since it was first recognized as a severe problem.

9.14 EFFECTIVE METHODS TO REDUCE PLASTIC WASTE POLLUTION

Solving this problem will require widespread participation from both individuals and institutions in adopting measures that significantly reduce waste production. Figure 9.23a and b explain the solution to plastic waste management and the recycling process.

The management techniques for PW are broadly classified into six categories, namely landfills, recycling, pyrolysis, liquefaction, road construction and tar, and concrete production, as displayed in Figure 9.23b.

Some recommendations for reducing one's use of plastic are:

Store-Friendly: Many reusable bags can be folded down to almost nothing, making plastic bags, formerly a modern convenience now obsolete. Simply multiply the frequency of your grocery shopping trips by the number of bags you typically carry home to get an estimate. Indeed, a plastic mountain has risen! Keep a bag handy at all times, and recycle plastic bags whenever possible.

FIGURE 9.22 Plastic pollution effects on land (https://en.wikipedia.org/wiki/Plastic_pollution#/media/File:Pollution_plastique_2.jpg. Credit: Mouenthias. CC BY-SA 4.0.)

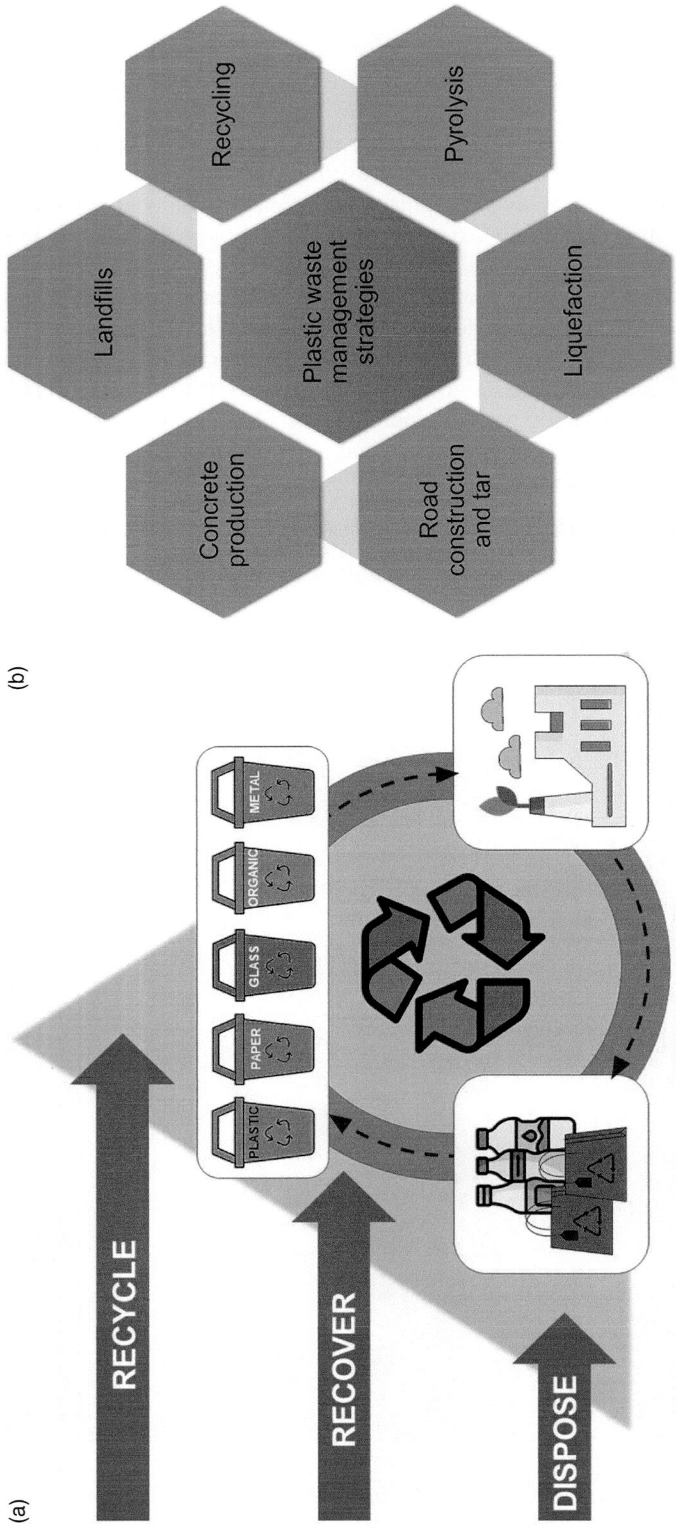

FIGURE 9.23a Waste management strategies and waste-processing cycle (Wiśniewska et al., 2023.)

FIGURE 9.23b Plastic Waste Management (Huang et al., 2022. https://doi.org/10.3390/ijerph19084556.)

Give up the bottled water: Plastic water bottles have become a common and practical solution for consumers to fulfill their daily water intake recommendations. Most of these can only be used once before being thrown away, which means that every time a bottle is empty, it gets thrown away. Polyethylene terephthalate (PET), the material generally used to make plastic bottles, has a biodegradation time of more than 400 years. Nowadays, many establishments provide reusable water bottles as an alternative to single-use ones. Having a metal water container on hand is your best bet.

Take-Out Containers: You might be surprised to learn how much plastic goes into making and shipping food containers. Just picture the paper cup you get at your favorite cafe. Most likely, it's insulating plastic (pour a cup of coffee on some cardboard and see what happens). A simple and efficient alternative to disposable plastic food containers, lids, and utensils is reusable containers.

Nothing Should Be Thrown Away: If at all possible, choose products that are packaged in recycled and recyclable materials other than plastic. More and more things can be recycled now, so make sure to check before you toss anything out. Keep in mind that plastic doesn't biodegrade easily (if ever), so even after recycling, the material is still plastic. Therefore, even after recycling, you aren't lowering plastic use or exposure.

Using No Plastic Products: Some examples of the many single-use plastic things we buy and then toss away are grocery bags, plastic wrap, disposable cutlery, straws, and coffee cup lids. It takes only a few repetitions before it becomes second nature to bring your own reusable bags to the store, silverware to the office, or a travel cup and a metal reusable drinking bottle. As if the health benefits of home-cooked meals weren't enough, eating in also negates the need for disposable containers and doggie bags. Avoid using disposable plastic cutlery and containers by informing eateries that you will be eating in.

Enhance Your Domestic Habits: Small changes you make at home can have a big influence on the environment, and it's crucial to abandon your throw-away attitude. Choose products that use less plastic in their packaging. Avoid using cosmetics and personal care items that include microbeads, the tiny plastic particles found in some face cleansers and toothpaste. Recent research has shown how dangerous microbeads may be to aquatic life and human health. Avoid wearing microfiber synthetics of any kind. The tiny threads that come off in the wash end up in the ocean and eventually in the stomachs of fish and other marine animals. The concentrations of pollutants on microplastics in marine and coastal environments show spatial differences. Concentrations of pollutants on microplastics from remote areas are lower than those in big cities. The concentrations of HCHs and DDTs on microplastics are lower than PAHs and PCBs in global environments (Guo and Wang, 2019). Nanotechnology for smart packaging, genetic engineering of natural products, and the utilization of bio-based, biodegradable polymers are all promising directions for development.

9.15 PRIME CAUSES OF PLASTIC WASTE

First, plastic is inexpensive, accessible, and widely used. Plastic is used for so many things since it is cheap and long-lasting. This includes things like packaging, bottles, straws, bags, and so on. The perpetual cycle of plastic production and disposal will persist unless businesses switch to using more sustainable materials (like paper).

Second, urbanization is on the rise along with the global population (https://cleanstreets.westminster.gov.uk/plastic-waste-complete-guide/#7). Simply put, as the global population rises, so too does the demand for low-cost resources, such as plastic. In the first decade of this century, for example, more plastic than ever before was manufactured in response to surging urbanization and consumer demand.

Thirdly, we treat plastic with a throwaway attitude. Consider the short useful lives of common plastics like grocery bags, water bottles, straws, and storage containers for dry goods and wet foods. Also, we don't bother keeping individual pieces because they're so inexpensive to produce.

Furthermore, plastic waste is often improperly disposed of, leading once again to accumulations in landfills. Plastic is durable because of the chemical links that hold it together. Depending on the specific plastic, the rate of degradation can be anything from 50 to 600 years. A depressing notion for all of us is that, according to the US EPA (Environmental Protection Agency in the United States), virtually all of the plastic ever manufactured and disposed of in landfills or the environment still remains.

The cycle continues as long as new plastic products are produced daily. Commercial fishing and maritime transportation are related industries that use ships to transport fish and other goods. Plastic waste and pollution, especially in the oceans, are also caused by the shipping and fishing sectors. Ships and fishing nets, which are typically made of plastic, are a major contributor to plastic pollution of coastal areas. In addition to contributing to environmental pollution, plastic fishing gear poses a direct threat to marine life by entangling or ingesting marine life.

9.16 CONNECTIONS AMONG THE SUSTAINABLE DEVELOPMENT GOALS (SDGS), WASTE MANAGEMENT, AND PLASTIC POLLUTION

Both terrestrial and aquatic ecosystems are plagued by plastic waste. When plastic garbage enters the ecosystem, it causes issues and poses a serious threat to all living things. Because of careless use, insufficient recycling, and landfill deposits, plastic is accumulating in the environment at a rate never seen before. Because of their lengthy environmental persistence and resistance to degradation, plastics are a geological indicator of the Anthropocene Age and have recently emerged as a threat to the environment. Although plastic is a fantastic substance that contributes to economic development and synthetic modernity, it is known as plastic litter when it is disposed of carelessly and unethically in any environment. The objectives of the SDGs are mainly to take the collective decision against the most critical multi-dimensional socio-economic-environmental global threats (Figure 9.24).

The intricate interdependency of plastic use's economic and hazardous aspects is ingrained in the current lifestyle. Irrational production, improper landfill disposal, and poor recycling management turn plastics into waste. The extraordinary rate at which plastic garbage is leaking into the environment, including terrestrial and aquatic ecosystems, presents serious problems for waste management for expanding populations, particularly in emerging nations. Limiting the amount of plastic waste that is dumped in landfills and deposited in marine environments should be a priority. The UN seeks to mitigate and treat the effects of ocean acidification, manage and protect marine and coastal ecosystems sustainably, and prevent major negative impacts by enhancing their resilience to achieve sustainability and take action against climate change.

Plastic pollution is a significant environmental challenge that affects every aspect of human life, from public health to marine ecosystems. The management of plastic waste is closely linked to all 17 Sustainable Development Goals (SDGs), and the interlinkages between plastics pollution, management, and each SDG are complex and multi-dimensional, as shown in Figure 9.24.

Interlinkages between Plastic Pollution, Waste Management, and Sustainable Development Goals

To attain sustainability, many countries have adopted the Sustainable Development Goals (SDGs), which were introduced by the UN in September 2015 as an extension of the Millennium Development Goals. The SDGs aim to address the most pressing multifaceted socio-economic-environmental concerns on a global scale by promoting collaborative action. Only one of the 17 Sustainable Development Goals (SDGs) and 169 targets has a direct connection to plastic pollution. SDG 14: Life Below Water, with target 14.1.b as an indicator, aims to reduce marine (micro)plastic debris burdens by 2025, especially from land activities. The nature and behavior of plastics in the environment, including MPs pollution monitoring and management, are difficulties that each nation faces at the national, subnational, or supranational levels, as shown by SDG indicators. Thus, the

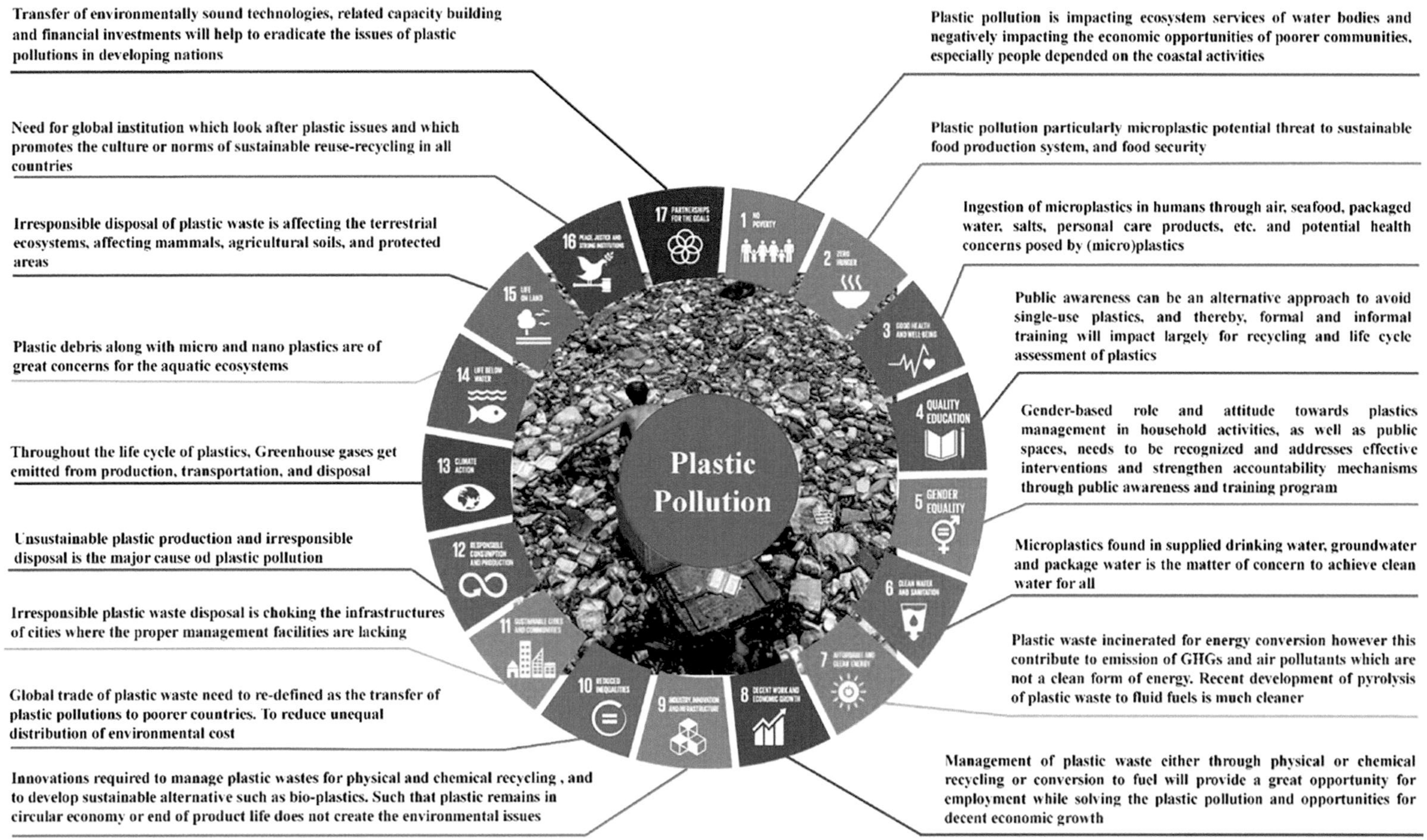

FIGURE 9.24 Perspectives on interlinkages and relationships between plastics pollution, management, and each SDG (Kumar et al. (2021), https://doi.org/10.3390/su13179963.)

connections between waste management and plastic pollution and each of the SDGs are underlined in the figure and covered in the ensuing subsections. Nearly 320 million tons of plastic garbage are produced each year, and by 2050, that amount is predicted to triple. By exposing plastics to the environment and keeping an eye on the detrimental effects on humans and aquatic life, a number of government agencies, research groups, and institutions have coordinated mitigation and management efforts to safeguard the ecosystem. The handling of plastic garbage remains appalling, particularly in developing nations, despite the current understanding of the effects of plastic contamination. For instance, various lifeforms are at risk for health problems due to the widespread use of open burning, inappropriate landfilling, and natural disposal methods like cremation. The existing linear paradigm of managing plastic garbage must be replaced immediately with a circular model that incorporates chemical or physical recycling.

Improving waste management systems and encouraging low-income nations to embrace ecologically friendly technologies from affluent nations are two basic ways to address the issue of plastic pollution. Whether or not households utilize bottled water depends in large part on climate factors. Lack of access to potable water or unreliable water quality in various locations speeds up the use of bottled water and contributes to plastic pollution, both of which are closely related to climate change. Formal and informal training may have a significant impact on the recycling and life cycle assessment of non-biodegradable and biodegradable plastics, since public awareness might be an alternative strategy to reduce single-use plastics. In the Anthropocene, plastic pollution has become an unavoidable part of human civilization.

9.17 CONCLUSION

Microplastics are a contaminant that can have negative effects on ecosystems and wildlife. Microplastics, like plastics of all sizes, pose a threat because they do not easily decompose into innocuous molecules. Microplastics appear as small, colorful plastic fragments in the sand on beaches. Microplastic pollution is a major problem in the water, and marine animals frequently ingest it. Some of this environmental pollution is the consequence of littering, but much of it is the result of plastic—both whole objects and microplastics—being carried into our oceans by storms, water runoff, and winds. While scientists have devoted a great deal of time and energy to studying microplastics in the marine environment, little is known about the nature and amount of microplastic pollution in continental habitats such as rivers, lakes, soil, and the air. The effects of microplastics are emblematic of a global problem that affects virtually every marine ecosystem, no matter how remote it may be from possible sources of pollution. Microplastics (MPs) can be found in the digestive systems (intestines, tissues) of many marine organisms. Since plastic particles cling to and can penetrate epithelial cells, it follows that epithelial cells play a crucial role in the entry of plastic into fish. Therefore, it is plausible that MPs could learn about the human diet by eating these gills and skin surfaces.

More people are becoming aware of environmental microplastics and the potential harm they may cause to biodiversity, ecosystem services, water security, and human health and welfare. Ecosystems, economies, and public health are all at risk due to the unsustainable use and disposal of plastic items in modern society. Despite efforts to reduce the negative effects of plastic pollution, present cleanup methods are unable to keep up with the increasing volume of plastic entering the environment. Therefore, reducing plastic inputs to the environment must be a top priority using a global, interdisciplinary strategy. Ten recommendations for stakeholders to reduce plastic pollution include: regulating production and consumption; eco-design; increasing demand for recycled plastics; reducing plastic use; using renewable energy for recycling; extending producer responsibility for waste; upgrading waste collection systems; prioritizing recycling; and using bio-based and biodegradable plastics.

REFERENCES

Abbasi, S. 2021. Microplastics washout from the atmosphere during a monsoon rain event. *J. Hazardous Materials Adv.* 4: 100035. https://doi.org/10.1016/j.hazadv.2021.100035

Adelugba, A. and C.Emenike., 2024. Comparative review of instrumental techniques and methods for the analysis of microplastics in agricultural matrices. *Microplastics.* 3: 1–21. https://doi.org/10.3390/microplastics3010001

Ahmad, M., J-L. Li, P-D. Wang, W.N. Hozzein and W-J. Li. 2020. Environmental perspectives of microplastic pollution in the aquatic environment: A review. *Mar. Life Sci. Technol.* 2: 414–430. https://doi.org/10.1007/s42995-020-00056-w

Alberghini, L., A. Truant, S. Santonicola, G. Colavita and V. Giaccone. 2023. Microplastics in fish and fishery products and risks for human health: A review. *Int. J. Environ. Res. Public Health* 20: 789. https://doi.org/10.3390/ijerph20010789

Allouzi, M.A.M., D.Y.Y. Tang, K.W. Chew, J. Rinklebe, N. Bolan, S.M.A. Allouzi and P.L. Show. 2021. Micro (nano) plastic pollution: The ecological influence on soil-plant system and human health. *Sci. Total Environ.* 788: 147815. https://doi.org/10.1016/j.scitotenv.2021.147815

Amato-Lourenço, L.F., R. Carvalho-Oliveira, G.R. Júnior, L. Dos Santos Galvão, R.A. Ando and T. Mauad. 2021. Presence of airborne microplastics in human lung tissue. *J. Hazard Mater.* 416: 126124. https://doi.org/10.1016/j.jhazmat.2021.126124

Andrady, A.L. 2011. Microplastics in the marine environment. *Marine Poll. Bull.* 62(8): 1596–1605. https://doi.org/10.1016/j.marpolbul.2011.05.030

Athey, S.N., S.D. Albotra, C.A. Gordon, B. Monteleone, P. Seaton, A.L. Andrady, A.R. Taylor and S.M. Brander. 2020. Trophic transfer of microplastics in an estuarine food chain and the effects of a sorbed legacy pollutant. *Limnol. Oceanogr. Lett.* 5: 154–162. https://doi.org/10.1002/lol2.10130

Bergami, E., S. Pugnalini, M.L. Vannuccini, L. Manfra, C. Falerii, F. Savorelli, K.A. Dawson and I. Corsi. 2017. Long-term toxicity of surface-charged polystyrene nanoplastics to marine planktonic species *Dunaliella tertiolecta* and *Artemia franciscana. Aquatic Toxicol.* 189: 159–169. https://doi.org/10.1016/j.aquatox.2017.06.008

Bhuyan, S.Md. 2022. Effects of microplastics on fish and in human health. *Front. Environ. Sci., 16, Sec. Toxicol. Poll. Environ.* 10. https://doi.org/10.3389/fenvs.2022.827289

Browne, M.A., P. Crump, S.J. Niven, E. Teuten, A. Tonkin and T. Galloway. 2011. Accumulation of microplastic on shorelines worldwide: Sources and sinks, *Environ. Sci. Technol.* 45: 9175–9179.

Campanale, C., C. Massarelli, I. Savino, V. Locaputo and V.F. Uricchio. 2020. A detailed review study on potential effects of microplastics and additives of concern on human health. *Int. J. Environ. Res. Public Health.* 17(4): 1212. https://doi.org/10.3390/ijerph17041212

Campanale, C., F. Stock, C. Massarelli, C. Kochleus, G. Bagnuolo, G. Reifferscheid and V. Uricchio. 2019. Microplastics and their possible sources: The example of Ofanto river in southeast Italy. *Environ. Pollut.* 258: 113284. https://doi.org/10.1016/j.envpol.2019.113284

Chatterjee, S. and S. Sharma. 2019. Microplastics in our oceans and marine health. *Field Actions Sci. Rep.* [Online]. Special Issue 19.

Chaukura, N., K.K. Kefeni, I. Chikurunhe, I.I. Nyambiya, W. Gwenzi, W. Moyo, T.T.I. Nkambule, B.B. Mamba and F.O. Abulude. 2021. Microplastics in the aquatic environment—The occurrence, sources, ecological impacts, fate, and remediation challenges. *Pollutants.* 1: 95–118. https://doi.org/10.3390/pollutants1020009

Cole, M., P.K. Lindeque, E. Fileman, Clark, C. Lewis, C. Halsband and T.S. Galloway. 2016. Microplastics alter the properties and sinking rates of zooplankton faecal pellets. *Environ. Sci. Technol.* 50(6): 3239–3246. https://pubs.acs.org/doi/10.1021/acs.est.5b05905

Cozar, A., F. Echevarria, J.I. Gonzalez-Gordillo and C.M. Duarte. 2014. Plastic debris in the open ocean. *Proc. Natl. Acad. Sci. USA* 111: 10239–10244. https://doi.org/10.1073/pnas.1314705111

Danopoulos, E., M. Twiddy and J.M. Rotchell. 2020. Microplastic contamination of drinking water: A systematic review. *PLoS ONE.* 15(7): e0236838. https://doi.org/10.1371/journal.pone.0236838

de Souza, M.A.A., W. Kolas Zarf, S. Hempe and M.C. Rilling. 2018. Microplastics as an emerging threat to terrestrial ecosystems. *Global Change Biol.* 24(4): 1405–1416. https://doi.org/10.1111/gcb

Dey, A., H. Choudhury, A. Mazumder, R. Ch. Bharali, K.K. Lal and D. Sarma. 2021. Spatial and temporal dynamics of fish species assemblage and distribution in the Kameng river, Northeast India. *Proc. Natl. Acad. Sci., India, Sect. B Biol. Sci.* 91: 123–130. https://doi.org/10.1007/s40011-020-01208-9

Dhiman, S., C. Sharma, A. Kumar, P. Pathak and S.D. Purohit. 2023. Microplastics in aquatic and food ecosystems: Remediation coupled with circular economy solutions to create resource from waste. *Sustainability*. 15: 14184. https://doi.org/10.3390/su151914184

Eerkes-Medrano, D., H.A. Leslie and B. Quinn. 2019. Microplastics in drinking water: A review and assessment. *Curr. Opinion Environ. Sci. Health*. 7: 69–75.

Evode, N., S. A. Qamar, M. Bilal, D. Barcelo´ and H. M. N. Iqbal. 2021. Plastic waste and its management strategies for environmental sustainability, Case Studies in Chemical and Environmental Engineering 4 , 100142. https://doi.org/10.1016/j.cscee.2021.100142

Fen, Q., J. Du, J. Gao, G. Liu, Y. Song, A. Yang, H. Wang, Y. Ding and Q. Wang. 2020. Bibliometric profile of global microplastics research from 2004 to 2019. *Int. J. Environ. Res. Public Health*. 17(16): 5639; https://doi.org/10.3390/ijerph17165639

Gambardella, C., S. Morgana, S. Ferrando, M. Bramini, V. Piazza, E. Costa, F. Garaventa and M. Faimali. 2017. Effects of polystyrene microbeads in marine planktonic crustaceans. *Ecotoxicol. Environ Saftey*. 145: 250–257. https://doi.org/10.1016/j.ecoenv.2017.07.036

Gela, S.M. and T.A. Aragaw. 2022. Abundance and characterization of microplastics in main urban ditches across the Bahir Dar City, Ethiopia. *Front. Environ. Sci*. 10. https://doi.org/10.3389/fenvs.2022.831417

Ghosh, S., J.K. Sinha, S. Ghosh, K. Vashisth, S. Han and R. Bhaskar. 2023. Microplastics as an emerging threat to the global environment and human health. *Sustainability*. 15: 10821. https://doi.org/10.3390/su15 1410821

Gómez, C. and H. Gallart-Ayala. 2018. Metabolomics: A tool to characterize the effect of phthalates and bisphenol A. *Environ. Rev*. 26(4): 351–357.

Guo, X. and J. Wang , 2019, The chemical behaviors of microplastics in marine environment: A review, Marine Pollution Bulletin, 142, 1-14. https://doi.org/10.1016/j.marpolbul.2019.03.019

Hamed, M., H.A.M. Soliman, Z. Eid., Y. Al Naggar and AE-DH, Sayed. 2022. Dietary feeding lycopene, citric acid, and chlorella alleviated the neurotoxicity of polyethylene microplastics in African Catfish (*Clarias gariepinus*). *Front. Environ. Sci*. 10 : 869727. https://doi.org/10.3389/fenvs.2022.869727

Hernandez, L.M., E.G. Xu, C.E.H. Larsson, R. Tahara, B.V. Maisuria and N. Tufenkji. 2019. Plastic teabags release billions of microparticles and nanoparticles into tea. *Environ. Sci. Tech*. 53(21): 12300–12310. https://doi.org/10.1021/acs.est.9b02540

Huang, S., H. Wang, W. Ahmad, A. Ahmad, N.I. Vatin, A.M. Mohamed, A.F. Deifalla and I. Mehmood. 2022. Plastic waste management strategies and their environmental aspects: A scientometric analysis and comprehensive review. *Int. J. Environ. Res. Public Health*. 19: 4556. https://doi.org/10.3390/ijerph19084556

Issac, M.N. and B. Kandasubramanian. 2021. Effect of microplastics in water and aquatic systems. *Environ. Sci. Pollut. Res*. 28: 19544–19562. https://doi.org/10.1007/s11356-021-13184-2

Jiang, B., A.E. Kauffman, L. Li, W. McFee, B. Cai, J. Weinstein and S. Xiao. 2020. Health impacts of environmental contamination of micro and nanoplastics: A review. *Environ. Health Preventive Med*. 25(1): 1–15. https://doi.org/10.1186/s12199-020-00870-9

Karthik R., R.S. Robin, R. Purvaja, D. Ganguly, I. Anandavelu, R. Raghuraman, G. Hariharan, A. Ramakrishna and R. Ramesh. 2018. Microplastics along the beaches of southeast coast of India. *Sci. Total Environ*. 645: 1388–1399. https://doi.org/10.1016/j.scitotenv.2018.07.242

Khuyen, V.T.K., D.V. Le, A.R. Fischer and C. Dornack. 2021. Comparison of microplastic pollution in beach sediment and seawater at UNESCO Can Gio Mangrove Biosphere Reserve. *Glob. Chall*. 5(11): 2100044. https://doi.org/10.1002/gch2.202100044

Koelmans, A.A., Besseling, E. and E.M. Foekema. 2014. Leaching of plastic additives to marine organisms. *Environ. Pollut*. 187: 49–54.

Kumar, R., A. Verma, A. Shome, R. Sinha, S. Sinha, P.K. Jha, R. Kumar, P. Kumar, S. Das, P. Sharma and P.V. Vara Prasad. 2021. Impacts of plastic pollution on ecosystem services, sustainable development goals, and need to focus on circular economy and policy interventions. *Sustainability*. 13: 9963. https://doi.org/10.3390/su13179963

Lai, H., X. Liu. and M. Qu. 2022. Nanoplastics and human health: Hazard identification and bio interface. *Nanomaterials*. 12: 1298. https://doi.org/10.3390/nano12081298

Lear, G., J.M. Kingsbury, S. Franchini, V. Gambarini, S.D.M. Maday, L. Waver and O. Pantos. 2021. Plastics and the microbiome: Impacts and solutions. *Environ. Microbiome.* 16: 2. https://doi.org/10.1186/s40793-020-00371-w

Lin, L., H. Yang and X. Xu. 2022. Effects of microplastics on fish and in human health. *Front. Environ. Sci., Sec. Water Wastewater Manage.* 10. https://doi.org/10.3389/fenvs.2022.880246

Lin, V.S. 2016. Research highlights: Impacts of microplastics on plankton. *Environ. Sci.: Processes Impacts* 18: 160–163. https://doi.org/10.1039/c6em90004f

Loganathan, Y. and M.P.J. Kizhakedathil, 2023. A review on microplastics – an indelible ubiquitous pollutant. *Biointerface Res. Appl. Chem.* 3(2): 126. https://doi.org/10.33263/BRIAC132.126

Lujan-Vega, C., J.L. Ortega-Alfaro, J. Cossaboon, S. Acuña and S.J. Teh. 2021. How are micro plastics invading the world? *Front. Young Minds.* 9: 606974. https://doi.org/10.3389/frym.2021.606974

Lumio, R.T., M.A. Tan and H.D. Magpantay, H.D. 2021. Biotechnology-based microbial degradation of plastic additives. *3 Biotech.* 11: 350. https://doi.org/10.1007/s13205-021-02884-8

Masó, M., J.M. Fortunio, S. de Juan and M. Demestre. 2016. Microfouling communities from pelagic and benthic marine matter to microplastic addition in Chinese loess soil. *Chemosphere.* 185: 907–917.

Mattsson, K., M.T. Ekvall, L. Hansson, S. Linse, A. Malmendal and T. Cedervall. 2015. Altered behavior, physiology, and metabolism in fish exposed to polystyrene nanoparticles. *Environ. Sci. Tech.* 49(1): 553–561. https://doi.org/10.1021/es5053655

Müller, C. 2021. Not as bad as it seems? A literature review on the case of microplastic uptake in fish. *Front. Mar. Sci., Sec. Marine Pollution.* 8. https://doi.org/10.3389/fmars.2021.672768

Osman, A.I., M. Hosny, A.S. Eltaweil, O. Sara, E.M. Ahmed, F. Mohamed, Y. Pow-Seng, W. Yuan-Seng, N. Saraswathi, B. Kalaivani, C.B. Gopinath Subash, J. Oliver Dean, S. Mahendran, S. Trideep, K. Puvanan, H.M. Hayrie Mohd and A.K. Adedamola. 2023. Microplastic sources, formation, toxicity and remediation: A review. *Environ. Chem. Lett.* 21: 2129–2169. https://doi.org/10.1007/s10311-023-01593-3

Padervand, M., E. Lighthouse, D. Robert and C. Wang. 2020. Removal of microplastics from the environment. A review. *Environ. Chem. Lett.* 18(3): 807–828. https://doi.org/10.1007/s10311-020-00983-1

Pan, Y., S.H. Gao, C. Ge, Q. Gao, S. Huang, Y. Kang, G. Luo, Z. Zhang, L. Fan, Y. Zhu, A.J and Wang. 2023. Removing microplastics from aquatic environments: A critical review. *Environ. Sci. Ecotech.* 13: 100222. https://doi.org/10.1016/j.ese.2022.100222

Parashar, N. and S. Haitet. 2023. Plastic rain—Atmospheric microplastics deposition in urban and peri-urban areas of Patna City, Bihar, India: Distribution, characteristics, transport, and source analysis. *J. Hazard Mater.* 458: 131883 https://doi.org/10.1016/j.jhazmat.2023.131883

Patil, P.B., S. Maity and A. Sarkar. 2022. Potential human health risk assessment of microplastic exposure: Current scenario and future perspectives. *Environ. Monit. Assess.* 194: 898. https://doi.org/10.1007/s10661-022-10539-1

Pauly, J.L., S.J. Stegmeier, H.A. Allard, R.T. Cheney, P.J. Zhang, A.G. Mayer and R.J. Streck.1998. Inhaled cellulosic and plastic fibers found in human lung tissue. *Cancer Epidemiol. Prevent. Biomark.* 7(5): 419–428.

Paul, M.B., V. Stock, J. Cara-Carmona, E. Lisicki, S. Shopova, V. Fessard, A. Braeuning, H. Sieg and L. B¨ohmert. 2020. Micro- and nanoplastics– current state of knowledge with the focus on oral uptake and toxicity. *Nanoscale Adv.* 2: 4350.

Peng, L., Mehmood, R. Bao, Z. Wang and D. Fu. 2022. An overview of micro (nano)plastics in the environment: Sampling, identification, risk assessment and control. *Sustainability.* 14: 14338. https://doi.org/10.3390/su142114338

Pironti, C., M. Ricciardi., O. Motta., Y. Miele., A. Proto. and L. Montano. 2021. Micro plastics in the Environment: Intake through the food web, human exposure and toxicological effects. *Toxics.* 9: 224. https://doi.org/10.3390/toxics.9090224

Prata, J.C. 2018. Airborne microplastics: Consequences to human health. *Environ. Pollute.* 234: 115–126. https://doi.org/10.1016/j.envpol.2017.11.043

Qin, F., J. Du, J. Gao, G. Liu, Y. Song, A. Yang, H. Wang, Y. Ding and Q. Wang. 2020. Bibliometric profile of global microplastics research from 2004 to 2019. *Int. J. Environ. Res. Public Health.* 17(16): 5639. https://doi.org/10.3390/ijerph17165639

Revel, M., A. Châtel and C. Mouneyrac. 2018. Micro(nano)plastics: A threat to human health. *Curr. Opin. Environ. Sci. Health.* 1: 17–23.

Ritchie, H. 2023. How much plastic waste ends up in the ocean? *OurWorldinData.org*. https://ourworldindata.org/how-much-plastic-waste-ends-up-in-the-ocean

Rochman, C.M., B.T. Hentschel and S.J. Teh. 2014. Long-term sorption of metals is similar among plastic types: Implications for plastic debris in aquatic environments. *PLoS ONE.* 9(1): e85433. https://doi.org/10.1371/journal.pone.0085433

Roslan, N.S., Y. Lee, Y.S. Ibrahim, S.T. Anuar, K.M.K.K. Yusof, L.A. Lai and T. Brentnall. 2024. Detection of microplastics in human tissues and organs: A scoping review. *J. Glob. Health.* 14: 04179. https://creativecommons.org/licenses/by/4.0/

Sayed, A.E.-D.H., M. Hamed, A.E.A. Badrey, R.F. Ismail, Y.A.A. Osman, A.G.M. Osman and H.A.M. Soliman. 2021. Microplastic distribution, abundance, and composition in the sediments, water, and fishes of the Red and Mediterranean Seas, Egypt. *Mar. Pollut. Bull.* 173: 112966. https://doi.org/10.1016/j.marpolbul.112966

Sayed, A. El-Di. H., M. H.M. Hamed and R. F. Ismail, 2022. Natural antioxidants can improve micro plastics-induced male reproductive impairment in the African Catfish (*Clarias gariepinus*), *Front. Environ. Sci.,* 2022, Sec. Toxicology, Pollution and the Environment, 9, 2021 . https://doi.org/10.3389/fenvs.2021.811466

Sharma, S and S. Chatterjee. 2017. Microplastic pollution, a threat to marine ecosystem and human health: A short review. *Environ. Sci. Pollut. Res.* 24(27): 21530–21547. https://doi.org/10.1007/s11356-017-9910-8

Singh, S., M. Kalyanasundaram and V. Diwan. 2021. Removal of microplastics from wastewater: Available techniques and way forward. *Water Sci. Tech.* 84(1): 3689. https://doi.org/10.2166/wst.2021.472

Silva, C.J.M., A.L.P. Silva, C. Gravato and J.L.T. Pestana. 2019. Ingestion of small-sized and irregularly shaped polyethylene microplastics affect *Chironomus riparius* life-history traits. *Sci. Total Environ.* 672: 862–868. https://doi.org/10.1016/j.scitotenv.2019.04.017

Smith, M., D.C. Love, C.M. Rochman and R.A. Neff. 2018. Microplastics in seafood and the implications for human health. *Curr. Envir. Health.* 5: 375–386. https://doi.org/10.1007/s40572-018-0206-z

Srivastava, R. and S. Godara. 2017. Use of polycarbonate plastic products and human health. *Int. J. Basic Clin. Pharmacol.* 2(1): 12–17.

Su, L., H. Cai, P. Kandasamy, C. Wu, C.M. Rochman and H. Shi. 2018. Using the Asian clam as an indicator of microplastic pollution in freshwater ecosystems. *Environ Pollut.* 234: 347–355.

Van Cauwenberghe, L., A. Vanreusel, J. Mees and C.R. Janssen, 2013. Microplastic pollution in deep-sea sediments. *Environ. Pollut.* 182: 495–499.

Varg, J.E., C. Bergvall and R. Svanbäck. 2022. The stressful effects of microplastics associated with chromium (vi) on the microbiota of *Daphnia Magna. Front. Environ. Sci.* 10: 875512. https://doi.org/10.3389/fenvs.2022.875512

Venâncio, C., I. Ferreira, M.A. Martins, A.M.V.M. Soares, I. Lopes and M. Oliveira. 2019. The effects of Nano plastics on marine plankton: A case study with polymethylmethacrylate. *Ecotoxicol. Environ. Safety.* 184: 109632. https://doi.org/10.1016/j.ecoenv.2019.109632

Vo, H.C. and M.H. Pham. 2021. Eco toxicological effects of micro plastics on aquatic organisms: A review. *Environ. Sci. Pollut. Res.* 28: 44716–44725. https://doi.org/10.1007/s11356-021-14982-4

Wagner, M. and S. Lambert. 2018. *Freshwater Micro Plastics: Emerging Environmental Contaminants.* Springer.

Wang, J., Z. Tan, J. Peng, Q. Qiu, and M. Li. 2016. The behaviors of micro plastics in the marine environment. *Mar. Environ. Res.* 113: 7–17.

Weber, A., C. Scherer, N. Brennholt, G. Reifferscheid and M. Wagner. 2018. PET micro plastics do not negatively affect the survival, development, metabolism and feeding activity of the freshwater invertebrate *Gammarus pulex. Environ. Pollut.* 234: 181–189. https://doi.org/10.1016/j.envpol.2017.11.014

Westphalen, H and A.A. Rasoul. 2018. *Challenges and Treatment of Microplastics in Water.* Intech. https://doi.org/10.5772/intechopen.71494

WHO (World Health Organization) 2019. *Microplastics in Drinking-Water.* IGO, World Health Organization. www.who.int/publications/i/item/9789241516198

Wisniewski, P., A. Hejna and M.R. Saeb. 2023. Recycling and processing of waste materials. *Materials.* 16: 508. https://doi.org/10.3390/ma16020508

Wright, S.L. and F.J. Kelly. 2017. Plastic and human health: A micro issue. *Environ. Sci. Tech.* 51(12): 6634–6647. https://doi.org/10.1021/acs.est.7b00423

Yee, M.S.-L., L.-W, Hii., C.K Looi., W.-M, Lim., S.-F, Wong., Y.-Y, Kok., B.-K, Tan., C.-Y, Wong. and C.-O Leong. 2021. Impact of micro plastics and nano plastics on human health. *Nanomaterials*. 11: 496. https://doi.org/10.3390/nano11020496

Zhao, S., Zhu and Li, D. 2016. Microscopic anthropogenic litter in terrestrial birds from Shanghai, China: Not only plastics but also natural fibers. *Sci. Total Environ.* 550: 1110–1115. https://doi.org/10.1016/j.scitotenv

Ziani, K., C.-B, Ioniță-Mîndrican, M. Mititelu., S.M. Neacşu., C. Negrei., E. Moroşan., D. Drăgănescu. and O.-T, Preda. 2023. Micro plastics: A real global threat for environment and food safety: A state-of-the-art review. *Nutrients*. 15: 617. https://doi.org/10.3390/nu15030617

10 Power Generation's Impact on Pollution and Associated Health Risks

10.1 INTRODUCTION

Energy use is central to human activity for preparing food, warming homes, powering travel, and producing goods, among many other purposes. Energy is engaged in all life cycles and is equally important in agriculture as it is in all other productive activities. A rudimentary food chain demonstrates the requirement for energy: crops require energy from solar radiation to grow, harvesting requires energy from the human body at work, and cooking requires energy from biomass in a fire. Food supplies energy to the human body. Energy is described as the ability to perform work, which in biology can be thought of as the ability to effect change. Energy can take many different forms, such as light, heat, and electricity. Energy generation is the primary driver of air pollution and global warming emissions. Energy availability is also associated with health at the household level, reflected in the terms' energy security and energy poverty (Smith et al., 2013).

Energy production is the generation of energy from renewable sources such as bioenergy, solar energy, geothermal energy, wind energy, and hybrid energy systems, that do not rely on fossil fuels and help maintain ecosystem stability and preservation. Energy is used in all life cycles and is just as crucial in agriculture as it is in any other productive activity. A simple food chain highlights the need for energy: crops take energy from sunlight to grow, harvesting requires energy from the human body at work, and cooking requires energy from biomass on a fire. Food provides energy for the human body. Energy is defined as the ability to do labor, which in biology can be interpreted as the ability to effect change. Energy can take numerous forms, including light, heat, and electricity.

Energy generation is the principal source of air pollution and greenhouse gas emissions. Energy production is the generation of energy from renewable sources such as bioenergy, solar energy, geothermal energy, wind energy, and hybrid energy systems, which do not use fossil fuels and contribute to ecosystem stability and preservation (www.planete-energies.com/en/media/article/energy-all-its-forms). Energy use and production can have a significant impact on human health (https://envirn.org/energy-and-health/). From the local impacts of mining and drilling for fossil fuels to the climate impacts from burning fossil fuels, there can be negative impacts to health and the environment all across the fossil fuel lifecycle. Fossil fuel energy sources include oil, coal, and natural gas, and are the world's primary energy source. Due mainly to particulate matter, sulfur dioxide, and nitrogen oxide emissions, power generation—especially that which comes from fossil fuels like coal and oil—contributes significantly to air pollution, which can result in several health risks, such as respiratory issues like asthma, heart disease, lung cancer, stroke, and early death. Large, densely populated populations are at serious risk for health problems due to the numerous fine particulate matter (PM2.5, PM $\leq$ 2.5 μm in diameter) precursors found in emissions from power plants in China and India.

DOI: 10.1201/9781003618164-10

10.1.1 Concepts and Types of Energy

Our body gets the energy to play, think, and do work from the food we eat. Similarly, non-living things like cars, buses, airplanes, fans, TV, etc., also need energy to work, but from different kinds of energy sources like petrol, diesel, or electricity. Energy exists in different forms (www.turito.com/learn/science/energy,www.thoughtco.com/main-energy-forms-and-examples-609254). Chemical energy is stored inside the bonds between molecules. Natural gas, gasoline, coal, and batteries are all energy sources. Even the food we eat has chemical energy. Electrical energy is derived from tiny charged particles known as electrons. Lightning bolts are one type of electrical energy. Humans generate the electricity that we use in our homes. Gravitational energy is connected with a gravitational field, such as the one surrounding the Earth. If you've ever fallen, you've felt the force of gravitational energy. Gravitational energy is the reason that riding your bike downhill is faster than riding it uphill.

Anything that moves uses kinetic energy. Examples include running, cycling, climbing, and even swiping your finger across your smartphone! Wind turbines convert the kinetic energy in wind into mechanical energy. Mechanical energy is stored in objects through tension. When tension is removed, motion happens. Both a compressed spring and a stretched rubber band possess mechanical energy. Mechanical energy is energy stored in objects and is made up of two other energy sources: kinetic and potential. Potential energy is defined as the energy held in stationary things as well as the energy associated with position. As the name implies, it is a potential kind of energy that emerges only when turned into kinetic energy. For example, when a ball is lifted, it gains potential energy (from gravity) that is only revealed when it falls. Nuclear energy is held within small atoms that are invisible but comprise the constituents of the entire cosmos. Nuclear energy is released when atoms combine (fusion) or separate (fission). The fusion reaction in the sun produces heat and light, but the fission reaction at a nuclear power plant generates enough energy to power vast cities. Solar (radiant) energy is energy derived from the movement of light. Sound energy is produced when an item is made to vibrate, resulting in sound. Your voice and musical instruments both generate sound energy. Thermal (heat) energy is derived from moving molecules. A fire produces thermal energy.

It is critical to understand the several definitions of energy. Energy, as defined in physics, chemistry, and nature, has many forms, all of which require the ability to perform labor. In these sciences, energy is a scalar quantity that exists as a property of objects and systems and is naturally preserved. All known natural phenomena can be explained using a variety of energy types, including kinetic energy, potential energy, thermal energy, gravitational energy, electromagnetic radiation energy, chemical energy, and nuclear energy.

10.1.2 Conservation of Energy

Energy can be converted from one form to another, but it cannot be created or destroyed. The law of conservation of energy, originally proposed in the early 19th century, applies to any isolated system. The total energy of a system remains constant across time, but its value varies depending on the frame of reference. For example, a seated passenger in a moving car has no kinetic energy relative to the vehicle but does have kinetic energy relative to the ground.

10.2 UNDERSTANDING ENERGY PRODUCTION AND USE: A HISTORICAL PERSPECTIVE

Upon the establishment of agricultural societies and across millennia, several civilizations built managerial norms in defense, economy, and education that enabled them to endure and prosper, leaving their mark on history. However, numerous communities have disintegrated over history. Civilizations that made their imprint were always led by leaders who excelled at strategic survival abilities, which have evolved as human understanding progressed. Similarly, the demise of large civilizations may have been caused by retaining status quo techniques that formerly aided their existence but failed to meet new obstacles. What is significant is that human knowledge advances as a result of constant change and increases in our understanding of the universal laws of nature, which are governed by the dynamics of natural forces, energy, and events.

Energy is required for even the most fundamental activities, such as eating, bathing, working, and sleeping (Kostic, 2004, 2007). It's also a must if you want to stay healthy. There is a degree of public health risk, occupational danger, and environmental risk associated with any energy source. Energy is the capacity to exert force, to cause motion, or to induce a transformation in a material medium. It's the ability to generate mechanical motion or generate thermal energy from the combustion of fossil fuels. When we talk about something's "energy," we're talking about its potential to make changes and provide the means for a better quality of life. Energy also facilitates change; it helps us accomplish goals. It gets vehicles along the road and ships over the water. The oven can create a cake, while the freezer can maintain ice cubes. It illuminates our homes and plays our favorite tunes on the radio. Energy is essential for physical development and mental function. Energy, according to the scientific community, is "the capacity to do labor" (Smith et al., 2013).

10.3 ENERGY GENERATION - SOURCES OF POLLUTION

Energy generation is the primary driver of air pollution and global warming emissions (Malanima, 2014). Energy production and use can have an impact on the environment as well as human health (https://en.wikipedia.org/wiki/File:Pathways_linking_energy_and_health.jpg). Pollution is the introduction of harmful substances into the environment (www.sciencedirect.com/topics/engineering/power-generation). They can also be caused by human activities, such as industrial waste or runoff. Pollution is the process of making land, water, air, or other aspects of the environment contaminated and unsafe to use. Pollutants degrade the quality of air, water, and land. All living organisms, from single-celled microbes to blue whales, rely on the Earth's air and water supplies (https://education.nationalgeographic.org/resource/pollution). When these resources become polluted, all forms of life are endangered. Air and water currents transport pollution. Heavy metals, pesticides, herbicides, industrial chemicals, and poisons are some of the substances found in illegally discarded garbage. Carbon monoxide, nitrogen oxides, and sulfur dioxide are all examples of pollutants. Ocean currents and migrating fish spread marine contaminants far and wide. Winds can pick up radioactive material that has been accidentally spilled from a nuclear plant and spread it over the world. Smoke from a plant in one country flows into another.

The sources of electricity production, such as coal, oil, and natural gas, have contributed to one-third of global greenhouse gas emissions (Kumar and Majid, 2020). Coal emits more pollutants than any other energy source. While coal generates only 44% of US electricity, it contributes 80% of power plant carbon emissions. Burning coal produces soot, pollution, acid rain, global warming, and carbon emissions. Natural gas, like coal, is a fossil fuel that emits pollutants and poses environmental and health dangers. Natural gas emits fewer greenhouse gases than coal and oil, but emissions are only one aspect of the issue.

Nuclear energy is one of the world's largest sources of renewable energy, although it is not completely clean. Nuclear energy is emission-free and saves around 2.4 billion tons of carbon emissions per year compared to coal, but it produces radioactive, high-level waste. Nuclear waste, unlike other types of industrial waste, becomes less hazardous with time. Waste from nuclear energy generation must be handled safely.

Geothermal energy is a clean energy source that extracts energy from beneath the Earth. High temperatures beneath the Earth's crust occur from the gradual disintegration of radioactive particles. Hot rocks beneath the surface heat water, generating steam, which is captured and used to power turbines. Rotating turbines provide power to generators.

Solar panels gather solar energy from the sun, which can then be converted into electricity. Many households now have solar panels that provide at least some of the home's electricity, and some corporations are even getting on board. Kohl's is a leader in solar panel use among retailers in the United States. Kohl's has solar panels at many of its stores, including solar trees.

Renewable energy sources (RES) meet 14% of the global energy demand. RES encompasses biomass, hydropower, geothermal, solar, wind, and marine energy. Renewable energy sources are primary, domestic, and clean, or endless. Large-scale hydropower generates 20% of global electricity. Wind power in coastal and other windy areas is a promising source of energy. Renewable energy sources will play a critical part in the world's future. Renewable energy sources are resources that can be utilized to generate energy repeatedly, such as solar energy, wind energy, biomass energy, geothermal energy, and so on. They are also known as alternative sources of energy. Pollution is the leading environmental cause of illness and premature death. Pollution causes around 9 million premature deaths, the bulk of which are due to air pollution. This is several times more deaths than from AIDS, TB, and malaria combined. Global health crises, as the COVID-19 pandemic, serve as reminders of the deep links between the environment and health, as well as the importance of addressing these ties consistently.

Every energy source has an impact on the environment. Most studies suggest that fossil fuels (coal, oil, and natural gas) cause much more harm than renewable energy sources in terms of air and water pollution, public health impact, wildlife and habitat loss, water use, land use, and global warming emissions. By increasingly relying on renewable energy sources like wind and solar power, we can reap a wide variety of benefits, including reduced air pollution, lower greenhouse gas emission levels, decreased healthcare costs, a more resilient electrical grid, and job generation. Renewable energy sources–which are available in abundance all around us, provided by the sun, wind, water, waste, and heat from the Earth–are replenished by nature. Photovoltaic (PV) building-integrated systems do not influence the environment or landscape. Larger photovoltaic facilities are usually built on undeveloped or agricultural land. In that case, their impact on the land is represented as flora and wildlife. There are also indirect consequences on wildlife. All forms of electricity generation have an environmental impact on our air, water, and land, but the extent varies. Producing and consuming energy more efficiently reduces both the amount of fuel consumed and the emissions of greenhouse gases and other air pollutants. Because no fuels are utilized, power generated by renewable resources like solar, geothermal, and wind has little impact on climate change or local air pollution.

10.4 ENERGY AND HEALTH RELATIONSHIP

Natural resources provide us with the energy we need to keep our homes illuminated, comfortable, productive, and mobile. Geothermal energy is generated by heat deep within the Earth. Tides are caused by the moon's and the sun's gravitational pull. Light (electromagnetic radiation) from the sun drives the hydrologic cycle and makes photosynthesis possible, while also generating wind. Solar energy is used by plants, algae, and cyanobacteria to develop biomass that can be converted into fuels like wood, biodiesel, and bioethanol. Biomass from photosynthetic organisms can be transformed into energy-rich fossil fuels via the geologic process of burial and transformation via heat and pressure over the course of millions of years. Multiple fields of study have found connections between energy use and production as key direct and indirect drivers of air pollution. Air pollution (Singh et al., 2024), climate change, water pollution, thermal pollution, and solid waste disposal are only some of the environmental issues that may be traced back to human activity in the energy sector. Most urban air pollution comes from vehicles and industries burning fossil fuels. The total amount of energy consumed is correlated with economic growth and population expansion, but the efficiency with which different cultures utilize energy varies widely. Higher economic output is typically related to the use of more efficient and versatile energy sources (liquid fuels and, especially, electricity).

Energy and health pathways are depicted in Figure 10.1 by delineating between primary and secondary energy forms (https://en.wikipedia.org/wiki/File:Pathways_linking_energy_and_health.jpg), as well as between intermediate secondary energy forms like electricity and final end-use energy services like transportation (Wilkinson et al., 2007), and the health risks increase with each successive stage.

One can classify each of these sources of power as either renewable or nonrenewable. Renewing renewable energy sources is possible in human timescales. Solar, wind, and biomass are just a few examples. Solar panels, wind turbines, and hydroelectric dams are just a few examples of the modern environmental technology that have allowed us to harness this naturally occurring energy and convert it into electricity or usable heat (Smith et al., 2013). The supply of nonrenewable energy will run out eventually, and it will take too long for humans to replace it. Nuclear power and fossil fuels are two examples of long-term energy sources. There is a disparity in the allocation of energy between countries, and all energy sources have some sort of negative impact on the environment and human health. Comparatively, Costa Rica generates more than 99 percent of its electricity from renewable sources, whereas the United States generates less than 10 percent.

Natural and human ecosystems are both impacted by the fuel cycle, which includes the extraction, refining, distribution, and use of energy sources like coal, biomass, and hydropower. Pollutants may be released regularly or accidentally from energy systems. The same sectors that contribute to GHG (greenhouse gases) and air pollutant emissions also contribute to transportation, industry, commerce, and households. All of these factors rely on the generation, transmission, and consumption of energy to function normally. When contaminants enter the environment, they alter its natural state in ways that disrupt our daily routines. According to a recent study (https://energy.economicti

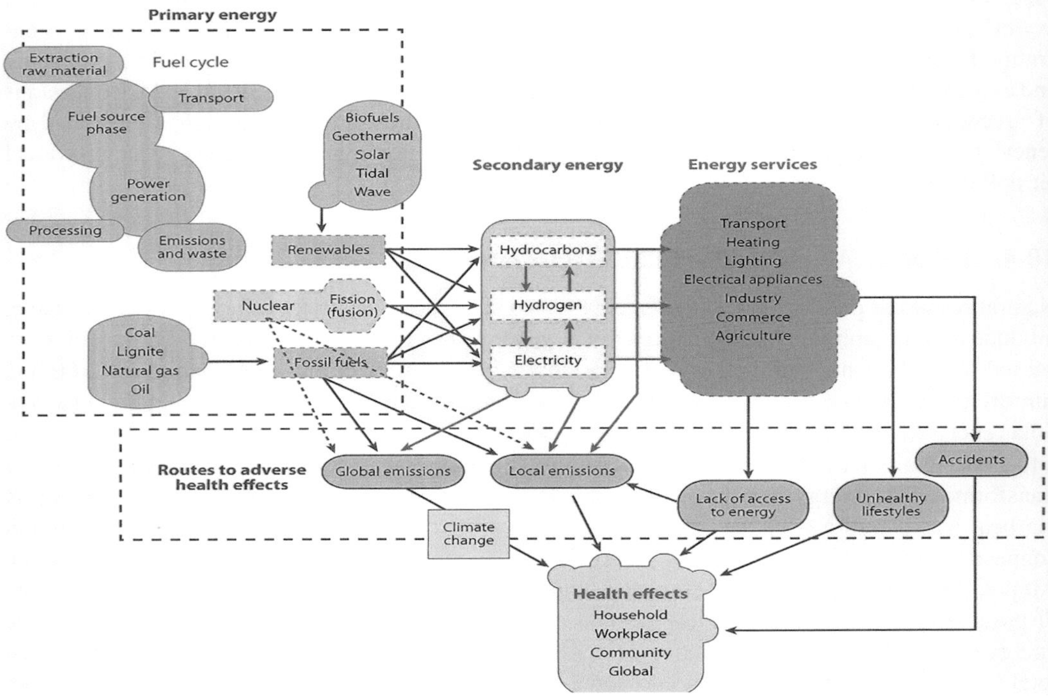

FIGURE 10.1 Pathways linking energy and health (https://en.wikipedia.org/wiki/File:Pathways_linking_energy_and_health.jpg) Under the Creative Commons Attribution-Share Alike 4.0 International license.

mes.indiatimes.com/news/power/air-pollution-leads-to-increase-in-electricity-consumption-study/78327617), pollutants, or waste products, are the primary causes of pollution. Our ecosystem and the delicate balance it depends on are threatened by pollution. Our increasing reliance on contemporary conveniences has resulted in record levels of pollution, which in turn has exacerbated climate change and made people sick.

Pollutants are released into the air when fossil fuels are burned to produce electricity. The majority of the United States' sulphur dioxide emissions, 13% of its nitrogen oxide emissions, and 40% of its anthropogenic carbon dioxide emissions come from fossil fuel-fired power plants. Smog, acid rain, and haze can result from these emissions. Moreover, these emissions from power plants raise the possibility of global warming (www.deq.virginia.gov/get-involved/pollution-prevention/virginia-information-source-for-energy/energy-and-the-environment).

Environmental governance, energy restructuring, and the impact of pollutant emissions on public health have all received significant global attention in recent years in response to climate change and the worsening of environmental quality (Xing et al., 2019). Rapid economic growth not only causes increased consumption of fossil fuels but also increases emissions of several pollutants. Numerous studies have revealed that some pollutants (e.g., PM10, PM2.5, and SO_2) emitted as a result of the usage of fossil fuels pose a threat to human health, both physically and mentally. Pollutant emissions are causing an uptick in diseases in developed nations, as determined by an investigation into the connection between air quality and public health conducted by Sharma et al. (2018). Environmental pollution poses serious hazards to public health because it has been shown to raise the likelihood of psychiatric illnesses in the United States and Denmark (Khan et al., 2019). According to the findings of Perera (2018), the use of fossil fuels leads to pollution emissions and climate change, both of which pose serious risks to the health of children.

There is no single cause for the damage that energy systems do to the environment. When fuels are burned, pollution and potential environmental harm result immediately; however, when electricity is used, there are no such immediate results. However, when fuels are burned to generate power, pollutants such as carbon dioxide and nitrogen oxides are released into the atmosphere, which can contribute to warming and acidification. These emissions can be prevented if the electricity is generated by wind turbines or solar panels (Hu et al., 2022). However, environmental emissions are linked to the manufacturing of wind turbines and solar cells due to the high volumes of materials required. So, it's not just the act of using the energy carrier itself that needs to be considered when weighing its environmental effects; rather, the entire chain of events beginning with primary energy extraction from the environment and ending with waste management must be included. The environmental implications of items like electricity and petroleum can be evaluated using the "cradle-to-grave" method of environmental life cycle assessment. From the time raw materials and fuels are extracted from the environment until the time those same materials are returned to the environment, the cradle-to-grave method considers the entire process (www.sciencedirect.com/topics/earth-and-planetary-sciences/environmental-impact-of-energy). One of the most frequently mentioned negative effects of energy production on local ecosystems is acid rain. Most acid rain come from power stations that burn fossil fuels, but hydrogen sulphide emissions from geothermal units contribute a minor amount too. The problem is global and regional in scale because acid rain can be carried great distances by the wind. Impacts on plants and forests due to things like reduced frost hardiness; deterioration of materials like buildings and fabrics; health effects; acidification of lakes, streams, and groundwater, which harms fish and aquatic life; plant toxicity due to acidic conditions and the release of heavy metals; these are just some of the consequences of acid rain.

At first glance, it may be difficult to make the connection between energy and health policy. However, the unintended consequences of using fossil fuels are easy to spot. According to the World Health Organization (WHO), both climate change and air pollution are major threats to human health as of January 2019. Infrastructure, availability, and quality of energy supply and healthcare, as well as access to both, are inextricably interwoven with each other and with other policy domains.

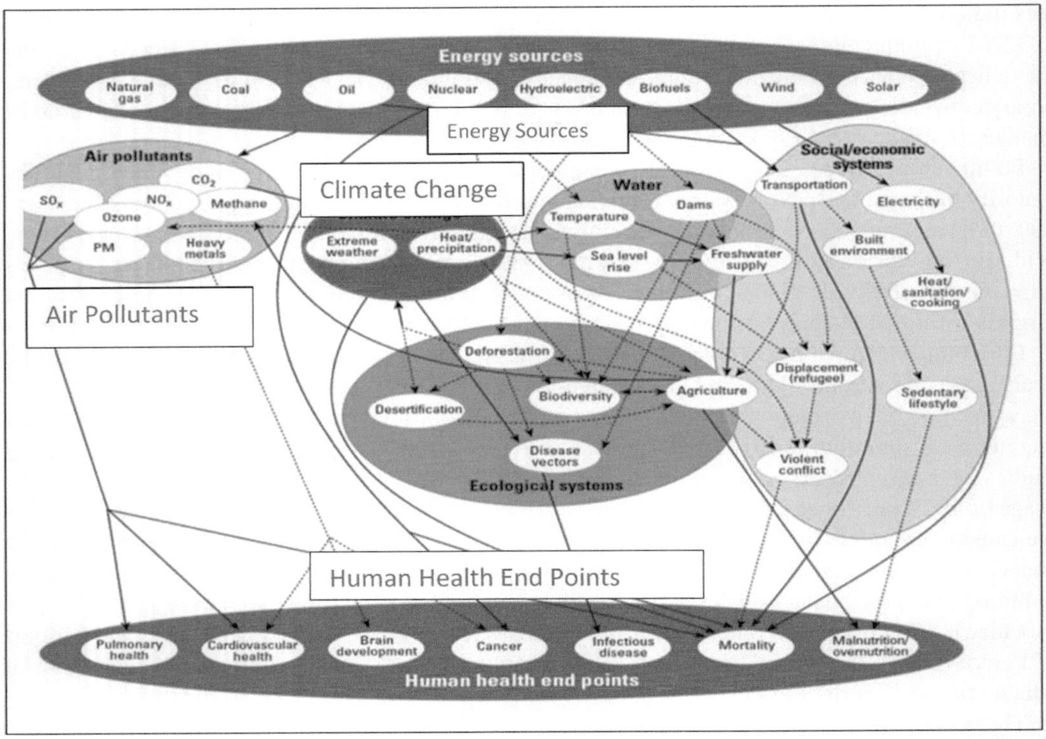

FIGURE 10.2 Energy sources' effect on human health. (Gohlke et al. (2008). https://creativecommons.org/licenses/by/4.0/.)

Cold chains are crucial for the safe storage of vaccines, and modern hospitals cannot function without reliable electricity. Both sectors of policymaking must be conceived and linked in a way that takes into account planetary limits and contributes to human security beyond national borders, in accordance with the United Nations' Sustainable Development Goals (SDGs). To reap the benefits of both, health and energy must be brought together. This would be a huge step in the right direction for the SDGs (www.swp-berlin.org/10.18449/2019C14/) to be put into action more quickly.

There is a correlation between a nation's energy use and its level of health security. For instance, people in nations with low or intermediate incomes may be at risk of being exposed to pollutants due to the inability to reliably generate clean energy. In addition, air pollution, safe water shortages, and inadequate medical care infrastructure are all ways in which people's health can be negatively impacted by energy use. The health-environmental security of a nation is threatened when its citizens rely on energy sources that increase the risks to human health, the future costs associated with climate change, and other forms of environmental harm. Due to insufficient medical facilities, inefficient energy use, and a larger share of overall energy consumption from fossil fuels, this problem is exacerbated in low- and middle-income nations. Energy sources may have both immediate and long-term effects on human health (Gohlke et al., 2008; von Hildebrand et al., 2009) (Figure 10.2).

10.5 ENERGY AND ITS IMPACT ON THE ENVIRONMENT

Environmental Impacts of Energy Technologies: Every energy source has an impact on the environment. Most studies suggest that fossil fuels (coal, oil, and natural gas) cause much more harm than renewable energy sources in terms of air and water pollution, public health impact, wildlife and habitat loss, water use, land use, and global warming emissions (Seddighi et al., 2023). Renewable energy sources, such as wind, solar, geothermal, biomass, and hydropower, have significant environmental consequences. Photovoltaic (PV) building-integrated systems do not influence the environment or landscape. Larger photovoltaic facilities are usually built on undeveloped or agricultural land. In that case, their impact on the land is represented as flora and wildlife. There are also indirect consequences on wildlife (such as habitat modification and food availability). All forms of electricity generation have an environmental impact on our air, water, and land, but the extent varies. Producing and consuming energy more efficiently reduces both the amount of fuel consumed and the emissions of greenhouse gases and other air pollutants. Because no fuels are utilized, power is generated by renewable resources like solar, geothermal, and wind, etc.

Energy efficiency and renewable energy regulations can lessen the need for and the amount of energy produced by fossil fuel power plants (such as those that run on natural gas, oil, and coal). The power grid, emissions, public health, and the economy can all benefit from energy efficiency and renewable energy thanks to the electricity savings and renewable energy generation that underpin these estimates. Asthma, other respiratory diseases, cardiovascular illness, and premature mortality can all be prevented by improved energy efficiency and the use of renewable energy sources. While there is always the chance of a catastrophic disaster with nuclear power, that risk is quite low. There is a high risk of lung cancer for uranium miners, and chronic low-level radiation exposure is a concern for everyone working in the nuclear industry. Even the slim likelihood that nuclear waste could be obtained to make a "dirty bomb" is cause for a serious security alarm. Radiation is a concern at every stage of the nuclear power process, from uranium mining and processing to transportation and storage at nuclear power plants. Underground mining for uranium sometimes involves considerable exposure to radon, a radioactive gas released from uranium, making it a significant occupational health concern in nuclear power generation. Lung cancer is made more likely by radon exposure. Even though uranium ore has a relatively high energy content, meaning fewer health consequences per unit of power produced, miners nevertheless face the same risks in the workplace as those mentioned above. Before the Chernobyl disaster, radiation exposure among nuclear power plant workers was uncommon. However, forty workers perished in that tragedy, either from the fire itself or from the acute radiation exposure they suffered. Cancer and other radiation-related impacts may have long-lasting consequences for personnel exposed during the accident, but it is too soon to tell (https://user.iiasa.ac.at/gruebler/Lectures/Graz-04/wea chapter3.pdf). Sulfur dioxide (SO_2) and nitrogen oxide (NOx) emissions from human activities are driven mostly by the energy sector, accounting for 82% and 76% respectively. Non-methane volatile organic molecules (56%), carbon monoxide (46%), and methane (34%) all serve secondary roles to energy processes.

Renewable energy can enhance the quality of life and save lives by lowering air pollution. By curbing greenhouse gas emissions, renewable sources help stabilize climate patterns, thereby preventing health emergencies related to extreme climatic conditions and safeguarding vulnerable populations from climate-induced health risks. No type of electricity generation is completely without negative impacts. However, surveys of the scientific literature have shown that the negative impacts of wind energy are significantly less than those of fossil fuel and nuclear generation. Properly planned and sited, wind energy can reduce the negative impacts of electricity generation. Policy, legal, and financial decision-making should take into account the health co-benefits of renewable energy sources (RESs) when evaluating their economic viability. Energy availability, energy efficiency, and the replacement of fossil fuels by renewable energy sources are key factors in the global development of sustainable energy. In many countries with limited fossil fuel resources, the sustainable development of renewable energy sources is an important tool in reducing dependence on imported fuels.

The most prominent types of multiple benefits of energy efficiency improvements and Multiple impacts of a clean energy system are shown in Figure 10.3a and Figure 10.3b.

(a)

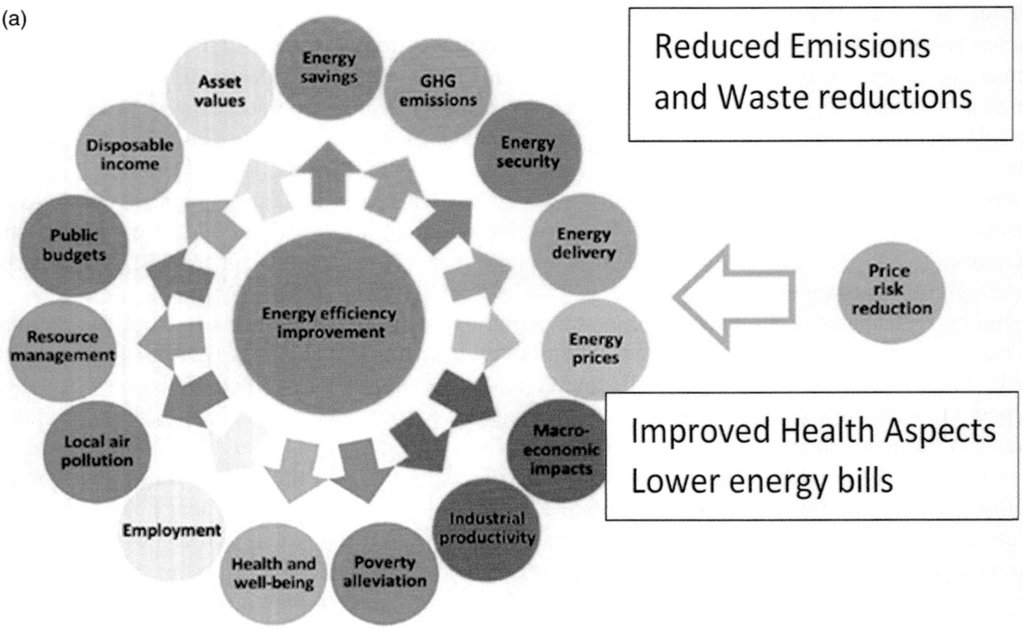

(b)

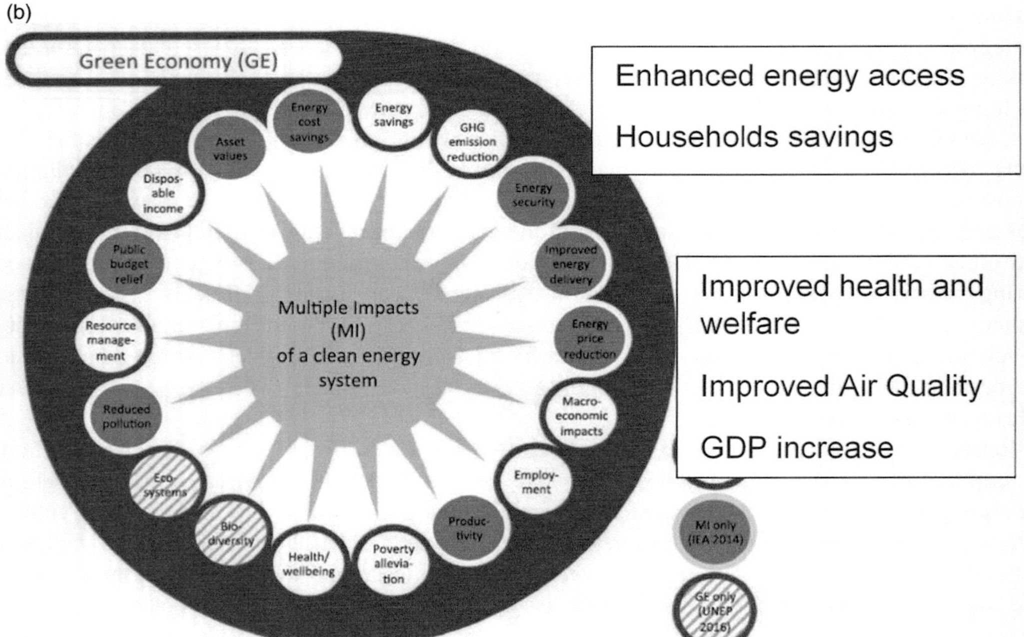

FIGURE 10.3a The most prominent types of multiple benefits of energy efficiency improvements (Tuominen and Seppänen (2017). Open access. https://creativecommons.org/licenses/by/4.0/.)

FIGURE 10.3b Concepts of multiple benefits: IEA concept and relation to green growth (Ringel et al. (2019). https://doi.org/10.3390/en12081537. Open access.)

Energy efficiency is a transformative force that extends well beyond its initial promise of saving energy and money. Its impact influences various aspects of our lives and the world at large. Localized advantages, such as energy affordability, social development, and enhanced health and well-being; sectoral benefits, such as increased industrial productivity, higher asset values, and less environmental damage; and international benefits, such as those listed above. Energy is a necessity in our lives because it enables us to do the things we need to do. Energy is required for nearly all that we do each day. Various types of energy can be found on Earth. The origin of the negative effects that all forms of energy have on the environment is varied.

The use of energy can affect communities for better or for worse. It is to humanity's advantage to have ready access to energy that is all of the above. However, the process of energy extraction, transportation, and consumption can have severe effects on people's health, the environment, and the economy. Furthermore, a nation's security may be compromised by its reliance on energy imports. Everyone feels the effects of energy choices differently. Because of their lower adaptive capacity and possibly weaker bargaining position, poor or marginalized cultures are more likely to bear the negative effects of energy decisions. Therefore, advances in energy accessibility, safety, or price can have a significant impact on marginalized communities.

Some of the most pressing problems facing our planet today are given below. When compared to renewable energy sources, the environmental damage caused by the use of non-renewable fuels like coal and petroleum is greater. Emissions of sulphur from coal burn trees. Coal mining causes acid mine drainage, which seeps into the ground and surface water and dissolves heavy metals. Erosion, ecological damage, and pollution may all arise from the process's potential for uprooting large amounts of topsoil (https://byjus.com/physics/energy-resources-environmental-impacts/).

When oil is extracted and burned, it releases dangerous carbon oxides into the atmosphere. When oil is accidentally released into a large body of water, it causes serious economic and environmental problems. Many marine organisms are in jeopardy because oil on the ocean's surface blocks sunlight and reduces the amount of dissolved oxygen. Feathers and fur lose their ability to insulate and repel water when exposed to crude oil, making oil-covered marine creatures and birds susceptible to hypothermia and death.

The environmental impact of even renewable resources is not negligible. Wind, solar, tidal, and other forms of renewable energy generation also have negative effects on the environment and biodiversity. Natural gas combustion releases nitrogen and even methane oxides, which are toxic to aquatic life and the humans and animals who drink from polluted water sources. When fossil fuels are burned, they release large amounts of carbon dioxide, which is a greenhouse gas. These gases contribute to global warming because of their propensity to hold heat in the atmosphere. Harmful gases are also produced due to the combustion of biomass. The ash produced when biomass is burned contains harmful metals like lead and cadmium, making it yet another waste product that presents disposal challenges.

10.6 HOW DO ALTERNATIVE ENERGY SOURCES CONTRIBUTE TO REDUCING GLOBAL WARMING?

It's practically difficult to generate, transport, or use energy without a major environmental impact (www.eea.europa.eu/help/glossary/eea-glossary/environmental-impact-of-energy); therefore, energy and environmental issues go hand in hand. Air pollution, climate change, water pollution, thermal pollution, and solid waste disposal are only some of the environmental issues that may be traced back to human activity in the energy sector. Most urban air pollution comes from vehicles and industries burning fossil fuels. The majority of greenhouse gases are released into the atmosphere due to the combustion of fossil fuels. The use of electricity results in a wide range of water pollution issues. Oil leaks are a major issue. There is always some risk of an oil spill occurring during any process involving petroleum. In addition to air pollution, water pollution is another potential consequence of coal mining.

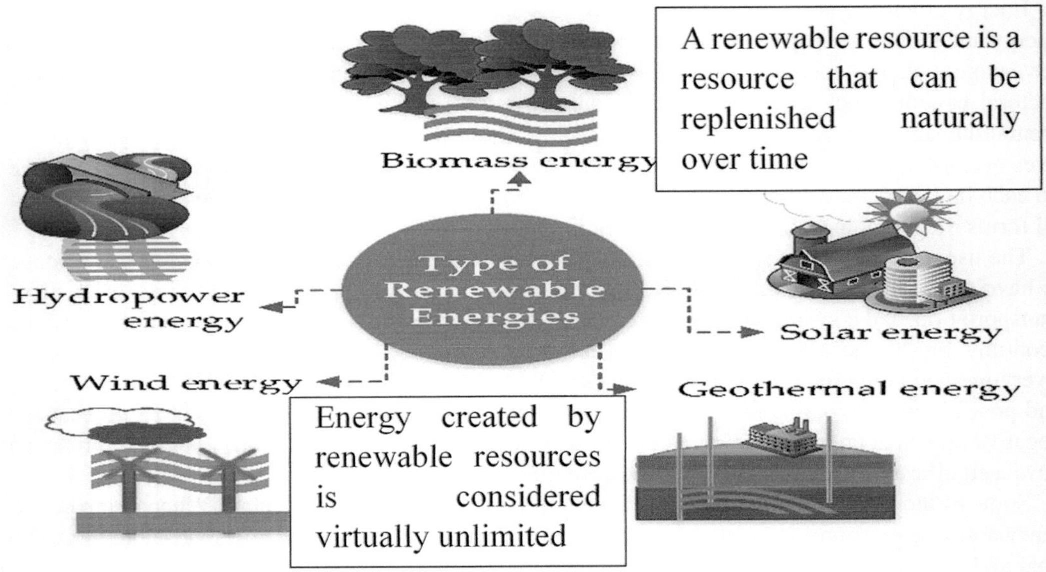

FIGURE 10.4 Types of renewable energy resources: biomass, solar, geothermal, wind, and hydropower (Avtar et al. (2019). http://creativecommons.org/licenses/by/4.0/.)

The effects of climate change on the world are worsening, and numerous factors are to blame. Research indicates that our reliance on fossil fuels is the primary cause of this problem. Fossil fuels, such as coal, natural gas, petroleum, bitumen, and shale oil, are the sources of nonrenewable energy. There are many climatic dangers associated with using fossil fuels, beginning with their extraction and ending with their consumption. Choosing renewable energy has the potential to improve the planet's environment and has many other good effects.

There is almost no pollution or greenhouse gas emissions from renewable energy generation. This has a net positive effect on the natural environment and reduces our carbon footprint. Greenhouse gas emissions from the burning of fossil fuels for energy are a major contributor to climate change. When considering the whole life cycle of the technologies, most renewable energy sources produce negligible or even zero emissions. The inherent purity of renewable energy makes it crucial. They don't use polluting fossil fuels, which they avoid burning. This is renewable energy, meaning that the resources from which it is obtained will not run out in the foreseeable future. The types of renewable Energies are highlighted in Figure 10.4.

Fuel availability, energy production, and the physical durability of existing and future energy infrastructure are all directly impacted by climate change. Heatwaves and droughts are already straining current energy generation, so cutting back on fossil fuel use is crucial. Wind, solar, hydroelectric, and tidal power all have a positive impact on global warming (www.nhm.ac.uk/discover/renewable-energy.html). There is an initial investment required to set up the infrastructure, as well as continuing expenses to keep it operational. If wind farms are constructed in the ocean, they will increase much further. Poorly sited wind farms, for example, if they are located in the flight patterns of migrating birds or bats, can be harmful to certain species of wildlife. Solar array installation is expensive, and there are other costs associated with energy storage. Solar panels require more space than some other renewable energy options, and their efficiency is sun-dependent. Panel production can cause environmental impact due to the mining and processing of minerals required to produce them.

Hydroelectric facilities need a lot of room and can't be built in every climate. They're vulnerable to dry conditions. The construction of artificial water reservoirs has the potential to reduce biodiversity in natural water systems by obstructing the migration of migratory fish populations and reducing the supply of nutrients. The decomposition of plants in the flooded area causes these reservoirs to produce methane, a greenhouse gas. Dams require massive quantities of cement for their construction. Carbon dioxide is released in enormous quantities during the production of this substance.

Setting up and maintaining offshore infrastructure is costly, and there are few good locations for tidal power plants across the world. They also pose a threat to marine ecosystems and the local flora and fauna. Since the production of biofuels necessitates the use of both land and water, the increasing demand for these fuels poses a threat to biodiversity. Without carbon capture, biomass combustion produces greenhouse gas emissions.

Better public health; coal and natural gas are major contributors to air and water pollution, which has been connected to various significant diseases. Clean energy technologies do not contribute to the same levels of air and water pollution that are responsible for the majority of these adverse health effects. In contrast to conventional power plants, renewable energy sources like wind turbines, solar panels, and hydroelectric dams do not release any harmful greenhouse gases into the atmosphere when producing electricity. Furthermore, wind and solar energy require very little water for operation, thus they do not contribute to water pollution or put pressure on water supplies by competing with agricultural, municipal, or industrial uses.

Multiple massive human activities are contributing to an alarming rate of global warming that threatens society. Large-scale deforestation for cattle "production," for example, is a serious issue that contributes to global warming just as much as the burning of fossil fuels for energy. To say that renewable energy is useless because it can't solve the problem on its own is to miss the point (https://cleantechnica.com/2018/01/25/renewable-energy-critical-stopping-global-warming/). If we want to have any chance of averting catastrophic climate change, we need to seek renewable energy as aggressively and rapidly as possible, as it is one of the most crucial answers to global warming.

10.7 THE EFFECTS OF RENEWABLE ENERGY TECHNOLOGIES ON HUMAN AND ENVIRONMENTAL HEALTH

All energy sources have some impact on our environment. Fossil fuels—coal, oil, and natural gas—do substantially more harm than renewable energy sources by most measures, including air and water pollution, damage to public health, wildlife and habitat loss, water use, land use, and global warming emissions. However, renewable sources such as wind, solar, geothermal, biomass, and hydropower also have environmental impacts, some of which are significant. The exact type and intensity of environmental impacts vary depending on the specific technology used, the geographic location, and several other factors. By understanding the current and potential environmental issues associated with each renewable energy source (www.ucsusa.org/resources/environmental-impacts-renewable-energy-technologies), we can take steps to effectively avoid or minimize these impacts as they become a larger portion of our electric supply. Wind power generates electricity without toxic pollution or global warming emissions, but it does have some environmental impacts that should be recognized and mitigated. The potential environmental impacts associated with solar power depend on the technology, which includes two broad categories: photovoltaic solar cells and concentrating solar thermal plants. The environmental impact of geothermal energy vary depending on the technology used to generate electricity and the type of cooling system utilized. Hydroelectric power includes both massive hydroelectric dams and small run-of-the-river plants, both of which have associated environmental impacts. Renewable energy is generally acknowledged as the key to countering climate change and achieving global climate goals, with sustainable energy transition rising to the top of the UN's list of sustainable development priorities. Renewable energy produces far less greenhouse gas (GHG) than fossil fuel resources like coal, oil, and natural gas. In some cases, renewable sources produce no GHGs at all once they are up and running.

A research study analyzed the negative impacts of energy sources on humans using two parameters, namely: deaths caused by air pollution, and accidents that occur in the process of producing energy–from mining raw materials, to transportation, construction, operation, and maintenance (Ritchie, 2020). A study by Ritchie (2020) revealed that the human death rates from air pollution and accidents per terawatt-hour (TWh) of electricity for solar and wind energy are 0.02 and 0.04, respectively; compared to 1.3 for hydropower, and 4.63 for biomass. These rates are considerably lower than that of coal (24.62), implying that renewable energy is significantly safer than coal. All energy sources have negative effects, but they differ enormously in size: as we will see, fossil fuels are the dirtiest and most dangerous, while nuclear and modern renewable energy sources are vastly safer and cleaner. From the perspectives of both human health and climate change, it matters less whether we transition to nuclear power or renewable energy and more that we stop relying on fossil fuels.

10.8 HEALTH IMPACTS OF NUCLEAR ENERGY

Since nuclear power facilities are thought to emit fewer carbon emissions than fossil fuel power plants, they are frequently promoted as a solution to the climate crisis. This argument fails to take into account the fact that every stage of the generation of nuclear energy results in waste streams and radioactive and chemical emissions that can pollute food chains and drinking water, as well as injure people who live close to nuclear reactors (https://envirn.org/energy-and-health/). Concerns have also been raised over the health hazards of uranium mining, processing radioactive waste, and unintentional spills and discharges that have contaminated water and land for millennia. For instance, uranium mining, which is used to power nuclear reactors, has left a hazardous legacy of contamination, especially on Indigenous territory. Even though the mines have been shut down, the land is still contaminated, and there are still high radiation levels in residences and water sources. In comparison to wind and solar energy, nuclear energy is also costlier to build and operate the plant.

Nuclear reactors do not emit carbon dioxide or other air pollutants during operation, in contrast to fossil fuel-fired power facilities. However, uranium ore extraction, refinement, and reactor fuel production all involve energy-intensive processes (www.eia.gov/energyexplained/nuclear/nuclear-power-and-the-environment.php). Additionally, nuclear power plants use a lot of concrete and metal, both of which need a lot of energy to produce. The emissions from burning fossil fuels may be linked to the electricity produced by nuclear power plants if they are utilized in the mining and refinement of uranium ore or in the construction of the nuclear power plant. It should come as no surprise that nuclear energy offers both health benefits and drawbacks.

On the plus side, nuclear energy, like other energy sources, produces power that enables civilizations to sustain or grow contemporary economies that have good health effects, such as increased access to healthcare and greater living standards. On the downside, nuclear energy, like other sources, depends on processes and materials that can be harmful to the health of both industry employees and nearby residents. These hazards could show up in populations as harmful health outcomes. Human health is greatly impacted by nuclear power plant accidents, which can lead to issues with evacuation, behavioral and emotional issues, and elevated cancer rates. These catastrophes may have a significant negative influence on the ecosystem, particularly on species, and have unforeseen health effects on the surrounding populace.

10.9 COAL-FIRED POWER PLANTS AND HEALTH EFFECTS

Numerous studies have estimated the aggregate premature mortality burden of India's coal fleet by combining air quality models with concentration-response functions, as well as quasi-experimental methods, and their estimates range from just under 50,000 to approximately 100,000 deaths per year (Singh et al., 2024). Absolute annual mortality ranges from less than 1 to over 650 deaths/year across units, and the mortality intensity of generation varies from under 0.002 to 0.43 deaths/GWh. Understanding the total premature mortality attributable to all coal-fired power plants is valuable for estimating overall damages from the sector, but insufficient for developing targeted policies to improve public health. Coal-fired power plants influence climate via both the emission of long-lived carbon dioxide (CO_2) and short-lived ozone and aerosol precursors. Air pollution particles from coal-fired power plants are more harmful to human health than many experts realized, and they're more than twice as likely to contribute to premature deaths as air pollution particles from other sources, new research demonstrates. Chronic exposure to ambient particulate matter (PM) has been suggested to be associated with an increased risk of lung cancer, but the results were inconsistent. Cui et al. (2015) performed a systematic review and meta-analysis of prospective studies to assess the association between exposure to PM and the incidence and mortality of lung cancer in adults.

Air pollutants released from coal power plants (Sarath et al., 2014; Singh et al., 2024) were associated with nearly half a million premature deaths of elderly Americans from 1999 to 2020. A landmark study in the 1990s, known as the Harvard Six Cities Study, linked tiny airborne particles, called PM2.5, to increased risk of early death (Douglas et al., 1993). Other studies have since linked PM2.5 to lung and heart disease, cancer, dementia, and other diseases. Coal-fired power plants emit large amounts of CO_2, which constitutes one of the largest causes of global warming. The global climate challenge requires the stabilization of atmospheric CO_2 levels as a matter of urgency. Given the rising energy demand, this implies the need for a massive reduction in CO_2 emissions from fossil fuels.

Every existing source of energy is affected by climate variability. Renewable energy sources tend to be more sensitive to global warming, but fossil energy production is also affected by air and water temperatures. The most direct impact of climate change caused by global warming on electricity production in conventional thermal power plants using fossil fuels is reflected in the operating conditions of the cooling system. As anticipated, climate change will lead not only to an increase in air temperature, but also to a reduction in available water for cooling, all of which increases the effect of reducing the efficiency and reliability of electricity production in the near future.

Coal-fired power plants emit 66% of sulfur oxides, 40% of carbon dioxide, 33% of mercury, and 22% of nitrogen oxides in the U.S. and are linked as risk factors to respiratory diseases, cardiovascular diseases, and other ailments shown to impact environmental and human health (https://hsrc.himmelfarb.gwu.edu/cgi/viewcontent.cgi?article=1093&context=gw_research_days). Along with the injurious health effects that come with the presence of hazardous waste sites like coal-fired power plants, there are broader socioeconomic trends and outcomes related to their siting, especially affecting those who live near these facilities (Zhang et al., 2021).

Long-term exposure to PM2.5 is associated with shorter life expectancy and higher mortality risks from lung cancer and cardiovascular diseases. In fact, the International Agency for Research on Cancer (IARC) has listed several coal-fired power plant-related agents, including coal combustion, coal production, outdoor air pollution, and radon, as human carcinogens. While lung cancer is prevalent, the proportion of cases attributed to environmental factors such as air pollution varies by country and is difficult to estimate. Nonetheless, improved air quality has been correlated to better health, prompting many countries to implement regulations on air pollution. Lin et al. (2019) mentioned that their study may help address a key policy question about the externality cost of coal power plants and estimates of the global disease burden from preventable lung cancer attributable to coal-fired power plants.

Numerous epidemiologic studies carried out in China have discovered links between exposure to coal smoke and lung cancer. Due to the abnormally high rates of lung cancer among non-smoking women who use smoky coals in open fire pits, 1993 has been the subject of decades-long investigations into the relationship between coal smoke and lung cancer. A new meta-analysis by

Smith et al. (2004, 2006) summarizes the odds ratios (ORs) for lung cancer linked to indoor coal consumption. ORs were calculated both with and without correcting for two significant confounding variables for comparison. Analyses were either conducted exclusively on nonsmokers or adjusted for tobacco smoking status. Adjustment was also made for chronic respiratory conditions, such as emphysema, asthma, TB, and chronic bronchitis, since these conditions may raise the risk of lung cancer in later life.

Since some prior lung conditions may be on the intermediary path from exposure to lung cancer, these modifications may underestimate the odds of developing lung cancer; in this scenario, they are not confounders and should not be taken into account. For women, the 95% CI (confidence interval) ranged from 1.02 to 1.35, and the OR (odds ratio) estimate was 1.17. However, the OR estimate for women significantly increased to 1.94 (95% CI, 1.09–3.47) when the analysis was limited to trials that controlled for smoking and chronic respiratory disease. When smoking and chronic airway illness were taken into account, the OR estimate for men was 1.79 (95% CI, 1.18–2.72), which was somewhat lower (OR = 1.5, 95% CI, 0.97–2.46). Both without and with the two confounding factors taken into account, the total OR estimates for men and women combined were 1.86 (95% CI, 1.48–2.35) and 2.55 (95% CI, 1.58–4.10), respectively. Similar findings were made by Zhao et al. (2006) in a more recent review. Strong epidemiologic evidence that indoor coal smoke exposure dramatically raises the risk of lung cancer is presented by the meta-analyses. According to Straif et al. (2006), the International Agency for Research on Cancer (IARC) recently categorized indoor coal combustion pollutants as carcinogenic to people (Group 1) based on adequate evidence in both humans and animals.

Lung cancer symptoms usually occur when the cancer is in an advanced stage, and thus, chemotherapy may not be effective. The survival rate for lung cancer patients is at 25.4% which is low compared to other cancer types. The main reason is the late diagnosis of lung cancer compared to other cancers, such as breast cancer. Lung cancer is considered a leading cause of cancer-related death across the globe among both men and women (Ige et al., 2024). In 2020, Lung and Bronchus cancer held the grim distinction of being the third-leading cause of cancer-related deaths in the United States, with a mortality rate of 31.8 per 100,000 people (95% CI 31.7, 31.9 per 100,000 people). Additionally, it ranked as the third most common cancer in terms of new cases, with an incidence rate of 47.1 per 100,000 people (95% CI 46.9, 47.3 per 100,000 people), highlighting its significant impact on public health. In 2020, Kansas faced a notable lung cancer burden with an age-adjusted incidence rate of 52.5 per 100,000 people (95% CI 51.4, 53.6 per 100,000). This surpassed the national average, underscoring the significant impact of lung cancer in the state with 1,892 reported cancer cases. A study has shown that coal-fired power plant workers have an increased cancer risk due to airborne PAH inhalation and dermal contact (Wang et al., 2015). Also, it is well known that people living near power plants suffer from higher rates of premature mortality, respiratory diseases, lung cancer, cardiovascular disease, poorer child health, and higher infant mortality.

10.10 THE EFFECTS OF SOLAR AND WIND POWER ON THE ENVIRONMENT

There are two major categories of solar energy technologies: photovoltaics (PV) and concentrating solar power (CSP). PV cells convert solar radiation directly into electricity, whereas mirrors are used in CSP to reflect and focus sunlight to generate heat and convert it into power. Solar energy installations can be distributed or large-scale depending on the size and location. Distributed solar energy systems (typically <1 MW) are independent of the grid, and they are often building-integrated. Large-scale solar installations are geographically centralized and sometimes built far away from where the power is needed, e.g., in deserts where the solar irradiation is high and the population density is low. Both concentrated solar thermal and photovoltaic solar plants have an impact on the environment. This impact varies depending on the size and type of the solar plants.

Large-scale solar energy infrastructures may block the movement as well as seasonal migration of wildlife species, which may cause a gene migration disruption between different species.

This habitat loss and fragmentation could threaten biodiversity and disrupt the ecosystem balance. Concentrated solar systems could pose a marginal threat to insects and birds flying near the focus of the reflectors. In the case of PV (Photovoltaics) plants, the occupation of land has a major impact on wildlife and habitat. Despite the benefits and many advantages of renewable energy, in this case, wind and solar power, increased production comes with increased waste accumulation and other environmental impacts. Both wind and solar power are portrayed as clean and sustainable, and earn this title by being a fuel-free source of energy. They also create energy without generating the damaging pollutants that result from burning fossil fuels. Overall, wind turbines have less of an environmental impact compared to solar panels. With the present technological standards, the wind turbine technology is more sustainable than the photovoltaic technology over their entire life cycles; however, both still pose a threat when it comes to the disposal phase and developing a proper waste management system.

Assessing the environmental impacts of wind and solar energy allows for the identification of sustainability opportunities and areas for improved technology. Once installed, virtually no greenhouse gases are emitted as a result of wind and solar power generation, and they pay off the energy related to their manufacturing and construction within a matter of months (https://ember-energy.org/latest-insights/why-wind-and-solar-are-key-solutions-to-combat-climate-change). Their existence prevents the continuous burning of fossil fuels for decades. Throughout a typical 3-megawatt wind turbine's 25-year lifespan, 154,484 tons of coal would need to be burned to produce the equivalent amount of power. This would emit around 400,000 tons of CO_2 into the atmosphere–roughly the same as driving an average petrol car around the equator 50,000 times. Global power sector emissions would have been 20% higher in 2022 if all the electricity from wind and solar had instead come from fossil generation. Wind turbines can also be harmful to bats in a similar way to bird populations. When a wind energy project is being planned, the risk to bats should be assessed for the location being considered. Appropriate measures should be taken in planning and design to minimize impacts, such as locating them a minimum distance away from hedgerows and trees.

Solar energy technologies and power plants do not produce air pollution or greenhouse gases when operating. Using solar energy can have a positive, indirect effect on the environment when solar energy replaces or reduces the use of other energy sources that have larger effects on the environment. However, producing and using solar energy technologies may have some environmental effects. Solar power plants don't release any harmful gases or liquids during operation, and the photovoltaic array can even act as a thermal barrier for the structure they're installed on. Negative environmental impacts of solar energy are caused by the presence of toxic metals in solar panels.

Extreme wind events and gusts, icing of the blades, sea ice, permafrost, and air density are all factors that could be impacted by climate change, although they have received less attention than wind speed itself. Variables that affect these aspects are notoriously difficult to predict. Despite its enormous potential, wind power generation has a number of unfavorable environmental effects that must be acknowledged and countered (Fitzner, 2019). Aesthetic impacts include visual intrusion, noise, and interference to television and radio signals. Other impacts on humans and ecosystems include the killing of wildlife, especially birds and bats, by the turbine blades. Wind power plants have relatively little impact on the environment compared to fossil fuel power plants. Wind energy facilities have adverse environmental impacts, which include aesthetics, plus other effects on humans and ecosystems, while solar energy does not have such adverse environmental impacts. Solar energy affects the environment negatively because Solar panels contain toxic metals like lead, which can damage the nervous system, as well as chromium and cadmium, known carcinogens as pollutants (www.earth.com/news/environmental-impacts-solarwind-energy/). All three metals are known to leach out of existing e-waste dumps into drinking water supplies. The washed out of the fragments of solar panel modules over several months by rainwater could pose a threat to human health. Wind turbines annually cause the deaths of millions of birds and bats. Large areas of desert can be converted into solar farms, displacing a variety of animals from pronghorns and tortoises to coyotes and rattlesnakes. Solar and wind power can only do so much for the environment if energy use is optimized and kept within reasonable bounds. Wind turbines and solar panels still require an initial investment of money. Wildlife extinction is another issue. Wind turbines kill birds and bats, although the total impact on species is minimal.

10.11 EXPLORING RENEWABLE ENERGY RESOURCES USING REMOTE SENSING AND GIS

Techniques from Remote Sensing (RS) and Geographic Information Systems (GIS) are very interesting to the renewable energy sector. In order to investigate the five primary renewable energy sources associated with water, soil, and air resources, Avtar et al. (2019) examined extensive potential uses of remote sensing and GIS methodologies. Policymakers may find their findings useful in developing alternative incentives for accessible renewable energy supplies. A successful approach to finding pertinent material is emphasized. Second, a concise synopsis of the state of traditional approaches has been provided, along with an explanation of the function of GIS and remote sensing methods about renewable energy sources. Third, a number of case studies involving the use of GIS and remote sensing methods in the search for renewable energy sources have been examined. In a nutshell, a review and discussion of the many study approaches that scholars have employed to investigate renewable energy resources utilizing GIS and remote sensing techniques has been given. In summary, remote sensing techniques give multi-spatiotemporal information in addition to increasing speed, cost efficiency, precision, and timeliness for estimating various parameters used in various explorations of renewable energy resources.

10.12 CONCLUSION

Renewable power is booming, as innovation brings down costs and starts to deliver on the promise of a clean energy future. Energy efficiency and utilizing all energy sources are necessary to meet the world's increasing energy demand. Renewable energy sources are limitless and clean, but the sporadic nature of most types is their biggest drawback. To solve this issue, an HRES (hybrid renewable energy systems)—a combination of multiple energy sources—is created. Energy use negatively impacts health through air pollution, clean water scarcity, and medical treatment. Fossil fuels pose risks. Access to electricity is crucial for universal healthcare and Sustainable Development Goals. Renewable energy extends life and reduces death rates, preventing chronic diseases and reducing mortality.

Half of all renewable energy sources pose less risk to employees, with central station photovoltaics and passive solar heating being the safest options. Residential wind with no storage or wood pyrolysis poses the most uncertainty. Photovoltaic cells use trace metal-doped semiconductors, causing toxic pollutants. Unforeseen consequences include land-use disputes, exploitation, poor interstate cooperation, and conflict escalation. Renewable energy in logistics can improve environmental sustainability and boost national pride.

Switching to clean, renewable energy sources is crucial to reducing exposure to harmful substances from fossil fuel production. Health professionals can facilitate discussions, request health impact assessments, advocate for clean energy investments, and express health impacts to local and national government representatives, ultimately alleviating health burdens worldwide.

REFERENCES

Avtar, R., N. Sahu, A.K. Aggarwal, S. Chakraborty, A. Kharrazi, A.P. Yunus, J. Dou, and T.A. Kurniawan. 2019. Exploring renewable energy resources using remote sensing and GIS—A review. *Resources*. 8: 149. https://doi.org/10.3390/resources8030149

Cui, P., Y. Huang, J. Han, Song, and K. Chen. 2015. Ambient particulate matter and lung cancer incidence and mortality: A meta-analysis of prospective studies. *Eur. J. Public Health*. 25(2): 324–329.

Douglas, W., C. Dockery, A. Pope, X. Xu, J.D. Spengler, J.H. Ware, M.E. Fay, B.G. Ferris, Jr., and F.E. Speizer. 1993. An association between air pollution and mortality in six U.S. cities. *Engl. J. Med.* 329: 1753–1759.

Fitzner. 2019. The environmental impacts of solar and wind energy Earth.com staff writer. www.earth.com/news/environmental-impacts-solar-wind-energy/

Gohlke, J.M., S.H. Hrynkow, and C.J. Portier. 2008. Health, economy, and environment: Sustainable energy choices for a nation. *Environ. Health Perspect.* 116(6): A236–A237.

Hu, H., X. Zha, Z. Li, and R. Lv. 2022. Preparation and performance study of solar pavement panel based on transparent resin-concrete. *Sustain. Energy Technol. Assessments*. 52(Part B): 102169.

Ige, O., Ratnayake, I. Martinez, J. Pepper, S. Alsup, A. McGuirk, and M.D.P Mudaranthakam. 2024. A regional study to evaluate the impact of coal-fired power plants on lung cancer incident rates. *Preventive Oncol. Epidemiol.* 2(1). https://doi.org/10.1080/28322134.2024.2348469

Khan, A., O. Plana-Ripoll, S. Antonsen, J. Brandt, C. Geels, H. Landecker, P.F. Sullivan, C.B. Pedersen, and A. Rzhetsky. 2019. Environmental pollution is associated with increased risk of psychiatric disorders in the US and Denmark. *PLoS Biol.* 17(8): e3000353. https://doi.org/10.1371/journal.pbio.3000353

Kostic, M. 2004. Work, power, and energy. In: Cleveland, C.J. (ed.) *Encyclopedia of Energy* (Vol. 6, pp. 527–538). Elsevier.

Kostic, M.M. 2007. Energy: Global and historical background. In: Anwar, S. (ed.) *Encyclopedia of Energy Engineering*. Taylor & Francis. https://doi.org/10.1081/E-EEE-120042341

Kumar, J.C.R., and M.A. Majid. 2020. Renewable energy for sustainable development in India: Current status, future prospects, challenges, employment, and investment opportunities. *Energy. Sustain. Soc.* 10: 2. https://doi.org/10.1186/s13705-019-0232-1

Lin, C-K., T.-T. Lin, T. Chen, C. Zigler, Y. Wei, and D.C. Christiani. 2019. A global perspective on coal-fired power plants and burden of lung cancer. *Environ. Health.* 18: 9. https://doi.org/10.1186/s12940-019-0448-8

Malanima, P. 2014. Energy in history. In: Agnoletti, M., and S. Neri Serneri (eds.) *The Basic Environmental History, Environmental History* 4. Springer International Publishing. https://doi.org/10.1007/978-3-319-09180-8_1

Perera, F. 2018. Pollution from fossil-fuel combustion is the leading environmental threat to global pediatric health and equity: Solutions exist. *Int. J. Environ. Res. Public Health.* 15: 16. https://doi.org/10.3390/ijerph15010016

Ritchie, H. 2020. What are the safest and cleanest sources of energy?. https://ourworldindata.org/safest-sources-of-energy

Ringel, M., R. Laidi, and D. Djenouri. 2019. Multiple benefits through smart home energy management solutions—A simulation-based case study of a single-family-house in Algeria and Germany. *Energies.* 12: 1537. https://doi.org/10.3390/en12081537

Sarath, K.G., and P. Jawahar. 2014. Atmospheric emissions and pollution from the coal-fired thermal power plants in India. *Atmos. Environ.* 92: 449–460. https://doi.org/10.1016/j.atmosenv.2014.04.057

Seddighi, S., E.J. Anthony, H. Seddighi, and F. Johnsson. 2023. The interplay between energy technologies and human health: Implications for energy transition. *Energy Rep.* 5592–5611. https://doi.org/10.1016/j.egyr.2023.04.351

Sharma, A.K., B. Palak, and K. Prashant. 2018. Air pollution and public health: The challenges for Delhi, India. *Rev. Environ. Health.* 33(1): 77–86. https://doi.org/10.1515/reveh-2017-0032

Singh, K., T. Peshin, S. Sengupta, S.K. Thakkar, C.W. Tessum, J.D. Hill, I.M.L. Azevedo and S.P Luby. 2024. Air pollution mortality from India's coal power plants: Unit-level estimates for targeted policy. *Environ. Res. Lett.* 19: 064016. https://doi.org/10.1088/1748-9326/ad472a

Smith, A.H, G. Marshall, Y. Yuan, C. Ferreccio, L.J.O. von Ehrenstein, C. Steinmaus, M.N. Bates, and S. Selvin. 2006. Increased mortality from lung cancer and bronchiectasis in young adults after exposure to arsenic in utero and in early childhood. *Environ. Health Prospect.* 114(8): 1293–1296. https://doi.org/10.1289/ehp.8832

Smith, K.R., S. Mehta, and M. Maeusezahl-Feuz. 2004. Indoor smoke from solid household fuels. In: Ezzati M., Rodgers A.D., Lopez A.D., Murray C.J.L. (eds) *Comparative Quantification of Health Risks* (Vol. 2). World Health Organization.

Smith, K.R., H. Frumkin, K. Balakrishnan, C.D. Butler, Z.A. Chafe, I. Fairlie, P. Kinney, T. Kjellstrom, D.L. Mauzerall, T.E. McKone, A.J. McMichael and S.H. Schneider. 2013. Energy and human health. *Annu. Rev. Public Health.* 34:159–188. https://doi.org/10.1146/annurev-publhealth-031912-114404

Straif, K., R. Baan, Y. Grosse, B. Secretan, F. El Ghissassi, and V. Cogliano. 2006. Carcinogenicity of household solid fuel combustion and of high-temperature frying. *Lancet Oncol.* 7: 977–978.

Tuominen, P., and T. Seppänen. 2017. Estimating the value of price risk reduction in energy efficiency investments in buildings. *Energies.* 10: 1545. https://doi.org/10.3390/en10101545

von Hildebrand, A., K.L. Ebi, J. Lapitan, and B. Menne. 2009. *Protecting Health from Climate Change: Global Research Priorities.* World Health Organization. www.researchgate.net/publication/315654872_Protecting_health_from_climate_change_global_research_priorities

Wang, R., G. Liu, and J. Zhang. 2015. Variations of emission characterization of PAHs emitted from different utility boilers of coal-fired power plants and risk assessment related to atmospheric PAHs. *Sci. Total Environ.* 538: 180–190. https://doi.org/10.1016/j.scitotenv.2015.08.043

Wilkinson, P., K.R. Smith, M. Joffe, and A. Haines. 2007. A global perspective on energy: Health effects and injustices. *Lancet.* 370: 965–978.

Xing, L., M. Xue, and M. Hu. 2019. Dynamic simulation and assessment of the coupling coordination degree of the economy–resource–environment system: Case of Wuhan City in China. *J. Environ. Manag.* 230: 474–487.

Zhang, H., Y. Da, X. Zhang and J-L Fan. 2021. The impacts of climate change on coal-fired power plants: Evidence from China. *Energy Environ. Sci.* 14: 4890–4902.

Zhao, Y., S. Wang, K. Aunan, H.M. Seip, and J. Hao. 2006. Air pollution and lung cancer risks in China—A meta-analysis. *Sci. Total Environ.* 366(2–3): 500–513.

Index

Note: Page numbers in *italic* refer to figures.

D

E

F

G

H

I

L

M

N

O

P

R

S

For Product Safety Concerns and Information please contact our EU representative GPSR@taylorandfrancis.com Taylor & Francis Verlag GmbH, Kaufingerstraße 24, 80331 München, Germany